On the Edge of Eternity

ON THE EDGE OF ETERNITY

The Antiquity of the Earth in
Medieval and Early Modern Europe

IVANO DAL PRETE

OXFORD

UNIVERSITY PRESS

OXFORD
UNIVERSITY PRESS

Oxford University Press is a department of the University of Oxford. It furthers
the University's objective of excellence in research, scholarship, and education
by publishing worldwide. Oxford is a registered trade mark of Oxford University
Press in the UK and certain other countries.

Published in the United States of America by Oxford University Press
198 Madison Avenue, New York, NY 10016, United States of America.

CIP data is on file at the Library of Congress

ISBN 978–0–19–067889–0

DOI: 10.1093/oso/9780190678890.001.0001

1 3 5 7 9 8 6 4 2

Printed by Sheridan Books, Inc., United States of America

This book was published with the assistance of
the Frederick W. Hilles Publication Fund of Yale University.

Arenam maris et pluviae guttas
Et dies saeculi quis dinumeravit?
"Who will count the sand of the sea, the water drops in the rain,
and the days of time?"

Ecclesiastes, 1:2

Contents

Acknowledgments

IN THE LONG years I spent researching and writing this book, I contracted many more debts than I will ever be able to repay. I was still a graduate student when Luca Ciancio encouraged me to investigate the cracks and inconsistencies in the conventional narrative of the history of deep time that I found so perplexing. The Huntington Library and the Italian Academy for Advanced Studies (Columbia University, New York) generously funded my early research and allowed me to become better acquainted with far more learned colleagues like Pamela Smith, Paula Findlen, William Connell, and Lawrence Principe, whose advice has been invaluable. The unwavering support I received from Tony Grafton, his patience in amending my naïve ramblings about Renaissance chronology, and what counts most, his warm friendship, still humble me. I owe special gratitude to him. Chapter 3 could not have been written without the Harvard/Villa i Tatti fellowship I spent in Florence in 2015–16, in the company and with the help of amazing historians of Renaissance art, literature, and technology. Among others, I am especially grateful to Holly Flora, Christian Kleinbub, Giulia Torello, Allen James Grieco, Francesca Borgo, and Pamela Long.

The referees who contributed so much to improving this book remain anonymous, but I can at least express my gratitude to the scholars who agreed to read early versions of my articles and chapters. Craig Martin, Philip Nothaft, Brian Ogilvie, Rita Librandi, Samuel Gilson, and Ronald Numbers gave me precious feedback. I gained countless insights and new perspectives from some colleagues who are also deeply engaged in reassessing the interactions of early modern Earth sciences, religion, and more, such as Lydia Barnett, J. B. Shank, and Hanna Romain. Mark Brandon of the Department of Earth and Planetary Sciences at Yale University helped me get a grasp of Leonardo da Vinci's geological notes, and Carson Koepke of the Medieval Studies Department assisted me in deciphering medieval manuscripts that were well beyond my competence. Colleagues and students in the History of Science and Medicine Program have been wonderful in their support of my project, but my gratitude goes in particular to Debbie Coen. While I have inconvenienced librarians and archivists in many cities and across many countries on two different continents, the staff of the Beinecke Library and of the Medical Historical Library at Yale University were especially helpful in satisfying my omnivorous curiosity. These years have not always been easy, and I tend to be a slow writer even in good times, but I always knew that Susan Ferber at Oxford University Press would wait patiently for my chapters to come along.

This book is for Paola, who believes in me even when I don't.

Note on Terminology

SUBJECTS RELATED TO geology, mineralogy, or Earth history remained part of the larger field of 'meteorology' until the late seventeenth century. They enjoyed no disciplinary autonomy, and no specific terms or categories existed to designate them. This book often uses "Earth history" for want of better terms, despite being anachronistic. Still, this term encompasses most of the related disciplines and captures the idea of the Earth as a dynamic and historical entity. "Geohistory" is also a popular term in recent literature, with the further advantage of breadth that includes the influence of geographical or geological factors on human history. The terms "geology" or "geological" are employed when the context warrants it, even though they were not in common use until the late eighteenth century in French or Italian (nor in English before 1800). In the late 1600s the so-called "History of the Earth" became an autonomous discipline, whose goal was to combine in a coherent picture natural evidence and biblical exegesis; "History of the Earth" capitalized is used in this specific sense.

Many crucial terms employed in medieval and Renaissance meteorology were fraught with semantic ambiguity. "Deluge" and "flood" could mean either a local or a universal inundation, but in this book "Deluge" or "Flood" capitalized designate specifically the latter. It should be noted that in medieval and Renaissance meteorological literature, "the Flood" was normally a cyclical phenomenon (not unlike "the Earthquake") rather than a specific event. The same sources make scarcely any distinction between "earth" as one of the four elements and the terraqueous globe. However, when not in direct quotes "earth" refers to the element and "Earth" to the globe we inhabit. "World" could likewise designate either the Earth or the universe in general. In Aristotelian philosophy, the difference was often immaterial since they were both assumed to have always existed. This was not true (or not necessarily true) in biblical, Platonic, or Epicurean cosmologies, but sources may refer to "the age of the world" even when a distinction between the Earth and the universe as a whole would have been meaningful. In philosophical jargon, "ancient" often meant "eternal" and "new" simply stood for "created" rather than "recently created." In this book, I use "Ancient Earth" in the sense of a created Earth, whose age is greater than that of the traditional biblical chronology.

I have consulted primary sources in Latin, French, Italian, and English in the original, but quotations come from authoritative English editions whenever possible. In a few cases, I have modified translations that proved incorrect or inadequate, as explained in the corresponding notes. When not stated otherwise, translations are mine, as is the sole responsibility of everything written in this book.

On the Edge of Eternity

Introduction

*If one considers the generation of mountains and islands, he
will realize that there was a time when they were made and
concealed by the sea; and that the time will come again,
when what is now covered by the sea will become inhabitable, and what is inhabited and cultivated must one day be
hidden by the ocean.*

GIROLAMO FRACASTORO, *Homocentrica*, 1538

Ascensions into Antiquity

In the early spring 1539, learned physician Torello Saraina (1485–1550) led
a party of local gentlemen to the discovery of the antiquities of Verona in
northeastern Italy. Renaissance humanists could hardly have conceived of
a worthier social activity. The old Roman city, nested in the bend of River
Adige, had plenty to offer to their conversation. As they crossed the river and
climbed the hill on the opposite bank, they encountered renowned painter
Giovanni Caroto attending to a tentative reconstruction of the ancient theater. The vast hemisphere, dug into the slope of the hill overlooking the city,
lay right under their feet but was encrusted with the buildings that had grown
over it. Its arches, in ruins or incorporated into more recent structures, led to
grottoes that were the hideout of lovers and artists.

Yet, the company knew that the Roman settlement had been erected over
the vestiges of even older antiquity. When they arrived at the other side of
the hill, they stopped by the massive ramparts erected years earlier to reinforce the thin medieval wall, made vulnerable by modern artillery. Saraina
could still recall his wonder when the construction works revealed a network

On the Edge of Eternity. Ivano Dal Prete, Oxford University Press. © Oxford University Press 2022.
DOI: 10.1093/oso/9780190678890.003.0001

of subterranean caverns, "as if the stone itself inhabited them as dwellings." He maintained that those caves were excavated by human beings in the first age of humankind, when—as classical authors like Vitruvius and Lucretius attested—men lived in the woods and sought refuge in the hollows of the earth. One of Saraina's friends remembered that those digs had revealed even greater wonders: mixed with ordinary rock, diggers often brought to the surface large amounts of stones resembling marine creatures. Sea urchins, sea shells, hermit crabs, starfish, and animals of many other species seemed to have been turned into stone and were somehow buried under that mountain seventy miles from the nearest seashore. Saraina observed, though, that such "petrifications" should not be surprising. As he read in the books of the ancient naturalists, wood, bones, and other substances can become stone "because of the great length of time." What really puzzled him was how these marine animals (assuming they used to be living creatures) ended up there. On this subject, he could relate what he heard from an illustrious fellow citizen, physician and philosopher Girolamo Fracastoro (1478–1553), when he had given him one of those petrified shells as a present.

On that occasion, Fracastoro had told him that three different theories circulated "among philosophers." According to the first, sea animals were thrown on land "in the time, when the dregs brought by waters overtook the mountains." Another opinion maintained that in certain mountains there are "salty juices" that can generate marine creatures within the stones; but sometimes the generation cannot be completed, and the stone carries the impression of the still-imperfect creature. Fracastoro thought little of both ideas. If it was a flood of the sea, then fossils should be found only on mountain tops; instead, they were common at every height and even inside the mountains. The second theory made even less sense to him. Apart from other considerations, if true, then even living sea animals should be discovered from time to time among the rocks, but this never happens. No, those shapes must once have been real animals, born and grown up in the sea. But in order to understand how they became part of a hill, one had to look into a past so remote as to overcome every human memory. As Fracastoro explained, every mountain formed from mounds of sand and debris accumulated underwater by the motions of the sea, thus incorporating the remains of marine creatures. Later, the sea slowly recedes, uncovering islands and hills, while other parts of the world are being submerged. This phenomenon was still taking place before their own eyes. According to the ancients, Egypt used to be a sea gulf; the Italian city of Ravenna was lying several miles from the sea, but in Roman times the shore was mere steps away. Saraina recalled that "Fracastoro said many more things on this topic, but this is a subject for philosophers. It is now time to resume our walk."[1]

For the most part, the book in which Saraina described his excursion among the antiquities of Verona remained relegated to the footnotes of local historians and antiquarians. The account of his conversation with Fracastoro, however, lived on as a famous episode in the history of paleontology. For more than 250 years, Fracastoro has been celebrated as one of the few minds who, in an age of acritical acceptance of the letter of Genesis, dared to challenge the religious orthodoxy and contemplate the dangerous idea of an Earth much older than a few thousand years. Already in 1757, the entry "fossil" in Diderot and d'Alembert's *Encyclopédie* remarked that after "centuries of ignorance" Fracastoro had recognized the origin of fossils from living creatures and "did not doubt that the sea carried them on the continent."[2] Eighty years later, the great geologist Charles Lyell lamented that the disregard of Fracastoro's "clear and philosophical views" had been a scientific tragedy:

And the talent and argumentative powers of the learned were doomed for three centuries to be wasted in the discussion of those two simple and preliminary questions: first, whether fossil remains had ever belonged to living creatures; and, secondly, whether, if this be admitted, all the phenomena could be explained by the Noachian deluge. It had been the consistent belief of the Christian world . . . that the origin of this planet was not more remote than a few thousand years; and that since the creation the deluge was the only catastrophe by which considerable change had been wrought on the Earth's surface."[3]

As the story goes, a deeply Christian society could not accept the ancient Earth implied by Fracastoro's theory. Until the secularization of the European culture began during the Enlightenment, dominant religious beliefs made it impossible to conceive—or at least to discuss openly—such a possibility. Given this premise, the biblical Flood remained the unavoidable *Deus ex machina*—quite literally—of the history of the Earth. Philosophers who invoked much longer timescales, and showed little concern for the creation story of the Bible, have been portrayed in histories of geology as lonely "precursors" or even "useless geniuses." They were the brave but scattered supporters of heterodox if not plainly heretical views, whose impact on the culture of their time remained negligible.[4] In his classic *The Dark Abyss of Time*, Paolo Rossi (1923–2012) argued that the eighteenth-century discovery of the depth of historical and geological time was a change as radical "in the position of man" as the far more celebrated Copernican revolution: "Men in Hooke's time (1635–1703) had a past of six thousand years; those of Kant's time were conscious of a past of millions of years."[5] In an age of "alternative facts," growing polarization, and increasing disconnect between the academy and much of society, the assumption that an ancient Earth is the recent product of a secularized science—for better or worse—remains a topic on which even professional geologists and Young Earth creationists can agree.

I first noticed oddities in Saraina's account, and soon after in the conventional history of the "discovery" of deep time, when I was invited to present a paper on Fracastoro's legacy in eighteenth-century astronomy and natural history. It was not what Saraina wrote that puzzled me, but rather what he did not write. What sources do not say can rarely be regarded as evidence for anything, but sometimes unexpected silence can provide clues to flaws and dissonances in a narrative. In this case, how could an author publish such an extraordinary conversation without any further remarks? If the ideas voiced by Fracastoro were so obviously scandalous and dangerous in his time, why did not the author try to distance himself from them? There was no attempt on Saraina's part to make clear that he did not share the philosopher's opinions or to warn the reader about their heterodoxy. Saraina and his companions did not even express surprise or discomfort at the philosopher's words. They just moved on, as if they had heard nothing particularly striking.

Upon further investigation into whether Saraina's book ever caused trouble to the author, or to the philosopher who dared disprove the diluvial origin of fossils and argue for an old Earth, the answer was unambiguous: it did not. Fracastoro was alive and well at the time, and he had already written influential medical and philosophical works.[6] A few years earlier, he had published a cosmological treatise that contained an exposition of the theory explained to Saraina.[7] Instead of being tried as a heretic, Fracastoro was appointed physician (with a hefty paycheck) to the Council of Trent, the assembly that should have reformed Christianity and recomposed its unity broken by the Protestant Reformation.[8] He was not even particularly original: much of what he published or said on the long-term changes of the Earth's surface could be found in Aristotle's *Meteorology*, composed eighteen centuries earlier. Aristotle's book had been read, commented on, and taught in universities all over Europe since the thirteenth century.[9] A survey of contemporary meteorological works showed that such supposedly heterodox ideas were commonplace among his colleagues and discussed with the same lack of concern. Compelling evidence emerged that the circulation of these doctrines was not even restricted to the learned elite who wrote in Latin.

Around 1540, they were apparently so common as to make their way into "popular science": translations, summaries, and compendia written in the vernacular and intended for the vast public of non-specialists created by the diffusion of the printing press. The authors of these books seemed to have no qualms sharing with the unlearned entirely secular theories on the evolution of the Earth and on its "eternity." They did so without any qualifications, as if discussing common-sense notions that were entirely uncontroversial. Instead of the monolithic orthodoxy deprecated by Lyell, a picture of Renaissance Europe emerged in which a plurality of views and approaches seemed to be the accepted social and cultural norm. Fracastoro's statements—or Leonardo da Vinci's geological notes, another example of alleged scientific and religious heterodoxy—no longer appeared as the surprising and isolated stunts of geniuses who managed to overcome the intellectual limitations of their time.

Rethinking the History of Deep Time

Authors who dealt with medieval and early modern Earth history have often stumbled over some of the same puzzling documents I encountered. Most historians of geology have focused on the last three hundred years, when their discipline has become recognizable as such; but even brief introductory surveys on the previous centuries yielded inconsistencies that were difficult to ignore.[10] In their 1965 *The Discovery of Time*, Stephen Toulmin and June Goodfield presented the conventional account of Western and Islamic cultures long constrained by scriptural literalism. According to Toulmin and Goodfield, even the age of the scientific revolution contributed little to the understanding of the "history of Nature." By 1730, "the traditional six-thousand-year-scale still retained much of its earlier authority."[11] Yet, the authors recognized that biblical diluvialism was in fact a recent construction. Its authority did not rest on "superstitious relics of medieval religion" but on "scientific theories built up during the seventeenth and eighteenth centuries by men who prided themselves on their own modernity."[12] In the early eighteenth century, the timescale of the Genesis appeared "more fixed" than that of medieval Europe.[13] The appreciation of diluvialism as a product of early modernity, rather than of medieval traditions, has since been fully accepted and thoroughly developed in later literature.

Only a few years after Toulmin and Goodfield's work, Martin Rudwick masterfully exposed some of the most evident cracks in the traditional narrative. His discussion of sixteenth-century theories on the origin of fossil seashells in *The Meaning of Fossils* (1972) uncovered a lively debate and a plurality of views and approaches. Rudwick pointed out that the adoption of an ancient Earth of indefinite age seems to have been the norm at Padua University in northeastern Italy, where Fracastoro also studied and taught. Apparently, the time scales it entailed were no "stumbling blocks" for Christian thinkers, as "a metaphorical exegesis of the 'six days' of the creation had been acceptable . . . since the times of Augustine [fifth century]." Rudwick concluded that, in the Paduan tradition at least, it was "not uncommon to accept an Aristotelian view of the ever-changing geography of the globe, with occasional local inundations of purely natural causation, of which the scriptural Flood had been but one."[14]

Scholars whose research extended further back in time realized that the antiquity of the Earth was hardly a sixteenth-century innovation or a local peculiarity. In fact, this has been known since Pierre Duhem's rediscovery of medieval science and geology in the early twentieth century. Later authors have mostly built on his legacy and relied on the documents he found in the French archives and published. In *A History of Geology* (1991), Gabriel Gohau concisely examined geological theories and relative timescales in the Islamic and Christian Middle Ages, none of which seemed to care much about the chronology of Genesis. After discussing the models expounded in the fourteenth century by the French philosopher Jean Buridan (1301–1358), who mentioned time scales of hundreds of billions of years, Gohau could not help noticing "at least briefly, without going into details, how independent this fourteenth-century scholar at the University of Paris was from the Church."[15] A more accurate statement would be that the Church was not enforcing the biblical chronology at all. As usual for a medieval scholar, Buridan was himself a clergyman, and in the following centuries he continued to be one of the most revered Masters of Arts in the history of that university. Likewise, geophysicist Pascal Richet commented on the medieval and early Renaissance "liberty to speculate," only later undermined by the repercussions of the Protestant Reformation and by the rise of biblical literalism.[16]

Historians of geology may have been amazed at the unproblematic acceptance of such deep antiquity, but no student of medieval and Renaissance philosophy could really be surprised. Following the introduction of Aristotle's works in thirteenth-century Europe, the indefinite age of the Earth became the normal assumption in natural philosophy and in university teaching. Frictions between philosophers and theologians concerned the problem of whether the world was created at all, rather than when. When the Creation took place was rarely regarded as a substantial issue and certainly not a matter of faith. Historian of Renaissance literature William Connell has remarked that "in Renaissance Italy it became common to discuss the possibility that the world is eternal." Historical thought, in particular, was deeply affected by the spread of Aristotelian eternalism. Increasingly mindful of the unfathomable depth of the past, Renaissance historians adopted an approach that "emphasized the fragility of human memory."[17] Rather than focus on the origin of the world and of humankind, they addressed the problem of "how, in an eternal world, human history might have lost track of the distant past. The logical answer lay in the potential of cataclysms, particularly catastrophic floods (including, but not limited to, that of Noah's), to eliminate most of the human race."[18] In turn, this solution was clearly indebted to the tradition of medieval and Renaissance "meteorology"—a field that included what would today be called geology, mineralogy, and Earth history.

In spite of the increasing awareness of the inconsistency and inadequacy of the current narrative, historians have not proposed any serious alternative. A few lines in a book or a digression of a couple of pages in another are perhaps sufficient to signal the existence of a problem, but they are hardly enough to develop a different approach. The purpose of this book is to take the first steps toward a new paradigm for the history of deep time in Western culture. It aims to replace the view of a relatively recent discovery of the "abyss" of geological time with one that accounts for the complexity, diversity, and social and cultural significance of pre-modern Earth history. It argues that, alongside literal interpretations of the creation story of Genesis and of its timeframe, medieval and early modern Western culture allowed for a multiplicity of theories and beliefs, often built around the assumption that the Earth was created in a distant and undetermined past. Such opinions were largely unproblematic, circulated freely even outside the learned elites, and were not necessarily deemed incompatible with each other and with the biblical account.

The core teachings transmitted by the book of Genesis, such as that the world was made by God with an act of free will and that a sinful mankind was exterminated by a flood in a not too distant past, were rarely questioned. On the contrary, they retained their moral value even as they became either incorporated into theories of a cyclical renovation of the Earth's surface and of humankind itself or were excluded from the investigation of the physical world, as supernatural miracles better left to theologians. The holders of those views were neither judged nor did they consider themselves lesser Christians just because they preferred a figurative reading of the Bible when it came to explaining the origins and history of the natural world. European culture was of course deeply permeated by the short time frame of the Bible;

in the sixteenth century, the editions of the most read world histories based on the traditional chronology vastly outnumbered philosophical and scientific popularizations. Still, alternative accounts that were often perceived as complementary rather than opposed to those of the Bible remained easily accessible even to poorly educated readers.

In fact, for centuries there was no such thing as a Bible-based Earth history, only a Mosaic account of the creation alongside theories of the evolution or cyclical change of the globe that built upon Greek philosophy and its Islamic interpreters. The construction of a "pious natural philosophy," as historian Ann Blair calls it, began in earnest only in the late sixteenth century following the rise of biblical literalism in both Protestant and Catholic Europe.[19] As biblical chronology and theories of the Earth centered on Noah's Flood gained influence and even preeminence, their more secular alternatives remained vital and debated. The alleged Enlightenment discovery of geological time may have never happened, because the idea that the Earth could be very old had always been common. The eighteenth century, though, buried the memory of the richness and complexity of pre-modern Earth history. On its grave, it erected the historiographical myth of a Christian tradition that uniformly rejected the antiquity of the world as an extraneous body, as opposed to a new secular science ready to welcome it. Largely unchallenged for almost three centuries, that account solidified over time into a still-dominant truism.

The supposedly recent discovery of deep time has been crucial to the narrative of the rise of modern scientific rationalism. Paolo Rossi was certainly right when he equated its significance to that of the Copernican revolution. Indeed, the argument that medieval and early modern Christians could be comfortable with an Earth of indefinite age may still sound unnatural and counter-intuitive. Yet, recent scholarship has made the traditional account unsustainable. Among the developments that helped lay the ground for this research, the profound revision of the historical relationship of science and Christianity is possibly the most important. The traditional narrative has long been brandished in support of the (in)famous "conflict thesis" popularized in the late nineteenth century by John W. Draper and Andrew D. White. The thesis postulates the unavoidable hostility and incompatibility of science and religion, particularly Catholic Christianity. One chapter of Draper's *History of the Conflict between Religion and Science* (1874) is entirely dedicated to the "controversy regarding the age of the Earth."[20] White's less confrontational and more thoroughly researched *A History of the Warfare of Science and Theology in Christendom* (1896) maintains nonetheless that until the late

1600s, "to doubt [that man was created from four to six thousand years before the Christian era], and even much less than this, was to risk damnation."[21] The conflict thesis still holds currency among the public, but scholars soon began to question its simplistic dualism and its documentary basis.[22] After John H. Brooke's seminal works in the early 1990s, the same viability of "science" and "religion" as ahistorical and rigidly defined categories has been increasingly questioned. In a context that emphasizes the complexity of the historical engagement of science and religion, the binary view of a pre-modern culture stubbornly clinging to the letter of the Bible, as opposed to the latter triumph of a scientific worldview, looks like an oddity ripe for debugging.

Together with the rejection of the conflict thesis, modern scholarship has reversed the old idea that Western science became independent from theology in the wake of the scientific revolution of the sixteenth and seventeenth centuries. On the contrary, the age of the scientific revolution is now regarded as the time when theology and the study of nature strictly entangled, rather than separate. From this perspective, it becomes easier to conceive that medieval and Renaissance authors and audiences could set aside the letter of the Genesis to consider the possibility of an ancient world. One of the central claims of this book is that the idea of an ancient Earth did not remain the precinct of restricted intellectual elites, familiar with scientific and philosophical literature in Latin. This approach builds upon the recent reorientation of history of science studies toward the social and cultural context of scientific research and the inclusion of low-class practitioners and manual workers. The same rise of modern experimental science, once regarded as the creation of exceptional individuals like Galileo, is widely seen as the appropriation by members of the "high culture" of methods and practices long cultivated in the workshops of chemists and artisans.[23] Other essential parts of this book are supported by studies in a range of different fields that have become available only in recent years. Sixteenth-century meteorological publications in the vernacular have been the starting point of this research. It would have been impossible to contextualize them properly without Joëlle Ducos's and Craig Martin's works.[24] Federico Barbierato has documented how easily even explicitly heretical beliefs spread among the lower classes of a major center of culture like Venice.[25] Barbierato's research lends further support to the contention that such notions like the antiquity of the Earth could circulate unhindered beyond the learned elites. The main competitor of an ancient Earth was of course the short chronology based on the Bible. Their reciprocal relation (or lack thereof) is a recurring theme of this book. Between the late sixteenth and the early eighteenth century, chronology grew into a highly technical yet immensely popular field.[26] By comparison, medieval chronology remained largely understudied until Philipp Nothaft began to unearth, analyze, and publish an impressive corpus of medieval literature on chronological and calendric subjects.[27] The Noetic Flood, however, remained an important theological and moral event for people who lived during the Middle Ages and in the early Renaissance, but largely irrelevant to the physical history of the Earth. Recent literature, particularly Lydia Barnett's *After the Flood* (2019), has explained its meteoric rise in the sixteenth and seventeenth centuries within the context of European colonial expansion. Only the reality of a universal inundation could guarantee at once biblical inerrancy and the descent of every human race from the same sinful couple, thus justifying their conversion to Christianity.[28]

The geographical and cultural focus of this book is Western Europe. For practical and methodological reasons, some chapters pay special attention to central and northern Italy. Simply stated, for much of the period under consideration, that context offers the most numerous sources, particularly when it comes to scientific culture elaborated and transmitted in a vernacular language. Yet, this work does speak to efforts to place the history of deep time within global and colonial perspectives. Among the most recent contributions, Pratik Chakrabarti's *Inscriptions of Nature* traces the European discovery of deep time in colonial history and in the interactions of "colonial antiquarianism, Orientalism, exploitation of natural resources, and the study of aboriginality and tribal mythologies," whose timescales often dwarfed those of Genesis.[29] While colonial encounters undoubtedly transformed the meanings and uses of deep time, those narratives still rest on the assumption that such a "discovery" did in fact take place. Within the broader chronological landscape examined in this book, the inclusion of literary and material sources from beyond Europe appears to have rekindled and reframed long-lasting controversies rather than opening up entirely new temporal horizons. Late medieval and early Renaissance natural philosophy and visual arts constructed a landscape that was already "historical," being the product of slow erosive processes and a witness of the passage of both human and geological time.[30] In many ways, pre-modern European understandings of human and Earth history always included a global perspective, since they relied heavily on Greek and Arabic materials that absorbed elements from a much wider world.

Saraina and his acolytes were not taken by surprise by the increasingly deep layers of antiquity they encountered in their ascension. From their perusal of classical and medieval authors, they assigned the ruins of Verona's Roman

Theater to the 40th or 42nd year of Emperor Augustus—almost exactly fifteen centuries earlier. As they kept delving into the past of their hometown, they knew that myth would soon encroach history and make every effort to tell one from the other tentative. The chronology of facts and events became more controversial and, for the eras preceding the biblical Flood, perhaps pointless. As Christians, they probably did not doubt that there was a creation in time, and like everyone they were familiar with the universal histories that placed it sixteen centuries or so before the Flood. But alongside the traditional chronology, many authors also presented accounts of far more ancient times. The petrified seashells they examined reminded them that the book of Genesis conveyed moral and theological truths, not physical ones, which were better left to observation, reason, and profane authorities. In front of the narrow shore of knowable history, sparsely littered with human and sacred documents, an unfathomable abyss may have extended with countless more deluges, raised and submerged continents, annihilated and resurgent humanities. That ocean may not have been without boundaries, but it certainly seemed to stretch well beyond the horizon.

* * *

Chapter 1, "Footprints in the Dust," introduces the reader to the long and complex medieval debate on the eternity of the world. Christian and Muslim thinkers tried hard to reconcile the Mosaic account of the creation with Greco-Roman natural philosophy, which never contemplated the creation of matter, space, and time from nothing. Their endeavors rested on the possibility, which was widespread and largely accepted, of a figurative reading of Genesis and in particular of the first "days" of creation. Not until 1215 did the Lateran Council IV proclaim that the world had been created in time, thus closing one of the many options available up till then. The conciliar decrees, though, aimed to establish a principle rather than its chronology: whether the world was created at all, not when, was the matter of contention.

Chapter 2, "The Medieval Earth," turns to medieval models of the physical changes of the globe, which often entailed the possibility or "probability" of an ancient Earth. The Bible was not the only source that described the destruction of humankind by a catastrophic flood: Plato, Aristotle, Ovid, and Seneca all posited that a physically and morally corrupted Earth was periodically renovated by "deluges" of water or fire, either universal or local, occasioned by particular astral conjunctions. Scholars like Jean Buridan and Albert of Saxony deployed impressive intellectual efforts to show that new mountains could replace aging ones as they were eroded, since the absence of such a mechanism would have denied the possibility of an ancient Earth. While they could not determine when the world began (nor was this important to either the Christian or to the philosopher), they went to great extents to demonstrate that its extreme antiquity was plausible from a physical point of view, as well as compatible with the faith.

Crucially, the diffusion of those doctrines was not restricted to the small intellectual elite that communicated in Latin. Chapter 3, "Vernacular Earths," focuses on medieval and early Renaissance Tuscany to show that the notion of an ancient Earth spread unhindered in vernacular culture and was easily appropriated by artists, craftsmen, and merchants. The rise of a prosperous mercantile economy in the 1200s was accompanied by the production of literary and scientific works meant to be accessible to those without a formal education. In the following centuries, the Earth and its strata were extensively studied for practical reasons by builders, hydraulic engineers, and water prospectors, while artists began to portray its surface as a landscape of natural ruins scarred by old age and by the passing of time. When the most famous of the Renaissance artists/engineers, Leonardo da Vinci, turned to the problem of the nature and evolution of the Earth in the early 1500s, the idea that the terraqueous globe might have been only a few thousand years old never crossed his mind.

Chapter 4, "A Christian History of the Earth?" details how the new natural, geographical, and ethnographical knowledge Europe's colonial expansion depended on shook the foundations of the medieval Earth. The radicalization of religious tensions, the Protestant and Catholic reformations, and the rise of biblical literalism shattered the barriers that for centuries guarded the autonomy of natural philosophy from theology, and vice versa. Restrictions to intellectual freedom all over Europe came with unprecedented calls for a Christianized science of nature. Yet, the new orientations often broadened the scope and breadth of traditional meteorology rather than outright replacing it. In Catholic Europe, authors were only required to state that the world had a beginning, not to uphold a recent one. A created world of undetermined age continued to inform much of the relevant literature.

Colonial expansion, and the demise of medieval cosmography, had even more crucial implications. Chapter 5, "The Rise of Diluvialism," expounds on the rise of the biblical Flood to become the ideological pillar of European imperialism, since it guaranteed both the universality of the Christian Revelation and the common origin of humankind from the same sinful couple. Once a mainly moral and theological story that natural philosophy could bypass entirely, the Noetic catastrophe solidified in the course of the seventeenth century into a major physical event that must have left considerable traces of its passage. As the historical truth of the Revelation increasingly depended on natural evidence for the Noachian Flood and its chronology, authors who defended an ancient Earth and rejected the diluvial origin of fossils began to find themselves associated with materialism and irreligion.

In the course of the eighteenth century, Earth history was eventually weaponized into a tool of cultural wars and political propaganda. Chapter 6, "The Invention of the History of Deep Time," deals with the works of immensely influential *philosophes* like Voltaire, Boulanger, and d'Holbach, who during the Enlightenment turned the recent association of diluvialism and Christianity into a paradigm of the alleged historical clash of religious prejudice and rational enquiry. As they rearranged sources and evidence in a way that suited their political and cultural agendas and their conception of history, they outlined for the first time what remains to this day the common understanding of pre-modern Earth history. Theories based on an ancient Earth were presented not in their uninterrupted continuity with medieval and Renaissance traditions, but as recent feats of a human reason freed from Christian obscurantism. Still, the *philosophes* may not have been so successful had their conservative opponents not adopted a similar framework. Recast as a tool of anti-Christian propaganda, the antiquity of the Earth came to be perceived in conservative milieus as a direct threat to the basis of the existing social order, rooted in the proverbial alliance of the throne and the altar.

Chapter 7, "Political Fossils," shows that fears of social subversion only increased in the decades leading to the French Revolution, powerfully contributing to the politicization and polarization of Earth history. Italy, and particularly Venice, provide once again an especially compelling case. The ruling elites of its state embraced diluvialism and a young Creation as instruments of social and political preservation, using their social clout to impose a diluvialist orthodoxy where none had existed before. Fossils, rocks, and other natural specimens were put into the service of a politically conservative science, which strove to demonstrate the reality of a recent inundation of the globe to uphold the historical accuracy of the Revelation. Ultimately, the identification of anti-diluvialism and of an ancient Earth with irreligion was instrumental in establishing a new narrative of the history of deep time. Conservatives bolstered the appeal and legitimacy of their positions by claiming, too, that the biblical chronology had been an unquestioned tenet of Christianity, only recently challenged by modern secularism.

The conclusion of this book bridges the gap between ca. 1800 and the present day by briefly discussing its implications for the relationship of science and religion in Western societies, and the rise of Young Earth creationism in America. As the French Revolution engendered not the dreaded egalitarian and democratic new order, but the socially conservative Napoleonic Empire, milder versions of diluvialism and an ancient Earth became again politically acceptable. In the following decades, however, growing evidence against the Flood and for Darwinian natural selection were instrumental in the revival of biblical literalism on the one hand, and on the other, of the "conflict thesis" that singled out religion—in particular, Catholic Christianity—as the major historical obstacle to the intellectual and material progress of humankind. Both sides built upon the by-then commonsensical narrative of the traditional opposition of Christianity to the idea of an ancient Earth. The same assumption underpinned the growth of Young Earth creationism, as the most conservative strands have found it tempting to undermine evolution by denying or casting doubt on the millions of years it requires. Often singled out as an outstanding example of religiously motivated backwardness, a Young Earth has nonetheless received support from prominent US politicians and state legislatures. It is a fascinating idiosyncrasy of our time that the one assumption regarded by every side as a basic fact has turned out to be a historiographical myth.

I

Footprints in the Dust

THE ETERNITY OF THE WORLD IN
THE MIDDLE AGES

*For, they say, if a foot had always been from eternity in the
dust, it would always have had a footprint under it, but no
one would doubt that the footprint was made by that tread,
and that neither preceded the other, even though one was
made by the other.*

AUGUSTINE OF HIPPO, *On the City of God* (413–427 CE)

IT IS PERHAPS ironic that so few firm dates are known about the life and
works of Eusebius of Caesarea. One of the greatest fathers of the Eastern
Church, he was born around 260–265 in the then-sprawling city of Caesarea,
not far from Haifa, and passed away between 337 and 340.[1] Eusebius has
always been considered if not the initiator, then at least the scholar who
solidified the great tradition of the Christian chronology: the almost obses-
sive effort, for at least fifteen centuries, to assign a precise time to the main
events of the history of humankind and of its salvation. The original Greek
text of Eusebius's *Chronicle*, which spanned from Adam to his own times,
has been lost, but it survives in an Armenian translation discovered in 1787.[2]
The second book, consisting of chronological tables that were revolutionary
at the time, was translated into Latin by St. Jerome around 380 and became
the foundation on which countless followers built.[3] Twelve centuries later the
work of Joseph Scaliger (1540–1609), a crucial figure in the development of
modern historiography, still took Eusebius's *Chronicle* as its starting point and
aimed to amend its inaccuracies.[4]

On the Edge of Eternity. Ivano Dal Prete, Oxford University Press. © Oxford University Press 2022.
DOI: 10.1093/oso/9780190678890.003.0002

Creation Without a Beginning?

All of the three Abrahamic religions devoted a lively and sophisticated interest to history and chronography. Yet, the drive to know and understand the dates of important events, especially when they directly concerned the history of salvation, was especially ingrained in the Christian tradition.[5] At a time when Christians had recently ceased to suffer from organized persecutions, and their apologists were still jousting with pagan opponents, the study of pro-fane chronology and its harmonization with that of the scriptures could be seen as a way to offer assurance about the historicity of the events described in the Christian revelation.[6] Of the innumerable problems raised by the dating of sacred and profane events, the exact year of Christ's birth and crucifixion or that of the biblical Flood certainly stood out, but none has received more enduring attention than the time of the Creation. Not only would that event seem to be the natural anchor of any Christian universal history, but its un-equivocal determination could also put an end to the constant debate about an uncreated or eternal world among ancient pagan philosophers. Following Eusebius's tables, founded in turn on the Septuagint Greek version of the Bible, St. Jerome (ca. 345–420) reckoned that Christ began his predication in the year 5228 since the creation of the world. For the following thirteen or fourteen centuries, chronologists quarreled about these dates, adding or subtracting a few centuries but without questioning the fundamental assump-tion that the Earth was created just a few thousand years earlier. Looking at the immense literature they produced, one may easily infer that this must have been the undisputed tenet of every good Christian. However, this was not always the case.

Eusebius himself seriously doubted that it would be possible to shed light on the earliest times of the Creation, those that unfolded from the very be-ginning to Adam's exile from the terrestrial Paradise. He certainly gave little credence to the pagan "fables" of the Chaldean priest Berossus, who in the early third century BC bragged about the 400,000 years of history recorded in the archives of Babylon.[7] He likewise turned down the fabulous stories claiming that gods and heroes had initially ruled Egypt for a little less than 16,000 years before ceding their throne to the pharaohs.[8] The Bible, on the contrary, seemed to offer a firmer and verifiable history down to the time "when our human race experienced mortality and from [the time of] our first ancestor who set out on that path."[9] Still, the account of even earlier times remained beyond his reach. Eusebius showed little inclination to take literally the seven days of the Creation, to try to determine the number of years Adam spent in Eden, or how and when exactly the Almighty "established heaven and earth":

> There is no way one could determine the length of time that [Adam] dwelled in the Paradise of God, as it is called. It seems to me that the marvelous Moses inspired by the Holy Spirit alludes to a better dwelling than the world we inhabit, a kind of godly, I say, abode of blissful life. [Moses] calls Paradise the first dwelling of the human race; Adam lived indeed in Paradise, enjoying a sweet life and full of happiness, above the human condition. And it has been assuredly found that [Moses] was talking of the whole of the human race.[10]

Eusebius's chronicle begins with Adam's expulsion from Eden rather than with the creation of the universe, "even though others believed it could be done."[11] As he put it: "The early [chronology], whatever its history was, will be kept distinct as unknowable from the continuity of subsequent times."[12] The library on which Eusebius formed his culture was largely put together by Origen of Alexandria (185–253), the first systematic Christian philosopher.[13] It was then augmented by Pamphilus, who became Eusebius's teacher in Caesarea, bequeathed his library to him, and instilled in his pupil a strong admiration for Origen's thought and style. Among the fathers of the Church, no one diverged more radically from the linear reading of the history of creation so eagerly embraced by St. Jerome. Origen did so not for lack of confidence in the authority of the Scripture, which was the fundamental source of all of his writings, but because he maintained that Scripture must often be interpreted allegorically if any sense is to be made of it.[14] While inspired by God, the human author maintains agency and freedom, and he makes conscious choices as to how to deliver his message. In an argument reminiscent of the Stoic philosophy, Origen posited that Genesis describes not the original creation from nothing, but only the present iteration of a succession of worlds that are destroyed and remade one after the other:

> Not then for the first time did God begin to work when He made this visible world; but as, after its destruction, there will be another world, so also we believe that others existed before the present world came into being.[15]

Isaiah's famous vision of "new heavens, and a new earth," or the "ages which have been before us" mentioned by the Ecclesiastes, refer for Origen to the world that will follow this one of ours and those that preceded it. The creative activity of an eternal principle can hardly find limits in a finite time. Origen was enormously influential during his lifetime and the following two or three centuries, but unlike other fathers of the Church he never became a saint. A strong backlash against his doctrines developed on separate occasions after ca. 400, leading in 543 to the condemnation of a number of propositions attributed to Origen and widely deemed heretical.[16] Contrary to a diffused belief, however, not even his enemies ever said that his succession of worlds was meant to be without beginning and without end. Origen himself disavowed the Stoic notion that every world would be exactly the same as the one that preceded it, forever repeating an identical chain of things, people, and events. Such a view left no room for human freedom, nor for the uniqueness of Christ's incarnation and death.[17]

Origen's particular reading of Genesis, and some of its possible implications for subjects such as the human soul (which he considered existing since the beginning, rather than newly created together with each human being) or the final judgment (apparently, there should be one at the end of each world, but even a final restoration), were destined to raise eyebrows. But the principle that the biblical description of the Creation cannot be taken at face value was shared even by authors who were never touched by the shadow of heresy. Augustine of Hippo (354–430), known as Saint Augustine by Christians, firmly established for the centuries to come the possibility or even plausibility of such an approach. Origen's succession of worlds was his answer to the question of what God was doing before the Creation. The underlying problem is how an eternal Being not subject to change comes to perform a certain action—in this case, the creation of the world—in time, which, according to Aristotle's classic definition, is measured by change. Augustine's reply was that this is an ill-devised question, because time itself was created together with matter and space. From this point of view, there was no such thing as a time before the Creation, or even a time when the universe did not exist. The world was not made in time, but with time. The moments that we perceive as a distinct succession are all present together in God's immutable eternity, which is more akin to a dimensionless point than to a line stretching without end in both directions.

Since there was no time when there was no created world, the Creation was an instantaneous act. Hence, the words of Genesis should not be taken "in a childish way, as if God exerted himself by working. For he spoke, not with an audible and temporal word, but with an intellectual and eternal word, and the things were done."[18] The story of the six days of the Creation is thus a "dramatic representation" of what took place at once.[19] Of course, not all things were made as we see them now. In the beginning God established the principles of all created things, but like seeds that are sown in the ground then will develop in due time according to the laws of nature.[20] For we do live in time; we can measure its passing, and we perceive the act of the Creation as unfolding in terms of successive events. Genesis can thus be considered a historical, though not a scientific, account of the sequence in which the creatures appeared on the Earth.[21] The human race itself came into being in time, and quite recently as well. As a rebuttal to those who believe that there was never a first man, Augustine maintained on scriptural grounds that "six thousand years have not yet passed since the institution of humankind." Just like Eusebius, however, Augustine refused to equate its age with that of the Earth or of the universe, preferring to profess his ignorance in this regard. The chronology of the patriarchs had the appearance of an accurate historical narration, the description of the six days of the Creation much less so. If they cannot be taken as actual human days, than there is no way to determine unequivocally "what ages passed before the human race was instituted."[22]

Augustine discussed the doctrine of the Creation in three commentaries on Genesis, as well as in *On the City of God*. Understandably, his position did not remain the same throughout and is sometimes unclear. In the later commentaries he tended to prefer a "literal" interpretation,[23] though this should not be confused with the plainest possible level of reading. For example, Genesis 1:3–5 talks of "light," "evening," and "morning" even before the creation of light, which comes later; Augustine reasoned that these terms are not describing the creation of physical light, but rather of the superior incorporeal world of the angels.[24] This is not a figurative reading but a literal one, just with a wider scope than that implied by a "merely corporeal signification of the words."[25]

The idea that matter, time, and space were created from nothing had been foreign to Greek philosophy.[26] If Origen could partially draw on the Stoic doctrine of successive worlds, Christian thinkers unanimously rejected with horror the infinite universe of Epicurus (341–270 BCE) and Lucretius (c. 99–55 BCE), where atoms eternally moved and combined randomly in the void. Even though both the Earth and humankind were perishable, they formed through purely material causes without any need for spiritual beings and a creator God. For a long time, "atomism" remained a synonym for "atheism" in the Christian West. Aristotle's universe instead was spatially limited; it admitted metaphysical entities and a "first cause" that one could even call God. The material universe, though, was equally eternal and uncreated, just like the Earth and the human species. The Supreme Being neither made nor cared about them. Their existence and characteristics derived from necessary laws of nature, not from the free choice of a loving Creator. Unlike Aristotle, Plato—whose views deeply permeated early Christianity—described the crafting of the universe by a rational "demiurge" that turned the primordial chaos into an ordered "cosmos"; but even Plato's divine artisan acted upon matter that pre-existed his creative work. There has been little consensus—in antiquity or since—on whether Plato assigned a beginning to time.[27] Nonetheless, his followers found a subtle solution that made this problem less critical and, eventually, became attractive even to Christian (as well as Muslim) thinkers. In a passage of *On the City of God*, Augustine related how Platonist philosophers came to think that the world may be at the same time created from nothing, yet without a temporal beginning. Since the sensible universe is the work of God, it depends on its maker for its existence. The act of creation, however, should be understood as an everlasting relationship, not as an event that took place in time:

> For, they say, if a foot had always been from eternity in the dust, it would always have had a footprint under it, but no one would doubt that the footprint was made by that tread, and that neither preceded the other, even though one was made by the other.[28]

There was never a time when the footprint was not present; yet, it is only there because of the foot. The opposite is not true, as the foot (God) does not need the footprint (the Creation) in order to exist. In spite of the obvious divergences, Christian philosophers were the pupils and heirs to their pagan predecessors. Whenever it seemed remotely possible, by and large they sought with determination to reconcile their views (in particular, Plato's and Aristotle's) with the teachings of the new religion. In fact, for a long time many of them appeared to be surprisingly at ease not only with a universe whose age could not be determined, but even with one that did not have a beginning. After the collapse of the ancient world between the fifth and the sixth centuries, the most relevant authors who remained continuously known in Western Europe were Plato, Augustine, and the late-Roman philosopher Severinus Boethius (ca. 480–524). What was transmitted of the first was actually limited to part of Plato's *Timaeus*, in the Latin version translated and extensively commented by Calcidius in the fourth century CE.[29] Because of the obscurity and ambiguity of the original text, one could hardly overestimate the significance of Calcidius's comment for the reception of Plato's work. The commentator emphasized both that "the origin and beginning of the works of God are incomprehensible" and that, as in the example of the eternal footprint, the "origin" of the world is "causative, not temporal."

> Nothing is known for sure; there is no indication of the time at which it began to be, only perhaps the cause—and even this is scarcely understood—why which of these things exist and for what cause, since it is certain that nothing was made by God without a cause.[30]

Boethius's legacy stands out even more prominently. Calcidius, about whom not much is known, could have been a Christian, but Boethius definitely was.[31] As the last Latin author with a firm command of the philosophical literature of antiquity, he handed down to the Middle Ages well-developed and sophisticated views on time and eternity, discussed and elaborated in a way that he considered compatible with the Christian revelation.[32] It is then significant that in *The Consolation of Philosophy*—a crucial text for the medieval civilization—Boethius effectively made a case for a world without a beginning. The philosopher distinguished between two kinds of eternity: that of God, which is "possession all at once of illimitable life," and that of the Creation, which he rather called "perpetuity" and defined as indefinite duration through time.[33]

> Whatever is subject to the condition of time . . . is nevertheless not such a thing as the eternal is rightly considered to be. For it does not grasp and include the whole of infinite life all at once; it does not yet possess the future which is yet to come. . . . For it is one thing to lead an interminable life, which Plato attributed to the world, and another thing to embrace the whole of interminable life as if equally present, which is clearly the property of the divine mind.[34]

In other words, one should not worry that a perpetual world would have the same status as God. As historian of medieval philosophy Richard Dales points out, Boethius does not show any sign of having a problem with a beginningless creation.[35] This kind of literature and its subject, however, did not lend themselves to unequivocal readings. Moreover, in the following centuries certain key terms underwent semantic shifts that contributed to generate confusion and misconceptions; among the most important, "perpetual" came to signify things that have a beginning but not an end. The result was that both Boethius and Plato were often considered supporters of the idea that time and the world actually had a beginning. According to Dales's survey of medieval discussions on the eternity of the world, half of the authors who wrote on the subject misunderstood Boethius's original intent.[36] On the other hand, half of them did not: his distinction between "eternal" and "perpetual" remained thus crucial to the possibility of a created, yet perennial, world.

While the option of an indefinitely old Creation continued to be explored, the idea of a recent world firmly took hold of Christian Europe's historical literature and of its very sense of history. Isidore (ca. 560–636), for three decades bishop of Seville and the author of an immensely popular digest of Roman sources known as the *Etymologies*, closely followed St. Jerome in reckoning 5,196 years from the Creation to Christ. An immeasurably old universe was certainly frowned upon for a number of reasons. One was that it would cast doubt on the belief, so widespread among early Christians, that its final days must be close. True, Isidore admonished that "the time remaining for the world cannot be ascertained by human investigation," since the Father had reserved this knowledge for himself;[37] he would probably be surprised, though, to find the Earth still in existence fifteen centuries later. Even disagreement on the exact time of the Creation could be perceived as dangerous, as it would weaken confidence in the possibility of an unequivocal reading of the Scripture. This would encourage in turn a merely allegorical interpretation of the first chapters of Genesis and, with that, the possibility of a perpetual universe. Indeed, Bishop Philastrius of Brescia (ca. 330–397) included the "uncertainty on the years of the world" in his notorious list of heresies:

> There is another heresy which says that the number of years [elapsed] since the origin of the world is uncertain, and that men ignore the course of time: while instead from Adam to the flood we have ten generations, and 2242 years.[38]

Considering the wide disagreement on the age of the world not only between sacred and profane sources but even between the Hebrew Bible and the Greek version of the Seventy, Philastrius's move seems rash, to say the least. There is little evidence that it was taken seriously in the following centuries.[39] In 708, a few monks went before the local bishop to accuse the English scholar Bede of heresy in his absence. Bede's sin consisted of the adjustment of St. Jerome's computation of the interval between the Creation and Christ's advent, which following the Hebrew Bible he reduced to only 3,952 years; it was then said that he denied that Christ incarnated in the Sixth Age of the world. Upon learning of the episode, Bede was incensed at the ignorance of his opponents and at the fact that the bishop had not defended him openly. He immediately wrote a letter to a correspondent, asking that it be read in front of the bishop, and that was the end of the affair. Bede did not suffer any consequence from the episode; allegedly, the monks were drunk when they spoke out against him.[40] Even in a sober state, they might have been equally horrified—but equally powerless—at John Scotus Eriugena's (ca. 815–877) leaning toward a universe that was both eternal and made. Under the influence of Augustine, Boethius, and Calcidius, as well as some Greek fathers he could read in the original, Eriugena argued that while the universe in its present state had a beginning, in a certain sense it is also eternal because it had always existed in God's eternal mind:

> And so, after having considered these arguments, who but a very slow or very contentious person would not concede that all the things which are from God are at once eternal and made?"[41]

"*The Incoherence of the Philosophers*"

The possibility of a world of eternal or indefinite age existing alongside a more common tradition that placed the beginning in a recent past was hardly an exclusive feature of Christian Europe. Medieval Islam shared with Christianity many of the same sources on the creation and the first ages of the world, since the first five books of the Bible have been traditionally regarded by Muslim theologians as divinely inspired. Islamic scholars retained a lasting attention to chronology, even though they did not develop the sophisticated astronomical techniques for the dating of past events later adopted in Europe. Abbasid astrologer Abū Maʿshar (787–886; Albumasar in Latin translations) managed nonetheless to provide the "astonishingly precise" figure of 2,236 years, 1 month, 23 days, and 4 hours from the creation of the world to the beginning of Noah's Flood.[42] The widespread attention toward chronology did not prevent many of the most influential Islamic thinkers from holding a very different opinion on the actual age of the universe. Compared to their colleagues in Christian Europe, they had early access to a wide range of philosophical and scientific works from classic and late antiquity, translated into Arabic mostly during the ninth century in Damascus and Baghdad. They included in particular the whole corpus of Aristotle's writings and later commentators, which became the main inspirer of the *falsafa* ("philosophy") tradition of Islamic learning; the eternity of the world was one of its main tenets. The revision and strengthening of arguments for the eternity of the world was indeed a major project for many medieval philosophers writing in Arabic.[43]

There is no question that the most authoritative representative of the philosophical tradition, the great Persian physician and philosopher Ibn Sina (Avicenna, 980–1037), believed the universe to be without a beginning. This was not a simple possibility for him, but the reality of the Creation. Time, to him, was akin to a line that stretches back without end. He discussed this thesis at length in his *Physics* and *Metaphysics*, and he employed in its defense a number of proofs—though they are mainly derived from those already given by Greek philosophers.[44] His considerations on the formation of mountains and layered rocks are based on the assumption that the Earth is immeasurably old. This does not mean that the universe was not made from nothing, but even for Avicenna the Creation consisted in a relation of dependence on the Creator, not in an event that took place in time.[45] Ibn Rushd (Averroes, 1126–1198), born in southern Spain at the other end of the Arabic-speaking area, shared a fundamentally similar approach to the issue of the age of the world, even though he preferred to liken time to a circumference where no point can be said to be the "beginning" of the circle.[46] The position of the "philosophers" was fiercely contested by more conservative authors closer to the other great stream of Islamic thought, the *kalam* (speculative theology), whose reaction became particularly effective in the twelfth century on the wake of Al-Ghazali's (ca. 1056–1111) work. The first, longest, and most substantial discussion in Al-Ghazali's *Incoherence of the Philosophers* addresses specifically the philosophers' view that the world is eternal.[47] Nonetheless, Averroes—whose ideas on the nature of the human soul were historically much more controversial than those on the age of the world—rejected allegations of irreligiosity. He argued that the revelation has an "apparent meaning" that is taught to the masses, but those who can pursue a rational investigation of God and nature should be allowed to take this more direct and superior path.[48]

Muslim thinkers in the *falsafa* tradition were hugely influential in medieval Christianity. In the same decades when Averroes wrote his replies to Al-Ghazali, Western Europe's economic, military, and intellectual recovery was in full swing. Italian trading powers like Venice, Genoa, and Pisa had already wrestled from Muslim navies control over the Mediterranean. Taking advantage of the Crusades, which were largely enabled by their naval supremacy, they began to establish commercial outposts on its eastern shores from Constantinople to Egypt. In Muslim Spain, the military pressure of the Christian states on the northern frontier became unbearable; less than two decades after Averroes's death, Muslim armies suffered crushing defeats followed by the fall of Cordoba in 1236. As Christian Europe expanded its presence toward the southern and eastern borders of the Mediterranean, learning flowed mostly in the opposite direction. The European appetite for natural knowledge, bolstered by the foundation of the first universities in the course of the twelfth and early thirteenth centuries, was nourished but not appeased by a wave of translations of Arabic philosophic and scientific works into Latin. Most of those writings came from areas where substantial Christian and Islamic (as well as Jewish) communities had coexisted or interacted for a long a time, like Spain,[49] or to a lesser extent Sicily, long disputed between Muslim emirates, Greek-orthodox Byzantium, and Western Catholics. After the Christian conquest of Toledo (1085), scholars from Italy and Great Britain flocked to its rich libraries, learned Arabic and sometimes Greek, and flooded the rest of the continent with the treasures of the classical tradition and of its Islamic commentators.[50] By 1230, most of the Aristotelian corpus had been translated several times into Latin from both Greek and Arabic and was already causing a shift to Aristotle as the main authority in natural philosophy. The change, though, did not happen overnight.

While Aristotle's work was already making its way into late twelfth-century scientific texts, they still relied for the most part on the usual, well-established sources. Cosmological views remained largely based on Plato's *Timaeus*, which was not a systematic treatise but a myth that needed decoding in order to unveil the underlying truths. Unlike the *Timaeus*, the Bible was of course divinely inspired; yet, its account of the origin and age of the world seemed every bit as nebulous and clearly required interpretation. Medieval readers were deeply accustomed to the exposition of truths as allegorical tales, because allegory itself had become a fundamental category of medieval culture; Bernard Silvestris's *Cosmographia*, a treatise on the formation of the cosmos written between ca. 1143–1148, was likewise composed as an allegorical narration.[51] A number of authors were convinced that the Platonic myth was compatible with the Bible, and the reconciliation of the two authorities emerged as a major cultural project. In his prefatory letter to Thierry of Chartre's *Treatise of the Work of the Six Days*, Clarembald of Arras (ca. 1110–1187) hoped that his work "will be found to have above all reconciled many of the opinions of the philosophers with Christian truth, so that Holy Scripture is strengthened and fortified even by its enemies."[52] The form that this reconciliation should take varied greatly from one author to another, but no twelfth-century author describing an ancient or beginningless universe felt himself to be teaching anything contrary to the faith.[53]

This was certainly the case for the mentioned treatise by Thierry of Chartres (composed ca. 1151–1155). While primarily a theologian, the French scholar did not hesitate to interpret the biblical narrative in light of his understanding of the physical evolution of the universe.[54] Following Augustine of Hippo, he explained that God made simultaneously the four elements "from nothing" in the beginning; from that moment on, however, the unfolding of the creation described in the six days of Genesis proceeded according to exclusively natural laws. Since light was created before the Sun, and the Sun itself only appeared on the fourth day, one is left of course with the problem of how to define a "day." Thierry of Chartres identified the "days" of the Creation with a corresponding number of rotations of the "highest and lightest" celestial sphere; whether this is the same sphere that, in medieval cosmology, imparts to the whole of the heavens its daily motion of 24 hours is not clear.[55] As a matter of fact, the author further differentiated between "natural days" going from sunrise to sunrise, and the "illumination of the air by the heavens" as opposed to "the darkness that is called night," warning that the Holy Scripture uses the word in both ways. The wording of some passages suggests that the first "days," at least, should not be intended as conventional "from sunrise to sunrise" days:

> Therefore, in the first rotation the fire illuminated the air, and the interval of the rotation was the first day. And the second rotation of the same fire, by means of the air, warmed the water and *placed the firmament* between water and water. And the interval of this rotation was called the *second day*.[56]

Authors less involved in theology or less connected with the Church may not have been particularly interested in explaining Scripture in physical terms; in fact, their writings are often rather ambiguous even when it comes to whether the world began. William of Conches (ca. 1090–1155), whose physical and astronomical theories deeply influenced Thierry of Chartres's treatise,[57] accepted Augustine's teaching that the universe was made together with time rather than *in* time; yet he seems to have seriously considered that the material universe had a beginning but not an end.[58] The anonymous treatise *On the Elements*, composed around 1160, is often mentioned as another example of the independence of medieval science from a literal reading of the Scripture. The author heavily draws on *Timaeus* and possibly Thierry de Chartres,[59] but he also presents the universe as made of atoms (defined as the smallest particle of an element) moving in the void according to their nature, with fire being the fastest, and the earth almost immobile at the center of the world. God created the elements in form of atoms, and it is their motion that gives origin to the material universe as we see it. In early modern Europe, atomism was associated with Lucretius's materialism and atheism, but Lucretius's poem *On the Nature of Things*, composed in the first century BC, remained almost entirely unknown before its rediscovery in 1417.[60] On the contrary, a well-known Christian authority like Isidore of Seville seemed fine with atomism, only warning that "certain philosophers of the heathens have thought that trees are produced, and herbs and all fruits, and fire and water, and all things are made out of them" (that is, without any divine intervention).[61] Against Aristotle's idea of matter as a continuum, most followers of the Islamic *kalam* tradition adopted an atomistic cosmology that seems to have points of contacts with that found in *On the Elements*.[62] It may not be a coincidence that the treatise was likely composed in southern Italy, possibly in Sicily.[63] As R. Dales writes, it would even be hard to tell if the author was a Christian, a Muslim, or a Jew, if not for a reference to a Psalm of the Bible.[64] His declaration that "the origin of the elements is motion, which motion is from eternity,"[65] and further references in the treatise seem to indicate that he believed matter to be eternal, or at least that he did not feel any compelling need to define his position unequivocally:

> The origin of elements is motion, which motion is from eternity; indeed, since motion is a change of something into something else, which cannot happen without an alteration, its principle either goes back forever, or there is something stable and eternal from which motion proceeds.[66]

Like Thierry of Chartres, he probably did not perceive any fundamental incompatibility between the creation story of Genesis and that of the *Timaeus*.[67] Bernard Silvestris similarly conceived his work as a contribution to the harmonization of Greek philosophy and Judeo-Christian theology.[68] Even though his *Cosmographia* was an allegorical tale, some passages are transparent enough as to leave little doubt that he considered the universe to be created, yet without beginning and without an end. His only distinction between time and eternity is that time is characterized by motion and change, while the mode of existence of eternity is immutability. The material world continuously goes through never-ending cycles, because its cause is eternal too: "Beginning from eternity, time returns to the bosom of eternity, wearied by the long, long route. . . . And so time continues forever to trace and retrace these paths."[69]

Silvestris' was one the most significant and successful didactic poets of the Middle Ages, and his *Cosmographia* was one of the most widely read books of its time,[70] known across Europe to authors as diverse as Vincent de Beauvais, Dante Alighieri, and Geoffrey Chaucer; Thierry of Chartres was the dedicatee.[71] His point of view on the age of the universe may not have been shared by every reader, but it was certainly familiar to the public of those who, in the late Middle Ages, could read and appreciate a literary work of that kind in Latin. In those decades an ancient or even beginningless universe was broadly perceived as compatible with the Christian revelation. These writings did not raise any opposition in their time, even though there was no dearth of controversy on many other topics.[72] Yet, this is only surprising if one assumes that the medieval perception of time and of religious orthodoxy was necessarily locked in the mindset of a recent Creation. In the case of Silvestris, there is no indication that he was ever prosecuted or his work suppressed. His treatise survives in more than fifty manuscripts, many amply glossed; none of those examined by scholars appears to have been modified in order to delete passages of dubious theological orthodoxy.[73] A marginal note on one of the early manuscripts of the *Cosmographia* even suggests that the newly completed book was recited before Pope Eugen III, who was at the time traveling in France, and "it won his benevolent approval."[74]

The widespread acceptance of an ancient world coexisted, apparently without too many problems, with a chronological tradition of undiminished vigor. Historical writing increased enormously during the twelfth century, as did the number of universal histories that followed in the footsteps of Jerome and Eusebius.[75] At the same time, Christian chronologists began to put to good use the new astronomical sources (and instruments, like the astrolabe[76]) made available by translations from the Arabic and became better acquainted with Jewish calendric practices.[77] Chronology fully participated in the intellectual revolution brought about by the reception of Arabic science;[78] the result was a more sophisticated, more rigorous, and ultimately more authoritative discipline. The relation between a Bible-based chronology of the Creation and a philosophy of nature that tended to set no definite limits to the age of the universe (if not of the Earth itself), presented in the Middle Ages a complexity and a variety of layers that is hard to grasp for the modern mentality. In our epoch, the time scales of academic geology and of Young Earth creationism cohabit only in the sense that—generally speaking—they exist side by side in mutually exclusive spheres, but they are regarded as incompatible worldviews. In the twelfth century, however, the choice did not have to follow a binary logic, and it may not even be necessary to choose, as in the case of the anonymous author of *On the Elements*.

Medieval Christians could be comfortable with an ancient or even beginningless world because this did not imply in any way a rejection of their faith and of the idea of a divine Creation. On the other hand, there is no doubt that chronologists who spent their lives trying to determine the year, month, date, and, if possible, even the hour of this creation held a very different view of how the first chapters of Genesis could or should be read. Twelfth-century Christianity was flexible enough to accommodate these two extremes, in part because it had not hardened yet into the sets of dogmatic beliefs which came to constitute what we know as Christian religion(s).[79] Furthermore, there were many ways in which biblical chronology and natural philosophy could be perceived more as complementary than as mutually exclusive. Chronology accomplished a variety of crucial tasks in a Christian society: they ranged from establishing the historical accuracy of the events of the Bible to helping with the correct determination of important liturgical events. Formidable intellectual energies were devoted for centuries to the best method to determine the date of Easter, a preoccupation not necessarily limited to a handful specialists. People well read in natural philosophy may have doubted that it made sense to try to compute the date of the Creation, but their life and their perception of history would still be shaped by conventional chronology. A modern atheist, after all, will still count the years starting from the birth of Christ even though he may not even believe that Christ actually existed.

A history of the Earth based on Greek sources could easily accommodate a large inundation of the globe that took place only a few thousand years before, similar to that described in Genesis. There was no compelling reason why those relatively recent events, and their chronology deduced from the list of the patriarchs, could not be integrated into an ancient or timeless creation. The relation between natural philosophy and chronology was rarely confrontational, because they had different aims and methods; while tensions could occasionally surface, for the most part they remained two different fields that just ignored each other. Until the late sixteenth century, it is difficult to find authors who tried to merge natural philosophy and chronology. The only unavoidable overlap concerned the six days of the Creation, which even chronologists like Eusebius doubted could be read literally. The Platonic philosophy that was so popular among early Christian thinkers added a further layer, by assuming a distinction between pre-existing matter and the creation

of an ordered cosmos. Last but not least, nothing prevented those who
believed that both time and matter had a beginning to place this event into
a remote, instead of a recent, past. This remained true even after the Fourth
Lateran Council, summoned by Pope Innocent III in 1215, radically changed
the terms of the debate by proclaiming that the world had indeed a beginning.

Eternity Condemned

The decision to address an issue that raised so little controversy in the previous century was arguably precipitated by the rapid spread of Aristotle's natural philosophy in European universities.[80] There are various places in the vast
Aristotelian corpus where the philosopher seems to mention the eternity of
the world simply as a topic for discussion, but the vast majority of his writings
presents it as a definite teaching.[81] The notion itself was scarcely new of
course, yet Aristotle's version of eternalism featured numerous elements that
made it harder to accept. While Plato writes of a God-artisan endowed with
freedom and will, who crafts the universe from shapeless matter, Aristotle's
world and the Earth itself have always been—and always will be—more or
less as we see them today. No creative action is thus needed, nor was there
a first man, since animal species are also eternal. Plato's cosmogonic myth is
so vague that it could easily lend itself to a variety of Christianized readings;
on the other hand, Aristotle proceeds through cogent logical deductions,
necessary demonstrations, and inescapable conclusions. In face of the new
challenges, the teaching of Aristotle's books on natural philosophy, as well as
of their commentaries, was prohibited for the first time in Paris in 1210.[82] Five
years later, the confession of faith that opened the first canon of the Vatican
Council IV solemnly declared that "God by his almighty power at the beginning of time created from nothing both spiritual and corporeal creatures,
that is to say angelic and earthly, and then created human beings composed as
it were of both spirit and body in common."[83] By defining as a truth of faith
that time and matter had a beginning, the council's deliberation made it impossible thereafter to defend the idea of a universe *both* created and eternal.

Yet the canon said absolutely nothing on how and when the Creation took
place (apart from the assertion that "the human" was its last product), thus
leaving a door open to the possibility of a world of very old or undetermined
age. This was most likely a deliberate choice. It seems implausible that the conciliar fathers were not aware of the possibility of non-literal interpretations
of the six days of the Creation. It was considered essential to proclaim that
there was a temporal beginning; the "when" was not addressed. Historians

of medieval debates on the eternity of the world have given little attention to the first canon of the Lateran Council IV. This may be due to the fact that the decrees of the council seem to have done very little to stop conjectures and controversies, which simply took a different form. Philosophers were still free to conjure up every possible rational proof against the eternity of the world or to quarrel about whether such a demonstration is possible at all, thus making sure that the discussion would continue.[84] Moreover, if the decree was aimed in particular at Aristotle's natural philosophy then it was remarkably unsuccessful, as its penetration continued unhindered. At the University of Paris, the 1210 prohibition to teach Aristotle's natural books did not last long. A further injunction was issued in 1231, apparently with similar results. In 1254, professors in the Faculty of Arts were admonished that they should thoroughly teach a number of prescribed texts without taking any shortcuts. The list included Aristotle's *Physics, Metaphysics, On the Animals, On the Heavens, On the Soul, On Generation and Corruption,* and the first and fourth books of the *Meteorology.*[85] Even Averroes, who passed for a radical Aristotelian among Muslims and Christians alike, was known and quoted by the 1230s.[86] Aristotle had come to stay; somehow, accommodation had to be reached.[87]

Unfortunately, the eternity of the world is not a simple addition to the edifice of the Aristotelian universe but a massive cornerstone, whose removal would cause the collapse of the whole structure. Aristotle's philosophy can explain how the heavens and the Earth function, but not how they came to exist in their current form. Indeed, a popular reading of Aristotle's writings was that he had demonstrated not the eternity of the universe, but only that its generation cannot be explained by natural causes. Hence, at a certain point a supernatural intervention is needed. If so, Aristotle had taught nothing contrary to the faith. From Oxford University, Robert Grosseteste (1175–1253) protested energetically against those who wanted "to make Aristotle a Catholic . . . and by making Aristotle a Catholic, they make heretics of themselves."[88] He decried those who kept interrogating Genesis on the subject of the Creation, and then wondered "why is not the world older than Scripture says, and did not begin earlier."[89] Yet the great Dominican scholar Albertus Magnus, together with most other commentators, did not find anything dangerous in Aristotle's doctrines.[90] That the world had a beginning in time was to Albertus probable according to reason (though maybe not demonstrable), and anyway, "it must be held by faith":

But it is impossible that it began by motion and generation or that it might end through a motion toward another form of corruption into another matter. And this is all that Aristotle's arguments prove. Therefore, they conclude nothing contrary to faith.[91]

Another strategy for the incorporation of Aristotle's works in university teaching relied on the increasingly sharp distinction between theology and natural philosophy as academic fields. Natural philosophy was taught in the faculties of arts, where it served as an introduction to the more prestigious studies of theology, medicine, or law. Following the argument that Aristotle's works only deal with natural causes, professors of arts in Paris claimed that natural philosophy should be free to investigate the natural world using exclusively observation and reason, even though their conclusions may sometimes be different from what the superior light of Revelation teaches us. In the case of the age of the world, this means that philosophers acknowledged that the universe was created in time, as the Church proclaimed; however, they reserved the right to reason as if it were eternal—knowing full well that this is not actually true. In the philosophical jargon of the time, it was *absolutely* true that the universe had a beginning; its eternity was only a *relative* truth—that is, the conclusion reason would lead us to, in the absence of a more authoritative source. Natural philosophers should then be able to describe the universe as eternal, "philosophically speaking."[92] Boethius of Dacia, a leading figure in the faculty of arts in the 1270s, observed that "the creation of the world, or its production in being, is taught in no part of natural science, because their production is not natural and does not pertain to the natural."[93] Are the creation of the world in time, the generation of the first man, or the resurrection of dead bodies possible according to the (Aristotelian) laws of nature? Of course not. Therefore, they have no place in natural philosophy, even though they are believed by faith.[94]

The arguments of the masters of arts could seem dangerously subtle and prone to misunderstandings. Indeed, historians of medieval philosophy long held that those professors taught the existence of a "double truth," namely, that two contradictory statements (the universe is eternal, and it is not) can both be true in their own way. Of course this does not make any sense, and recent scholarship has entirely debunked this myth.[95] Even in the thirteenth century, though, they could create mistrust in minds not thoroughly trained in the new ways of the Aristotelian logic, whose introduction was still recent. This was the case with many conservative professors in the faculty of theology, who suspected that their colleagues were putting up a smokescreen to protect the teaching of patently heretical doctrines. Academic rivalries certainly entered the fray, as the soaring prestige of natural philosophy and of its professors threatened established hierarchies. In the late 1260s the clash between the two faculties turned white-hot, with professors of arts accused of holding heretical beliefs and of defending the theses of Averroes. In the attempt to quell the confrontation, the Dominican order sent to Paris their most prestigious theologian, Thomas Aquinas, who had studied and taught there a few decades earlier. The treatise *On the Eternity of the World*, which he composed in the French capital between 1270 and 1272, represented the culmination of his long-lasting concern about the issue of eternalism. While primarily a theological treatise on the power of God and the limits of human reason, Aquinas's work aimed to save the independence of natural philosophy in its own sphere.[96] In fact, his position proved remarkably close to that of "artists" like Boetius of Dacia. Building on the tradition that went back to the analogy of the footprint in the dust, he argued that a universe both created and eternal entailed no logical contradiction.[97] Still, certain knowledge of a beginning came from the Revelation and should not be part of natural philosophy.

In spite of these conciliatory efforts, conservative professors in the faculty of theology enjoyed the support of the archbishop of Paris, Etienne Tempier, and eventually gained the upper hand. In 1270 Tempier issued yet another condemnation of a number of Aristotelian theses—including the eternity of the world and that there was never a first man—forbidding any professor from holding them. Professors could have arguably replied that they only taught them "speaking as natural philosophers," not as "absolute" truths, and nothing changed. In early 1277, however, instructions from Pope John XXI to hold an investigation were exploited to conduct a much more sweeping repression. A commission of theologians summoned by Tempier quickly redacted a list of 219 theses that should not be taught, including any related to eternalism, and imposed heavy penalties on the disobedient. This time, they meant it. The condemnation effectively terminated the careers of some of the most visible professors in the faculty of arts, whose works disappeared, or at least were no longer quoted explicitly, not just in France but all across Europe.[98] Mere days after Tempier's condemnation, the archbishop of Canterbury convened a meeting of masters of every faculty at Oxford and issued his own censure, the first in the history of that university.[99] Two years later, the general chapter of the Franciscan friars in Assisi forbade its members to support the theses prohibited "by the bishop and the masters of Paris."[100] Traces of the condemnation can perhaps be found even in vernacular scientific treatises produced in those years in the area.

More than a century ago, Pierre Duhem famously hailed the 1277 condemnation as the action that freed medieval science from Aristotelian dogmatism and encouraged natural philosophers to examine possibilities which Aristotelian physics ruled out (such as the existence of vacuum in nature and of other worlds, or the motion of the Earth). Modern scholars, on the contrary, have downplayed its significance. Most of them regard it as an unwarranted measure, hastily put together with the inclusion of a number of statements obviously not heretical. Officially aimed at eradicating the doctrine of the "double truth," it may have been prompted instead by a mix of ignorance, academic feuds, or even outright dishonesty. As Dales summarized it:

> The alleged doctrine of the double truth was a fiction; the position that
> the world was actually eternal was taught by no one; and the assertion
> that its non-eternity is not demonstrable is certainly not heretical and
> was at least forty years old in the Parisian milieu by this time—hardly
> a dangerous novelty.[101]

The quarrel, Dales concludes, should not be considered a confrontation between Aristotelianism and Christianity.[102] In fact, Tempier's provisions quickly lost their strength. Displeasure toward the prohibitions was widely expressed in the 1290s, citing their negative impact on students and professors alike.[103] Thomas Aquinas's beatification in 1323 was a major blow to the residual possibilities of a strict enforcement. Less than two years later his doctrines, many of which were embarrassingly close to those of the masters of arts of the 1270s, were declared to be entirely orthodox.[104] Whether his measures were justified or not, all Tempier intended to do was to reinstate the principle that the world had a beginning. The time and modalities of the Creation were not a concern and remained outside the scope of his actions, exactly as in the 1215 decrees. Many theologians certainly thought that the alternative to a beginningless world should be a literal interpretation of Genesis, because a Creation of undetermined age would carry the risk of falling again into eternalism; to make of their opinion a general and generally enforced rule would be, however, an arbitrary extrapolation. The great controversies of the previous century left natural philosophers free to investigate the universe according to reason and to the aims of their discipline, provided that they acknowledged the "absolute" truth of the fundamental tenets of the Christian faith. Since 1215, the beginning of the world was one of them; a recent Creation was not. Neither should one be misled by the terminology used by medieval scholars, which is sometimes ambiguous: the adjective *novus*, for example, generally means "new" or "recent" but in this context it almost always stood for "created." On the contrary *antiquus*, "ancient," was often used in the sense of "eternal" and "uncreated."[105] Theories on the nature and mutations of the Earth had no need to set a specific limit to its age or, rather, to the length of its cyclical revolutions. Some may not have liked it, but it was certainly permissible, and it became indeed the norm. Until well into the 1500s natural philosophers just assumed that, for every practical and scientific purpose, the Earth could be considered eternal. "Philosophically speaking," of course.

2

The Medieval Earth

*I had the impression that William was not at all interested
in the truth. . . . On the contrary, he amused himself by
imagining how many possibilities were possible.*

UMBERTO ECO, *The Name of the Rose* (1980)

MEDIEVAL PHILOSOPHERS AND theologians were both capable and
willing to confront the possibility of an ancient world, even one without a
temporal beginning. The canons of the Fourth Lateran Council affirmed
in 1215 that a creation from nothing, rather than from pre-existing matter,
took place in time (or together with time); yet the conciliar fathers declined
to address the problem of how and when it happened, arguably because it
was irrelevant to the Christian faith. In the following centuries many continued to hold that, conciliar decrees notwithstanding, in the absence of a
firm date for the beginning, a door remained open to the ever-present danger
of eternalism. Their fears were not unfounded, but efforts to establish a chronology of the Creation—of which there was no shortage—failed to translate
into Church doctrine. The option of an ancient universe never trespassed the
precincts of orthodoxy, as long as a creation did happen. Still, there was no
requirement for natural philosophers to acknowledge a beginning. In fact, in
scholarly works it would have been inappropriate to even discuss the Creation
or other theological subjects when dealing with natural philosophy.

The controversy surrounding eternalism defined the boundaries within
which Earth history could move. The varied and sophisticated medieval
debates involving the nature and revolutions of the terraqueous globe remain
little known and insufficiently contextualized, despite the fact that a considerable amount of relevant sources and literature have been available for more
than a century. The extent to which many medieval natural philosophers
normalized the notion of an immensely old Earth, in particular, has rarely
been appreciated. This chapter and the following one seek to address the sort

On the Edge of Eternity. Ivano Dal Prete, Oxford University Press. © Oxford University Press 2022.
DOI: 10.1093/oso/9780190678890.003.0003

of cognitive blindness that befell medieval Earth history, obstructing the perception of the numerous and crucial continuities that connected it to later ages. Together, these chapters make it clear that the narrative of an Enlightenment discovery of the geological time, against the "religious prejudices" inherited from the Dark Ages, is untenable. In fact, many eighteenth-century theories of the Earth built on concepts that had been discussed for centuries in medieval and Renaissance Europe.

Deluges of Water and Fire

From antiquity to the late seventeenth century, the study of what is now called geology, mineralogy, or Earth history did not enjoy any disciplinary autonomy. Subjects related to the vicissitudes and productions of the Earth were instead part of the much broader field of pre-modern meteorology, which included of course atmospheric weather but also earthquakes, volcanic eruptions, mineralogy, and geological and climatic revolutions at both local and global levels. In practice, for much of the time period covered by this book, the field was defined by the range of topics discussed in Aristotle's treatise on the subject—namely, his immensely influential *Meteorology*.[1] There is a good reason why Aristotle gathered such disparate phenomena under the same roof. According to the Greek philosopher, a sharp distinction existed between the heavens and the earthly world. Composed of an almost divine "fifth element," the heavens were characterized by regular, cyclical, and predictable motions. On the other hand, the regions closer to the center of the universe were the places where the four traditional elements (earth, water, air, and fire) chaotically combined, recombined, and turned one into another, thus creating all of the natural phenomena that are part of daily experience. Paths of change were still discernible, but "with a regularity less than that of the primary element of material things" [the heavenly fifth element]."[2]

The alterations that affected the terraqueous globe and the surrounding region appeared chaotic or directional only because of the spatial and temporal limitations of human experience. In the long term and considering the Earth as a whole, Aristotle argued, change was as cyclical as that in the heavens, because it was the motion and heat of the Sun that transmitted movement to the lower regions of the universe. Celestial motions were thus responsible for the ignition of the underground fires that caused volcanic eruptions, or for the agitation of vapors and winds in vast subterranean caverns that led to earthquakes. Yet the most dramatic of the geological revolutions impelled by the heavens was also the least perceptible to human senses. In *Meteorology*

I, 14, Aristotle laid out the grandiose vision of an ocean that inexorably submerged and uncovered whole countries, slowly renovating the surface of the Earth:

> The same parts of the earth are not always moist or dry. . . . So also mainland and sea change places and one area does not remain earth, another sea, for all time, but sea replaces what was once dry land, and where there is now sea there is at another time land. This process must, however, be supposed to take place in an orderly cycle. . . . For wherever it has encroached on the land because the rivers have pushed it out, it must when it recedes leave behind it dry land: while wherever it has been filled and silted up by rivers and formed dry land, this must again be flooded. But these changes escape our observation because the whole natural process of the earth's growth takes place by slow degrees and over periods of time which are vast compared to the length of our life.[3]

Aristotle did not mention marine fossils, but none of his pupils and commentators ever doubted that this mechanism provided an elegant explanation for the existence of petrified sea creatures far from the sea: simply put, the ground we tread used to be the ocean's floor.[4] Heavenly motions were responsible for another outstanding phenomenon: the sudden floods that periodically sweep vast portions of the Earth, distinct from the progressive migration of the ocean. Just as winter occurs in a given season of the year, Aristotle wrote, "so in determined periods there comes a great winter of a great year and with it excess of rain. But this excess does not always occur in the same place."[5] The "great year" as previously defined by Plato in the *Timaeus*, was the length of time after which all celestial bodies, each one revolving around the center of the universe with its own speed and period, return to the same relative positions. Aristotle's Earth history could be characterized as change without evolution. Just like the heavens are eternal and uncreated, so are the Earth and the species of living things, every modification being purely local and cyclical. Mountains rise and collapse, rivers begin to flow and dry up, but on the whole, the Earth and humankind have always existed more or less the same.

While skimpy on details, instead focusing on causal explanations and general principles, Aristotle's *Meterology* remained by far the most systematic exposition of the subject produced in antiquity. However, it was by no means the only one. Some basic elements, such as the eternity of matter and time, the occurrence of periodical floods and droughts, and the influence of celestial bodies over meteorological phenomena, were more or less a given for every Greek philosopher. Most schools rejected, though, Aristotle's separation between celestial and sublunary physics, or the eternity of the Earth itself. Plato and his followers held that an intelligent being transformed the primeval matter into this ordered cosmos; while arguably much older than a few thousand years, the Earth definitely had a beginning, and its history was directional rather than cyclical. The Stoics maintained that the universe, ruled by an immanent rational principle, was continuously destroyed and recreated in never ending successions, every time reproducing exactly the same things, people, and events. For atomists like Epicurus or Lucretius, our Earth was just one among infinite that kept aggregating and dissolving in a boundless and eternal universe. Every form of life, including humans and superior animals, rose by spontaneous generation from the fertile mud that had once covered the young Earth. As its body grew older and less fecund, humans had to learn agriculture and build complex societies in order to obtain from the soil the crops it used to produce spontaneously.[6] Lucretius wondered why, "if there has been no first time for Earth and Heaven, and they have been always everlasting" have not "other poets sung other things beyond the Theban war and the ruin of Troy?" It must be acknowledged, the poet concluded, that "the world is young and new, and it is not been long since its beginning."[7]

The most serious threat to Aristotle's eternalism came from history rather than from natural philosophy; namely, from the apparent absence of any human documents or artifacts older than several thousand years. If the human race had existed eternally or even for a very long time, why were no memories left of those older eras? Why did many arts and inventions seem to have been acquired only recently? Aristotle's answer echoed that given by Plato a few decades earlier. While not an eternalist, Plato assigned to mankind a much greater antiquity than allowed by Greek (or Hebrew) historical records. In his *Timaeus*, a "very old" Egyptian priest explains that humankind is periodically decimated by floods of water and by their opposite, namely great droughts which recur "at long intervals" brought by the "shifting of the bodies in the heavens which move round the Earth."[8] Wars, famines, and plagues do the rest, leaving as sole survivors the illiterate herdsmen of the mountains and other remote places. The Greeks were therefore oblivious even of their own origin, because they repeatedly lost writing, which hands down the memory of past things.[9] After every catastrophe, civilization had to start over. To Plato's explanation, Aristotle added his own understanding of long-term climatic and geological change. The submersion or desertification of whole countries forces entire nations to leave their dwellings and, little by little, they settle elsewhere while losing every record of their previous existence.[10]

Pondering conflicting accounts of the origin of the world and of the human race, historian Diodorus of Sicily (ca. 90–30 BCE) recognized that he had no way to determine whether the humans who first established kingdoms and invented writing were generated by the loam of a newly created world or descended from the survivors of a distant cataclysm.[11] If periodical floods could submerge the whole of the Earth, as many argued, on its latest occurrence "the destruction of living things was complete." The renovated Earth then produced new animals and a new humanity, which bore no relation to the previous ones.[12] Assuming that there was some truth to Egyptian creation myths, then their civilization had begun between 10,000 and 23,000 years earlier.[13]

Medieval scholars were well aware that the book of Genesis could not claim any monopoly on stories of deluges of waters exterminating most of a wicked humanity.[14] The idea that the present world emerged from a catastrophic Flood, and that its physical decay went together with the moral and bodily decadence of the human race, was indeed pervasive in Greco-Roman histories of humans and nature. The Roman poet Ovid's (43 BCE–17 CE) *Metamorphoses*—one of the most avidly read texts in medieval and early modern Europe—explained in poetic language that, after "a God or Nature" made the universe, human beings passed through the proverbial ages of Gold, Silver, Bronze, and Iron, each time losing more of their original innocence. Jupiter finally decided to exterminate the whole of the earthly creatures, but not before promising the other gods that a new race of humans would replace the one soon to be extinguished. The only survivors, Deucalion and Phyrra, "the best of either sex," were too old to procreate; Themis, goddess of justice and good counsel, then suggested that they threw behind them the "bones of your mighty mother," which they understood to be the Earth and its rocks. The stones they cast became men and women, with a metamorphosis that mimicked Lucretius's verses on the generation of humanity from the soil of a new (or in this case, renewed) world.[15] In the fifteenth and last book of the poem, philosopher Pythagoras teaches in famous verses how the surface of the Earth is a witness to the perennial changes that affect the universe:

> I, myself, have seen the sea where previously there was very firm land.
> I have seen land made from the sea and sea shells very far away from
> the open sea and an old anchor has been found on the highest point of
> these mountains.[16]

One century later the Roman Stoic philosopher Seneca (4–65 CE) argued in his *Natural Questions* that "the fated day of the Flood" prepares a physically and morally corrupt Earth for its next cycle, so that "innocent people may be created afresh and no teacher of worse behavior may survive."[17] Like most other ancient authors, Seneca believed that great floods alternate with equally destructive "fires," so "whenever the world has decided on revolution, the sea is sent crashing down, just as heat and fire are when another form of extinction is approved."[18]

The advocacy by most early Christian authors of a recent Creation, and of the historical truth of Noah's Flood, should be seen in the wider context of Greco-Roman historical and philosophical traditions. A linear history of the Earth and of humankind, from its creation by an intelligent author to the final conflagration, conflicted with some pagan narratives but definitely not with all of them; Christian apologists were actually eager to appropriate and recast those that could be adapted to their worldview. It is little wonder that Eusebius of Caesarea dismissed as fanciful exaggerations the timescales of Egyptian or Babylonian chronicles, when even Diodorus questioned their reliability. If Moses had really been the human author of the first books of the Bible, than the *Pentateuch* could reasonably be regarded as the most ancient document of the early history of humankind. Philosophical and mythological accounts of past floods could easily be interpreted as the distorted echoes of real events, which a divinely inspired author transcribed more faithfully in the Bible. The existence of parallel narratives did not diminish that of Genesis. In fact, they reinforced it by showing that the catastrophe was still alive in the memory of nations other than the Hebrews. Even Lucretius, the flag-bearer of materialism and atheism, could be used to support a relatively recent origin of the Earth.

For late-ancient and medieval Christian readers, the core truths shared by these accounts were more meaningful than peripheral differences. Christian writers soon placed emphasis on fossil seashells as physical evidence of a past inundation of the globe, but that argument, too, was explicitly borrowed from pagan sources and could hardly have looked simplistic or implausible. In the early third century, Christian apologist Tertullian (ca. 155–240) wrote that "even now shells and buccinums from the sea wander as strangers on the mountains as if to confirm what Plato had said, that even the higher regions were flooded."[19] The same narrative was later adopted by Orosius (ca. 375–420) in his *Histories Against the Pagans*, and was finally passed down to the following centuries in a slightly more elaborated form by Isidore of Seville's *Etymologies*.[20] Isidore conflated Greek and biblical traditions in a short chapter

On Floods, whose title surprised historian of geology François Ellenberger because the Bible describes only one Deluge.[21] However, the bishop of Seville perceived pagan accounts as fundamentally compatible with that of the Bible, once purged of elements like the spontaneous generation of human beings.[22] Following Eusebius, he regarded the deluges of the Greek tradition as merely local events; evidence of the only truly universal one could still be observed "in the stones we are accustomed to see on remote mountains, stones that have hardened with shellfish and oysters in them."[23] The short chronology adopted by the author implied that slow geological forces had no time to produce any effects, but periodical upheavals could still enliven an otherwise directional history. Every change of the Earth's surface must then be brought by sudden and violent events such as earthquakes, volcanic eruptions, or local floods in addition to the universal one.[24]

After the collapse of the ancient world and the loss of much of its literary and philosophical production, early medieval scholars based their ideas of the physical history of the Earth on a limited number of sources. They included the relevant writings of the fathers of the Church, Ovid's *Metamorphoses*, Pliny the Elder's (23–79 CE) *Natural History*, Seneca's *Natural Questions* (though harder to find than other sources[25]), and Calcidius's commentary to Plato's *Timaeus*. While many of these works assumed a recent Earth, this idea did not necessarily contradict a more ancient (or even beginningless) creation, in the same way it did not contradict the eternal universe of the atomists or the cyclical one of the Stoics: whether co-eternal with God or not, matter and motion could pre-exist their organization in the current world. The most radical, coherent, and compelling advocacy of an eternal Earth, Aristotle's *Meteorology*, disappeared, together with the rest of his works, and exerted very little influence on early medieval culture. The ambiguity of various medieval authors reflected not heterodox views or fears of personal consequences, but the flexibility or vagueness of sources, theological doctrines, and interpretative approaches.

William of Conches's (ca. 1090–1155) *On the Philosophy of the World* is a good example of the attitudes that prevailed in Christian Europe around the middle of the thirteenth century and of the points that raised controversy.[26] Philosophical accounts of the formation of the Earth normally assumed that the globe had reached its present form through natural processes that carried out the divine plan for the Creation. In other words, they provided a physical counterpoint to the Genesis creation story. While the latter taught the "why," referring to images and metaphors adapted to the comprehension of common people, natural philosophy investigated the actual "how." When correctly understood, neither contradicted the other. In both cases, the Earth was shown to evolve from a disordered beginning to a state that made terrestrial life and the existence of human beings possible; this condition was not stable, though, as the physical and moral decadence of the world inexorably led toward its destruction or regeneration.

William of Conches warned that when philosophers mentioned the creation of the world, they did not mean its celestial or terrestrial bodies but only the four elements. Their motion then gave origin to everything else, without any need for further supernatural intervention.[27] Following their natural inclination, earth, water, air, and fire fell therefore toward the center of gravity of the universe, where they occupied four concentric spheres according to their specific weight. The skies were made of the same elements found on the Earth, though in different proportions. Philosophy thus agreed with Genesis that there was a time when "the earth was entirely covered with water, and that water was thicker and darker."[28] That said, William was not particularly happy with the prevalent opinion that the elements remained confused in a shapeless mass (for who knows how long) before the Divine craftsman began to arrange them. The most authoritative source for this claim, namely Plato, would have been misinterpreted: from the very moment space and matter existed, the elements could only move to the places that had been assigned to them in nature.[29] William also noted that it was the "common opinion of the philosophers, that terrestrial things are extinguished alternatively by the flood and by fire." They included not only the universal Deluge and the final arson of the Christian Revelation, which could only happen once, but even the more frequent local events of the Greek tradition.[30] The passage on the "floods" of water and fire that periodically scourge the surface of the Earth, and can even destroy it in their most devastating occurrences, was sourced directly from a chapter of Seneca's *Natural Questions*.

The Return of Aristotle

By the time William of Conches wrote his *Philosophy of the World*, many of Aristotle's works, including his *Meteorology*, had been available in Islamic countries for over three hundred years. In the early ninth century, as Western Europe struggled with external threats, political fragmentation, and cultural and economic isolation, Al-Bitriq composed a paraphrasing compendium of the *Meteorology* from a now lost version in Syriac.[31] In spite of its limits, the historical impact of this adaptation proved huge, because it was used by authors such as Ibn-Sina (Avicenna) or Ibn-Rushd (Averroes).

Another Arabic compendium was authored by Hunayn ibn Ishaq (d. 876), who also translated a paraphrase of the commentary on the *Meteorology* by the late-ancient Aristotelian philosopher Olympiodorus (ca. 500–570).[32] In the hands of its Islamic commentators, Aristotle's natural philosophy rose to a prominence never enjoyed in the ancient world. Its influence was already pervasive in the writings of the mysterious group of scholars known as "Brethren of Purity" active in Basra in the tenth century. Their *Epistles* merged Aristotelian meteorology, Ptolemy's astronomy, Platonic philosophy, and Islamic theology into a theory of geological change that filled some of the many gaps left by Aristotle.[33] In *Epistle 19*, the author(s) adopted the period of the precession of the equinoxes in Ptolemy's *Almagest* (36,000 years) as the duration of the "great year" that Aristotle left undetermined.[34] That length of time also corresponded to a complete geological cycle of the Earth, during which "mountains become deserts and waterless spaces; the places like deserts become sea, ponds, and rivers":[35]

> Know, O my brother, that . . . the flood of rains brings those rocks, stones, and sands down to the interior of wadis and rivers, and transports them, in the violence of its course, towards seas, ponds, and reed-beds. And the seas, by the violence of their waves . . . expand those sands, argil, and pebbles at their bottom, layer upon layer, in the duration of time and ages.[36]

When the ocean withdraws, "these mountains and these hills appear" with their layered structure and the remains of the sea creatures they incorporated. They will then be destroyed by erosion, while new ones accumulate on the sea floor and this "will continue throughout the ages." The Brethren of Purity maintained that the world did have a beginning, and they fought explicitly the "ignorance, confusion and doubts" that led others to state its "timeliness and eternity"; yet the Earth was "great in body, big in structure, long in life, prolonged in permanence, [and] it is not known when it was generated and when it will be corrupted."[37]

The age and revolutions of the Earth, the origin of mountains, and the formation and properties of rocks and minerals (an interest certainly fueled by the diffusion of alchemical practices) were often debated in the following centuries. The author of a *Book of the Elements* long attributed in Europe to Aristotle, but who was almost certainly an Arab, may have referred to the Brethren when he refuted the thesis of "certain people, among those who composed discourses," who believed that the sea changed its place on the surface of the terraqueous globe.[38] While the author agreed that such movements could only be caused by the revolutions of the celestial spheres, he noticed that none seemed slow enough. Even the precession of the equinoxes, whose period of 36,000 years corresponded to the slowest known celestial motion, would bring underwater one-quarter of the globe (that is, the whole of the emerged land according to medieval cosmography) in just 9,000 years.[39] Alterations on a vast scale should have occurred in historical times, but none was recorded. The conclusion, however, was not that the Earth must have been created recently or in a different state, but that the observed phenomena required different explanations.[40]

A number of possible such causes was put together by Avicenna in a small but crucial treatise on the formation of mountains, minerals, fossils, and geological strata he composed in 1022 as part of his *Book of Remedies*.[41] Avicenna first attributed the origin of mountains to earthquakes, which can either raise parts of the ground or create hollows in areas of softer soil; erosion by water and wind does the rest, leaving the toughest rocks as high ground or mountain peaks.[42] Earthquakes were often invoked in the Middle Ages as a cause for the formation of mountains, but usually not alone, and not for large ones; Avicenna was no exception. Building on the doctrine of the formation of rocks he had previously exposed,[43] he went on to explain in thoroughly Aristotelian terms that "this habitable world was in former days uninhabited and indeed, submerged beneath the ocean."[44] Little by little, it then became exposed, and the moist clay that used to be the sea bottom turned into stone; erosion later sculpted the newly hardened soil, creating mountains and valleys.[45] It is for this reason that "parts of aquatic animals, such as shells etc." were often found in rocks.[46] The Persian philosopher arguably did not mean that the Earth was once entirely covered by water: rather, the specific part now inhabitable used to be submerged, while other areas were dry. Having formed in ages of which we have no memory, most mountains were by his time "in the stage of decay and disintegration."[47] Given Avicenna's commitment to a beginningless creation and to a fundamentally Aristotelian Earth, there can be little doubt that the decadence of the current world prefigured the end of a cycle rather than its final demise.

Greek and Arab meteorological works began to flood Europe with the great wave of translations of the late twelfth century. Aristotle's *Meteorology* was turned into Latin from an Arabic compendium by Gerard of Cremona (1114–1187);[48] his so-called old translation was widely used until superseded, around 1260, by William of Moerbeke's "new translation" entirely from the Greek.[49] In the first half of the thirteenth century, Averroes's commentary on the *Meteorology* also became accessible to European readers.[50] Even though he shared some elements with the writings of the Brethren of Purity, Averroes remained faithful to the eternity of the Earth and to the Aristotelian displacement of the ocean across the globe, even citing marine fossils as evidence for this phenomenon.[51] One or two decades before the year 1200, Alfred of Sareshel, also known as Alfred the Englishman, wrote an abridged Latin version of Avicenna's treatise on mountains and minerals, later included at the end of his own commentary to the *Meteorology* as a separate chapter *On Minerals*.[52] Often attributed to Aristotle himself, Alfred's *On Minerals* was used by almost every European scholar with an interest in chemical and geological subjects. Indeed, it represented a crucial source for the most influential mineralogical treatise of the Middle Ages, Albertus Magnus's *Book of Minerals*.[53] The *Book of the Elements* began to circulate in Latin as well in the thirteenth century.

The injection of Greek and Arab meteorology into the circuit of the European culture was a powerful stimulus to a renewed attention to the nature of the Earth, its mutations, and its productions. This interest had important practical sides, embodied in the rediscovery of alchemy and in the lasting fascination with the healing and magical properties of stones and gems. Moreover, meteorology became a deeply contentious field, involved as it was in the crucial controversy on eternalism and on the relation between faith and natural knowledge. In the course of the thirteenth century, the traditional assumption that the Earth had an actual history, characterized by a development through different stages, was by and large replaced by the idea of change without evolution, in which even large-scale mutations were in the end local and cyclical. Aristotelian meteorology was the discipline of how the Earth works, not of how it came to be, since it never began in the first place. If so, the Earth was either without a beginning, or the Creator made it in the present form since its very inception. In both cases, the age of the Earth coincided with that of the universe, and the problem of its origin was better left to theology.

Whether such creation happened in recent times or in an unfathomable past remained an open issue, to which meteorology could offer crucial cues. On the one hand, the absence in historical times of dramatic geomorphological modifications seemed to deny that the precessions of the equinoxes, or other celestial motions of comparable duration, could set the pace for the renewal of its surface. The very existence of mountains in the face of unrelenting erosion was cited against the eternity or even the extreme antiquity of the Earth. These arguments were countered with the elaboration of bold and sophisticated theories on the origin of mountains, in which the replacement of worn-out elevations with new ones no longer depended on heavenly motions.

Unlike present-day geology, meteorology was not expected to deliver definitive evidence of the age of the Earth. This discipline, as opposed to other parts of Aristotelian philosophy, was built by induction upon the observation of physical phenomena rather than deduced through logical reasoning from established truths. In other words, it was not deemed possible in meteorology to arrive at conclusions as cogent as logical or mathematical demonstrations, which would constitute instead true and certain knowledge. As Aristotle himself remarked, since the direct cause of many meteorological phenomena is inaccessible to observation (earthquakes and volcanoes being a typical example, as they originate in the depths of the Earth), he would be satisfied with a plausible account that is "free from impossibilities."[54] The history of the Earth remained within the boundaries of the "probable," defined as a plausible opinion compatible with both reason and faith that could be legitimately entertained, even though others may disagree.[55] Like William of Baskerville in Umberto Eco's *The Name of the Rose*, medieval natural philosophers loved to explore how many possibilities were possible or, rather, probable. To the bafflement of modern readers, the resolution of a given problem may have required assumptions that were incompatible with those used in a different case, often producing mutually contradictory answers. Each of them would be "probable," provided the logical reasoning behind them proved correct. Yet William knew that not every possibility had the same degree of probability. Truth was one; the mysterious murders that aggrieved the abbey had a definite culprit; and an appropriate use of observation and reason might narrow down the many initial paths of enquiry to the most likely one. Likewise, medieval natural philosophers often insisted that meteorology could provide valid and dependable knowledge, though perhaps not with the same degree of certainty as geometrical demonstrations. By showing the physical plausibility of a created but ancient Earth, meteorological authors continued to provide a viable alternative to straight eternalism on the one hand, and to a literal reading of Genesis on the other.

Astrology and Earth History

Estimates of the length of the Earth's cycles did not involve the absolute or even relative dating of rock layers. This seemingly inexcusable oversight was caused not by lack of intellectual or observation powers, but by different assumptions about the ultimate origin of geological change. Since it depended on the motions and configurations of the heavens, it made sense to rely on the mathematical accuracy delivered by astronomy rather than on problematic geological evidence. The application of these disciplines to the history of the Earth was in fact an important, though often overlooked, aspect of the progressive "astrologization" of Aristotelian natural philosophy. Aristotle accepted the general principle of the influence of the celestial bodies into the lower world, but his natural philosophy was not properly "astrological".[56] Simply put, astrology in fourth-century BCE Greece was still a recent Mesopotamian import and a far cry from the pervasive system of thought and practices it became later. Greek astronomy itself was in its infancy and incapable of predicting with any accuracy planetary positions and conjunctions. Unsurprisingly, the Sun is the only celestial body with a clear role in Aristotle's system—including in the periodical "corruption" and regeneration of parts of the Earth.[57] His *Meteorology* does not provide any figures for the possible duration of a "great year," nor for the periodicity of great floods and draughts. Horoscopes are not documented in Greek culture, either, until the first century BCE. Ptolemy's codifications of mathematical astronomy (namely, how to compute the positions of the planets for a given date) and astrology (the practical application of the former) did not appear until ca. 150 CE, respectively in the *Almagest* and in the *Tetrabiblos*.

Islamic authors had already begun to merge Aristotle and Ptolemy in the ninth century. Building on their heritage, four centuries later European philosophers used astrology to weave mathematical disciplines such as astronomy and geometrical optics into the fabric of Aristotle's qualitative natural philosophy. In turn, the latter provided astrology with its philosophical foundations.[58] In other words, celestial influences became the force that bound Earth and heaven together; astrology was the discipline that made it possible to determine and quantify their relation. Periodical floods and arsons were thus attributed to conjunctions of planets in watery or fiery signs, whose epochs, periodicities, and even magnitudes could, in principle, be reckoned precisely (the more numerous the planets that gather in a certain sign, the bigger the event). Likewise, the "great year" became variously associated to heavenly motions of known or knowable duration, such as the precession of the equinoxes or the minimum common multiple of every planet's period. Albertus Magnus recognized that, of all the elements, earth was extraordinarily receptive to celestial influences, and this remained true at both the macro- and the microscopic level. Therefore, the effective preparation of talismans should take into account the appropriate material, but also the right astrological timing for their fabrication.[59]

Like in the Islamic tradition, astrology was often regarded as a potentially dangerous and illegitimate practice. Astrological determinism, in particular, would render Christianity moot by turning human beings into the stooges of a fate already written in the stars. Yet, towering authorities such as Robert Grosseteste, Albertus Magnus, Thomas Aquinas, and Roger Bacon insisted that the stars' influences extended only to matter, not souls: humankind thus retained full moral responsibility for their actions and choices. Mathematical astronomy, which computed the positions of stars and planets for a given epoch, might have been regarded as a potentially exact science; but astrology could only predict the probable outcome of their configurations.

The transformation of medieval Earth history did not happen overnight. In the late twelfth century, Alfred of Sareshel's commentary on the *Meteorology* still assumed that the Earth formed from pre-existing elements, rather than having always existed—a reading perhaps facilitated by the omission, in the "old translation," of the most explicit passages on the eternity of the world.[60] A few decades later, however, Albertus Magnus's meteorological works provided already an impressive measure of the pervasiveness of astrology, of competing interpretations of Aristotle's natural philosophy, of the renewed emphasis on empirical observation, and of a redefined relationship between faith and natural knowledge. Meteorology appeared to be a complex, mobile, and contentious field, as the problem of the nature of the Earth was related directly to that of the history of life and of humankind itself.

Albertus Magnus examined the nature of the Earth in his extensive commentary on the *Book of Elements*, in his *Meteorology*, and in the *Book of Minerals*—apparently composed in that order around the middle of the century.[61] Albert still took seriously the possibility of a directional Earth, in which dry land somehow emerged from the ocean that covered it entirely at the beginning of its history. A plausible physical explanation was provided by Alexander of Aphrodisias (ca. 200 CE), in a commentary to the *Meteorology* translated by William of Moerbeke in 1260.[62] According to his doctrine, the vapors raised by the Sun continuously turn water into air and fire, causing a progressive diminution of the sea level until part of the element of earth remained uncovered (the separation of water from land in Genesis). The process was still underway, and it would continue until the complete disappearance of water and the extinction of life on Earth (the end of the world).[63] Albert seemed tempted by this solution, which in various forms never ceased to circulate and made a spectacular comeback in the late seventeenth century.

In the end, though, he accepted the Aristotelian arguments for a cyclical Earth, noting that the generation of plants and animals must have been going on since the beginning of the world. This would be impossible if the Earth was once entirely covered with water.[64]

Still, he remained unconvinced that the Earth's surface could be renovated by the movement of the ocean across the globe. A crucial proponent of the astrologization of natural philosophy, Albertus Magnus fully embraced the criticism of that model expounded in the *Book of Elements*, and for the same reasons: such a phenomenon could be produced only by a celestial motion of the same period, but none was slow enough.[65] The advances and retreats of the seas that had occurred in historical times, and that he witnessed himself on the site of Belgian towns like Bruges, should therefore be considered "accidental" and due to local causes rather than to a global motion of the ocean.[66] The presence of marine fossils in locations well removed from the sea, that once again he had observed personally, was attributed to earthquakes which—he surmised—coastal areas are especially prone to;[67] afterwards, telluric forces happened to raise shallow sections of the sea bottom close to the shores.[68] Among lesser causes, he of course enumerated the catastrophic inundations of the sea that periodically scourge the Earth.

Albertus Magnus's discussion of "the Flood" (the singular denotes, as customary in meteorological literature, a recurrent natural phenomenon rather than a unique event) is exemplary of both the new importance of astrology and of the ongoing redefinition of the relation between theology and meteorology. The Revelation blended effortlessly into William of Conches's or Thierry of Chartres's natural science, which elucidated and confirmed the words of Genesis. Albertus Magnus did not perceive any contradiction either between the two, and he also believed that ancient myths like those transmitted by Ovid and Plato carried important truths; however, having been exposed to the rigor of Aristotelian philosophy, he considered it inappropriate to write about those matters in allegorical terms.[69] Furthermore, in his time the distinction between theology and natural philosophy was being drawn much more sharply as they became distinct disciplines and protected professions, whose boundaries must not be muddled. Albertus Magnus was careful not to mix the theological with the philosophical—not because he was wary of potential conflicts between faith and science, but for methodological reasons.[70] Of course, God was the root cause of everything, including Noah's universal Flood and others that similarly happened; some even said, Albert added, that "we should seek no cause for things of this sort other than the will of God." Yet, God normally acts in the world through natural causes, so that even the Deluge could and should be treated by philosophers exclusively as a natural occurrence.[71] As a natural phenomenon, a universal Flood depended on a rare concourse of celestial and terrestrial circumstances. However, Albert was quick to point out that "particular" floods, unleashed by smaller planetary conjunctions in a relevant sign or even by an earthquake, were a much more common occurrence.[72] Once again, floods of water implied the symmetrical existence of the "flood of fire."[73] Unlike William of Conches, he declined to specify that Noah's Flood was a unique event. On the contrary, his wording suggests that this may not have been the case at all.[74]

The possibility of other deluges had momentous implications, as it raised the issue of how terrestrial creatures come to be replaced after such events. As Albertus Magnus wrote, an *altercatio magna*—a great dissension—existed between Avicenna and Averroes on this subject. The Persian philosopher relied on the powers of spontaneous generation, in which the influence of the stars provided the equivalent of the male semen while the renovated earth acted as a womb; once recreated, animal species could then reproduce sexually. Averroes objected that, while this happened for lower creatures even under ordinary circumstances, large and complex animals had never been seen to rise from mud or decomposing organic matter. More faithful to Aristotle, the Cordoban philosopher maintained instead that a flood of waters can never be universal, so there will always be survivors.[75] Albert did not side decidedly with either authority, but he seemed persuaded by Averroes's criticism of the limits of spontaneous generation. Most importantly, he observed that human beings (and perhaps, even superior animals) could never be recreated naturally "because a rational soul is not educed from matter . . . and for this reason the first human *hypostases* [what constitutes the essential nature of human beings] were created and formed by God."[76] Still, having just explained how universal floods could happen naturally, Albert was reluctant to adopt Averroes's solution that only allowed for local ones. The problem of the resettlement of the Earth remained unresolved.

The Problem of Water and Earth

In the centuries between ca. 1300 and 1600, hypotheses about the renovation of the Earth's surface (or lack thereof) were normally framed within the overarching problem of how dry land could exist at all. In theory earth, water, air, and fire should settle around the center of the universe in this order, according to their specific weight. An earth completely surrounded by a lighter sphere of water would then be the ideal state of equilibrium of the world, with each element in its natural place—or as close as possible to it. Of course, experience showed that this was not the case.[77] Alexander of Aphrodisias's hypothesis of the progressive transformation of water into air and fire could work for a directional Earth, but not in a steady-state Aristotelian universe, whose stability required that, while the four elements keep changing into one another, their respective quantity must always be roughly the same. Yet, water is less dense than earth, meaning that it must occupy a much larger volume for a given mass, further complicating the problem. Indeed, it was a common tenet of medieval meteorology that the volume of the sphere of water could be up to ten times larger than that of the earth and its radius about twice as much.[78] Why some of it emerged was easily explained in teleological terms, as necessary for the generation of terrestrial animals in general and human beings in particular. How it managed to do so was the most debated problem of medieval meteorology. In the early 1300s the controversy resonated in universities, cathedral schools, princely courts, and artisanal workshops alike.[79]

A permanent miracle was an appealing solution, and it required only its temporary suspension to unleash a universal Flood.[80] Still, most maintained that God acted through the laws of nature and that its natural causes should be researched. Many authors postulated that the earth had a swelling on the inhabited side, so that some of it could pop out of the water. Giles of Rome (1247—1316), one of the most famous philosophers of the late thirteenth century, even reckoned the height of the earth's "hump" at about five-fourths of its radius. This protuberance could have been created directly by God in the beginning, but the effect over time of the magnet-like attraction of the northern stars (considered brighter and more powerful than those in the southern hemisphere) on the body of the earth was also credited.[81] As astronomer Robert the Englishman wrote in 1271: "We have evidence of this in the places that used to be underwater and are now dry, as we can see in certain locations in England."[82]

In 1319 the great Florentine poet Dante Alighieri (1265–1321) was involved in an animated controversy on this subject in Mantua, northern Italy. A few months later, he resolved to defend the thesis of the Earth's "hump" in a lecture, On the Problem of Water and Earth, delivered in a church of nearby Verona before "the whole Veronese clergy," using essentially the same arguments as Giles of Rome.[83] The contrast with the description of the Earth he gave in his Inferno, completed only a few years earlier, could not be more revealing of the medieval approach to these problems. In the religious poem, Dante conceived of Hell as a funnel-like cavity within the earth, whose base was close to the surface, right under the holy city of Jerusalem. The tip of the cone coincides

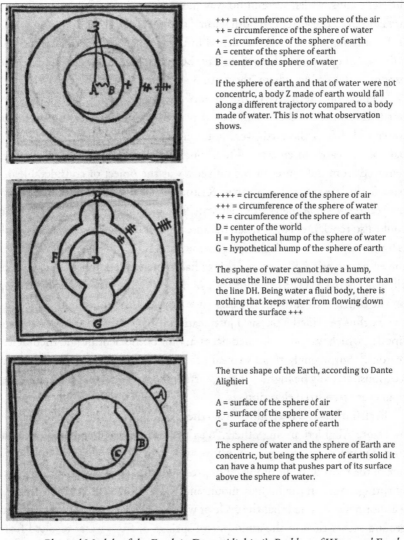

+++ = circumference of the sphere of the air
++ = circumference of the sphere of water
+ = circumference of the sphere of earth
A = center of the sphere of earth
B = center of the sphere of water

If the sphere of earth and that of water were not concentric, a body Z made of earth would fall along a different trajectory compared to a body made of water. This is not what observation shows.

++++ = circumference of the sphere of air
+++ = circumference of the sphere of water
++ = circumference of the sphere of earth
D = center of the world
H = hypothetical hump of the sphere of water
G = hypothetical hump of the sphere of earth

The sphere of water cannot have a hump, because the line DF would then be shorter than the line DH. Being water a fluid body, there is nothing that keeps water from flowing down toward the surface +++

The true shape of the Earth, according to Dante Alighieri

A = surface of the sphere of air
B = surface of the sphere of water
C = surface of the sphere of earth

The sphere of water and the sphere of Earth are concentric, but being the sphere of earth solid it can have a hump that pushes part of its surface above the sphere of water.

FIG. 2.1 Physical Models of the Earth in Dante Alighieri's *Problem of Water and Earth*
Dante Alighieri, *Questio florulenta ac perutilis de duobus elementis aquae & terrae etc.*, ed. Giovanni Benedetto Moncetto (Venice: per M. de Monteferrato, 1508).

with the center of the Earth and of the universe, where the huge body of Satan is stuck for eternity. The fall of the rebel angel caused the emergence of dry land, dug the cavity of Hell, and "maybe" even the mountain of Purgatory, at the antipodes of Jerusalem.[84] Dante's meteorological lecture did not even mention this cosmic geography; the poet-theologian and the natural philosopher simply described the same phenomenon in different terms and with different aims, which were in no way mutually exclusive. The emergence of dry land from the water had both natural and theological explanations, which in Dante's *Inferno* were also couched in the language of poetic fiction.[85] As Albertus Magnus argued, it would have been inappropriate to do so in a philosophical discussion, where "the truth is explained as it is."[86]

A common alternative to the "hump" theory held that some earth is dry not because the northern stars raised part of it, but because the sphere of water and that of the earth were not concentric: the resulting offset caused part of the earth to emerge. Which one of the two spheres had its center removed from the center of the universe was the object of considerable debate. In the late 1200s, Tuscan artist Restoro of Arezzo surmised in his vernacular *Composition of the World* that stars, just like magnets, could attract but could also repel. Dry land existed because the stars pushed away water (which is lighter and easier to displace) from one of its sides; the sphere of water was the eccentric one. This state of things had remarkable implications. First of all, in order to clear the inhabited side of the globe, the ocean must be so deep on the other as to rule out the existence of islands or continents there. The theory thus provided a physical justification for the inhabitability of the antipodes, which was often deemed necessary for both scientific and theological reasons.[87] Surprisingly, this also meant that the surface of the ocean on the far side must actually be higher (that is, farther from the center of the universe) than every terrestrial elevation.

In all fairness it was a controversial theory even in its time, attacked fiercely by Dante Alighieri among others. Yet the concept was much less irrational than it may look at a first glance. Above all, it provided an elegant solution for one of the most difficult problems of pre-modern meteorology: the existence of springs even on the highest mountains. Before the eighteenth century, it was normally assumed that the cycle of water took place mostly in the underground rather than in the atmosphere; springs of fresh water resulted from the filtering of sea brine running through the Earth in subterranean veins, but what force could rise the liquid to such heights? The internal heat of the Earth was not a viable candidate, as its sources were thought to be right under the crust. The center of the Earth was actually the coldest place of the universe, being the most remote from the Sun's heat.[88] However, the pressure exerted by a sphere of water whose surface was higher than every mountains could provide the desired mechanism. In the 1260s, Florentine scholar Brunetto Latini, an extremely well known author and popularizer, found it credible that from the Atlantic coast of northern Africa, the ocean appeared "much higher than the earth, yet it is self-contained so that it neither falls, nor runs over the earth."[89] The fact that this hypothesis was widely entertained, including in vernacular literature, reveals not faulty reasoning but intellectual sophistication. Medieval authors and readers were willing to accept the physical hypotheses that best explained the observed world, even when they seemed to defy common sense. The motion of the Earth several centuries later, or Einstein's relativity one hundred years ago, were every bit as counterintuitive in their times.

A different version of that theory maintained instead that the eccentric body was not that of water, which being a fluid would always form a perfectly spherical figure around the center of the universe, but that of the earth. Alexander of Aphrodisias argued that because of the great cavities within

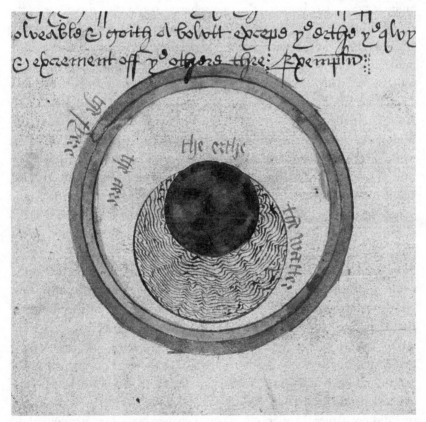

FIG. 2.2 **The spheres of earth and water.** The sphere of earth emerges from the sphere of water due to the offset between the geometrical centers of the two spheres. In this case, the center of the element of earth coincides with the center of the universe, while the sphere of water is displaced on one side.

Beinecke Library, Yale University, Ms. 337 (ca. 1526–1527).

the earth and its supposedly sponge-like structure, its density is not uniform and one side will always be lighter than the other. This means that, while the center of mass of the Earth coincides with the center of the universe as it should, its geometrical center does not; the resulting offset causes the emergence of part of the earth.[90] This theory had the advantage that dry land was higher than the ocean, as common sense seemed to dictate, while on the other side the ocean was still so deep as to rule out the existence of the antipodes.

Regardless of the physical reason that occasioned its emergence, medieval philosophers knew that as soon as dry land became exposed to the weather, erosion began its unremitting action: rains, rivers, floods, and winds washed away rocks and soil to the sea, shoveling matter from the emerged side to the bottom of the ocean until none of the former would be left. The earth's hump, or the offset between the two spheres, would then disappear as the elements inexorably regained their proper place. Some advocates of a young Creation cited indeed the existence of mountains as proof that the Earth must be recent. Thomas Aquinas argued that the beginning of the world was not rationally demonstrable, but perhaps astronomy and meteorology could succeed where logic, metaphysics, and chronology failed. In the early 1260s Dominican chronologist Giles of Lessines lined up geological and astronomical arguments to corroborate his evidence for the young age of the world. The astronomical proof had to do with a supposed variation over time of the point where the Sun is closest to the Earth (the Sun's orbit was circular in Ptolemy's system, but the Earth was not in its center to account for the unequal length of the seasons); the effect would be to bring major perturbations to the world's climate, though there were no written or empirical records of that.[91] If so, the heavens must have been revolving for no more than a few thousand years. On top of that, Giles observed that, "if this world had existed since eternity, there would be no mountains left by now, but they all would have been reduced to a plain because of the infinity of time."[92]

The notion of the "great year" also came under suspicion. The thesis that "when all celestial bodies have returned to the same point—which will happen in 36,000 years—the same effects now in operation will be repeated" was included in the list of propositions condemned in Paris in 1277. The censure did not target theories of geological change but rather astrological determinism, or the Stoic notion that each "great year" brings back exactly the same people and events: they both denied human freedom and personal responsibility, not to mention the uniqueness of Christ's incarnation.[93] Around 1280, the Franciscan monk Arlotto of Prato observed once again that if the "uninhabitable earth was once habitable," then "there is no written memory of this phenomenon, nor has anything relating to it come down to us."[94] Arlotto did not question that the motion of the highest sphere could bring vast geological mutations; he contended, though, that it must have been set in motion recently, so that it had completed only a mere fraction of a full 36,000 years revolution.

The Journey of a Rock Crystal

The "geological" objection to an ancient Earth could not be countered without a credible mechanism for the continuous creation of new high ground. The causes most commonly discussed by medieval authors, such as earthquakes or the accumulation underwater of mounds later exposed by the withdrawal of the ocean, seemed inadequate for anything larger than hills or small mountains. Even worse, they were tied to heavenly motions that appeared way too fast to account for the slowness of geological change, if such change was to be accepted as physically plausible.[95] Around the middle of the fourteenth century, Jean Buridan at the University of Paris finally decoupled the formation of mountains from the revolutions of the celestial spheres. The author of major innovations in logics and in the physics of motion, Buridan was long revered as one the most prestigious professors of the Faculty of Arts, where he spent the entirety of his academic career.[96] As a Christian, Buridan acknowledged that the universe had been made in time (or with time); as a natural philosopher, he espoused a cyclical Earth whose beginning could not be discerned through observation and reason and that could be treated as if it were eternal.[97] He did not discuss how dry land came to emerge, because the Earth had always existed more or less as we see it since its creation by God;[98] how its surface was renewed through the ages, however, must be explained without any supernatural intervention.

A rough sketch of Buridan's theory of geological change appeared first in his commentary *On the Heavens*, but a fuller account was given later where it belonged most: in the *Questions on the First Three Books of the Meteorology* (ca. 1352).[99] The French philosopher combined Alexander of Aphrodisias's theory, which postulated the eccentricity of the sphere of earth, with the fascination for concepts of balance and dynamic equilibrium that began to emerge in the second half of the thirteenth century.[100] Buridan noted that the emerged hemisphere must be lighter than the opposite, or there would be no distinction between the center of gravity of the earth and its geometrical center. The exposure to the air and to the heat of the Sun, which inflates the vapors in its pores and caverns, tends to make it even lighter. The difference is further increased by the continuous ablation of earth to the bottom of the ocean, where lack of heat and the pressure of water packs earth to much higher densities. The inhabited hemisphere then gets lighter and lighter, the opposite one heavier. Because of this process, the center of mass of the earth would tend to shift further away from the emerged side, but this can never happen because of its tendency to fall toward the center of the universe (just like any heavy object wants to fall down). As a consequence, the earth as a whole continuously crawls in the direction of the emerged hemisphere. As earth is washed down from the mountains to the sea, where it forms new layers of sediments, what used to be the ocean floor moves toward the center; on the other side new land emerges, on so forth *ad infinitum*. The rocks that are on the top of a mountain used to be on the bottom of the sea; then they passed through the whole body of the earth and finally emerged on the opposite side. "Nor is there any other way," Buridan concluded, "such mountains can be generated and maintained."[101]

Historians of science have often celebrated Buridan for repudiating astrological causes in his theory of the Earth, while embracing an essentially "mechanical" model.[102] This may not be entirely correct. While he downplayed the role of astrology in meteorological phenomena, the main result of his theory was to decouple the "great year" from the precession of the equinoxes, which made the hypothesis of an ancient Earth vulnerable to criticism on historical, astronomical, and geological grounds. In fact, heat and light from the Sun were not fundamentally different from other astrological influences. Far from rejecting them, he assumed that the largest possible flood, which marks the end of a "great year" and the beginning of the next (but is still not universal), requires a conjunction in Pisces of every single planet. This event, however, may be so rare as to take place "perhaps" once in "a hundred million years, so that in three or four thousand years no great movement would be perceived."[103] The duration of such cycles would be comparable to the advancement or withdrawal of the sea from the coastlines, a phenomenon that for Buridan was not needed at all to "save the generation of the highest mountains,"[104] and whose rate he estimated at ten leagues (ca. 20 miles) in the previous ten thousand years.[105] A rough calculation shows that a complete cycle would then take many millions of years at a minimum. Its pace was too slow even to determine whether coastlines changed as a result of a global movement of the ocean or were merely a local phenomenon.

For all of the boldness and brilliance of Buridan's construction, one may wonder whether it was intended to represent an actual possibility rather than just a thought experiment. After all, his model was based on the assumption that the world is "eternal," a proposition that he declared himself to be "absolutely" false. There is little doubt that he meant it. As the rector of the university in 1340, he signed the statute condemning philosopher Nicholas d'Autrecourt, who was proposing a form of actual eternalism.[106] Buridan may have stood on the edge of eternity, but he refused to cross that threshold. Yet his theory of the Earth does not seem to have been a purely intellectual exercise, devoid of any pretension to describe a process that might have been real. The eternity of the universe is of course just a philosophical convention; even the largest flood can never be universal, so the biblical one must be considered a miracle and outside the scope of natural philosophy; but almost everything else was admissible, though maybe not "certain." As he pointed out in a passage of his *Physics*: "There are some conclusions which are probable to me, which Aristotle and the Commentator [Averroes] have conceded, and which also need not be negated according to faith."[107] In his *Questions on the Metaphysics*, he insisted that conclusions of empirical nature were accepted as valid "because they have been observed to be true in many instances and to be false in none."[108]

"Probable" did not necessarily mean fictitious. In the 1270s, Siger of Brabant stated that "true notion" could be had of most meteorological phenomena, even though it was not the same kind of certitude attained in mathematics.[109] It was argued that their causes could be considered more secure than other kinds of knowledge not in spite of, but because they were based on sensory experience.[110] Buridan's meteorology emphasized personal observation, the description of real-world events located precisely in time and space, and even popular wisdom and the knowledge accumulated by crafts and trades.[111] The French philosopher was fully aware of the difficulties of induction, yet the regeneration of mountains was a "necessary" phenomenon, unlike those "transient" events (weather, earthquakes, or comets) of which we cannot have real "science."[112] In search of empirical evidence that could support his model of geological change and the immense age of the Earth it entailed, he turned to the origin of metals, rocks, and crystals. According to Avicenna and Albertus Magnus, they coalesce from mixtures of mainly earth and water through extreme and prolonged cold or heat. Buridan then argued that minerals that require extreme cold must originate close to the center of the Earth, considered the coldest place in the universe since it was the most distant from the Sun's heat. Yet they could be mined right under the surface we inhabit:[113]

> Ores of these minerals [exist] within the first, second or third layer of the habitable land, even though comparable cold or lack of heat does not exist there; indeed, these minerals formed in the course of a very long time at the center [of the earth], or very close to the center.[114] And even when their generation required instead vapors or hot gases, their ores can extend to great depths, where such gases or vapors do not penetrate. Minerals, indeed, come to this place from other regions.[115]

Rock crystals, Buridan argued, formed in the depths of the earth and were then lifted toward the surface by the global upward push. Likewise, metals that originated in the upper layer, where heat is available, had gone through an almost complete geological cycle and could now be found below that region. The pace of the earth's movement was set by the rate of erosion of the surface: since the mountains mentioned by the ancients (of negligible height compared to the radius of the Earth) were still there, and coastlines had since advanced or withdrawn by several miles at most, there was no way the journey of a crystal rock toward the surface could not have taken an unimaginable length of time.

Every Possible Earth

Buridan's doctrine was adopted, in different versions, by prominent scholars who were part of the same intellectual circles, such as Nicole Oresme (ca. 1320–1382), Albert of Saxony (d. 1390), Themon Judeus, and Marsilius of Inghen (ca. 1330–1396).[116] The doctrines of the eccentricity of the sphere of earth and of the global push that continuously replaces worn-out mountains were normally taught in Paris for at least a couple of centuries and became widely known across Europe. Some of Buridan's disciples or colleagues proposed non-trivial adjustments to his model. In his questions on the *Physics*, Albert of Saxony posited that earth and water should not be treated as separate spheres each with its own center, but rather as a single terraqueous globe (a conclusion that was reversed a few years later in the *Questions on Aristotle's on the Heavens*).[117] Above all, Albert of Saxony played a crucial role in the spread of Buridan's doctrine of the Earth, particularly in German-speaking countries and in Italy, where his commentary *On the Heavens* met with extraordinary success and was adopted at Bologna University as early as 1368.[118]

Buridan's commentaries *On the Heavens* and on the *Meteorology* were never printed (though there is ample evidence of their circulation in Paris as late as the early sixteenth century[119]). However, Albert of Saxony's *Questions* on the *Physics* had at least five editions, and those *On the Heavens* counted five Italian and three German editions between 1481 and 1534.[120]

The hypothesis of an eternal world became so common and established in Italian meteorological literature that previous precautions and distinctions became soon unnecessary. After 1350, it was no longer questioned that Aristotle taught the eternity of the world as a matter of fact.[121] In commentaries and treatises produced at Italian universities, locutions like "philosophically speaking" or "according to natural reasons," which circumscribed the value of potentially troublesome theses, became ritualistic formulas whose meaning the authors no longer cared to explain. In many cases they did not use them at all, while their texts carefully omitted any references to Scripture or to the Christian faith. Eventually, Italy proved to be a fertile ground for Averroes's radical interpretation of Aristotle. The teaching of heterodox or plainly heretical beliefs such as the mortality of the rational soul or the eternity of the world was normally permitted in the universities, provided the professors made it clear that the issue at hand was being discussed exclusively as a problem of Aristotelian natural philosophy. This was not always the case. In 1396 Biagio Pelacani of Parma, who taught astrology and natural philosophy in Bologna, Padua, Pavia, and Florence, was summoned by the bishop of Pavia (who was also the rector of the local university) for teaching the mortality of the rational soul and the eternity of the world as physical realities rather than philosophical conventions. In a passage of his *Questions on the Soul*, he also called the biblical story of Noah bringing wild beasts on the Ark "an old wives' tale," a story no educated person could believe.[122] Pelacani's teaching of *On the Heavens* was heavily based on Albert of Saxony's commentary, which, like Buridan's, ruled out the possibility of a universal Flood through natural causes.[123] Apparently, Pelacani was willing to go a step further and deny on philosophical grounds the historical accuracy of the biblical text. In doing so, he trespassed on the established boundaries between philosophy and theology. Yet it is likely that the discussion of the Flood generated little alarm, and that the whole affair was motivated, instead, by the suspicion that Pelacani upheld astrological determinism.[124] Having pleaded for a pardon and reassured the bishop that he had been speaking all the time "as a natural philosopher," he was immediately reinstated in his chair and stipend without further consequences.[125] His suspension lasted one day.

The popularity of Albert of Saxony's works did not mean that they were received without criticism.[126] The eccentricity of the sphere of earth was largely accepted, but Buridan's theory for the creation of new mountains met with staunch resistance. Among other issues, its adoption meant to renounce the Aristotelian tenet of the immobility of the Earth, since the whole of its body would move constantly in the direction of the emerged land. While these views were known and discussed, Italian professors remained generally committed to the Aristotelian orthodoxy as established in Averroes's commentaries. The most influential among them was probably Augustinian monk Paolo Nicoletti from Udine, also known as Paul of Venice (1369–1429). The author of major works on logic and metaphysics, he also composed a number of summaries and compendia that contributed significantly to his lasting fortune, and in the Renaissance they became convenient sources for vulgarizers of Aristotle's natural philosophy.[127] His *Exposition of Aristotle's Natural Books* (1408) was written mainly as a reference work for university students, who could find in a single book notions normally spread across many volumes.[128] After rejecting Buridan's doctrine on the creation of new mountains, Paul of Venice resorted to the combined action of winds, earthquakes, erosion, and the "motions of the sea"—most likely, the Aristotelian displacement of the ocean.[129] "If the sea will turn into dry land, this must be caused by some celestial motion"; the only candidate was of course the celestial sphere whose rotation produces the precession of the equinoxes. The problem with this thesis is that the sphere "completes its own revolution in 36,000 years, and in this time you cannot have but one permutation." Paul of Venice answered that, if so, then we must accept that these are the timescales of geological change.[130]

Because of the probabilistic nature of meteorology, the hypothetical character of its assumptions, and the ample (though not unconditional) tolerance of the ecclesiastic authorities, a whole gamut of options became available concerning the actual realism of those theories. At one end of the spectrum, more than a crack was left open to the infiltration of actual eternalism. A created world of undetermined duration was definitely less problematic and normally considered "probable." Finally, it was possible to stick to the traditional chronology of the Creation as to the true one, and regard the antiquity or eternity of the Earth as part of the exploration of what would be rationally and physically plausible if we did not have a higher Source. There is little doubt that this position was also extremely common, even in the faculties of arts. Nicole Oresme's French commentary on Aristotle's *On the Heavens* (ca. 1370–1377) offers an interesting counterpoint to those works that presented the extreme age of the world as the "most probable" case. It is often difficult (and perhaps pointless) to determine whether Oresme regarded his many different scenarios as anything more than pure mental experiments, meant to test the boundaries of the theologically and philosophically admissible. Exploring the possible cause of the universal Flood, Oresme observed that, "if God and nature caused the habitable portion of the earth to become as heavy as the other or caused the weight of the other part to diminish," then the emerged land would be plunged into the sea and "there could be a universal deluge without rain."[131] In the ensuing pages, he also suggested that an imbalance in the reciprocal transmutation of the elements (as mentioned in Seneca's *Natural Questions*) could cause the habitable land to wander across the globe "for thousands and thousands of years." Conceivably, the Antarctic pole would then take the place of the Artic one so that "the Sun would rise from the part we call West and would set in the opposite."[132] It was certainly possible to regard such speculations as intellectual exercises, suspend any judgment on their value as representations of the real world, or even think that the phenomenon was actually taking place without upholding *ipso facto* the possibility of an ancient Earth. Oresme seemed to choose the latter option when, quoting among others the Roman geographer Pomponius Mela (ca. 15–45 AD), he mentioned that the ancients claimed to have already witnessed the inversion of the celestial poles. However, he added that while this could happen in theory, in practice the Earth is not old enough:

> "[according to Pomponius] 'Letters preserved confirm that, while the Egyptians have been in existence, the sun has already set twice where it now rises.' . . . And although that is false, for the world has not existed sufficiently long so that this could have happened, nevertheless, if the movement of the heavens was perpetual, this would surely be possible."[133]

Still, theories that upheld a long duration of the world had to be taken seriously by their opponents. One of the most emblematic cases is that of Pierre d'Ailly (1351–1420), a high-ranking clergyman convinced that the end of the world would be a matter of centuries at most, and that its total duration could not be immensely longer. Eventually he turned to a form of Christian astrology that, he hoped, could help determine whether enough time was left for the conversion of the infidels and the healing of a divided and corrupted Church. His calculations, based on the intervals between the conjunctions of Jupiter and Saturn, reassured him that the Antichrist would not arrive before the year 1789.[134] Likewise, his *Concordance of Astrology with Theology*

(1414) established that 5,343 years had elapsed between the Creation and the birth of Christ.[135] The "great scandal," as he called it, of the disagreement concerning the time of the beginning of the world could thus be removed. This was a patent exaggeration: when the world exactly began was not a huge concern at all for the Church, even though chronologists, millenarians, and a few mathematically minded theologians may have thought otherwise.[136] His efforts to integrate the circular time of astrology within the linear chronology of the biblical tradition had to counter strong arguments for much longer timescales. In particular, he mentioned "some astrologers" who held that the total duration of the material world should span at least the 36,000 years required by one complete revolution of the "eight sphere." This would put off all his forecasts for an upcoming end of the world, or alternatively, his fix for the time of the creation.[137]

It may not be a coincidence that his compendium of the *Meteorology* (composed perhaps in the 1370s while he was lecturing at the Faculty of Arts in Paris) was littered with warnings, precautions, and scriptural references that it would be very hard to find in the comparable works of his Italian colleagues.[138] D'Ailly constantly emphasized that what he expounded was just Aristotle's view, and those teachings should be attributed solely to him.[139] He did not refrain either from rare but significant theological discussions, for example when he raised the oft-debated problem of why the rainbow did not appear before the biblical Flood, even though the necessary physical conditions must have existed already. His answer was that in this case, the reasons should be left to God alone: inscrutable divine will decided to reserve the apparition of that phenomenon to the time after the Flood, "so that through that particular cloud a sign of alliance or peace appeared, and we no longer have to fear that the world will die by another Flood."[140]

3

Vernacular Earths, 1250–1500

*And if in those parts [of the Earth where a great flood is
about to happen] there will be some wise man who knows
well the science of the stars, he will provide for himself and
his family, like that wise man Noah reportedly did.*

RESTORO OF AREZZO, *The Composition of the World* (1282)

THE ANTIQUITY OF the Earth, or at least its possibility, did not remain the
exclusive domain of professional philosophers and theologians and of their
Latin culture. Scientific subjects in general, and the nature and history of the
Earth in particular, raised interest in social contexts other than those of the
universities or the intellectual elites. In the course of the thirteenth century,
local vernaculars began to be used not only in translations and adaptations
of Latin or Arabic sources for the layperson, but even in original works. This
chapter examines a range of relevant scientific literature in the vernacular that
was produced between the late 1200s and the early 1500s in France and Italy.

Tuscany, in particular, offers by far the most numerous and most significant sources. While those works differed in terms of their approaches, intended readership, and reception, in general no constraints existed to the
dissemination of notions like the eternity or extreme antiquity of the Earth
across milieus as varied as aristocratic courts and artisanal workshops. Indeed,
some craftsmen consistently tried to merge their material expertise of the
Earth's surface with theoretical frameworks provided by natural philosophy,
proposing themselves as worthy interlocutors of the learned. The geological
writings of the most famous of them, Leonardo da Vinci, remain in popular
culture the feat of a unique genius centuries ahead of his time. In fact, his
elaborations built on centuries-old artisanal traditions, on the practices of
contemporary builders and water prospectors, and on models of the Earth
developed in the context of learned meteorology that found diffusion even in
the (overwhelmingly vernacular) world of the arts and trades.

On the Edge of Eternity. Ivano Dal Prete, Oxford University Press. © Oxford University Press 2022.
DOI: 10.1093/oso/9780190678890.003.0004

Earth History in Courts and Workshops

The first French version of Aristotle's *Meteorology* was completed around 1290 by a lecturer at the Faculty of Arts in Paris, Mahieu le Vilain, who wrote it at the behest of the Count of Eu in Normandy.[1] Epic novels, love poetry, religious literature, and even encyclopedias in the vernacular had long entertained the French courts, but proper philosophical subjects clearly represented something new for both authors and readers.[2] As le Vilain explained to the dedicatee, "I began to translate before you this book, word by word, as well as I can, in French language. But be aware: it is not possible to translate science in French as we do in Latin." Le Vilain picked meteorology exactly because it was considered the most approachable part of natural philosophy for a lay public, and one that could be "reduced more properly into French."[3] Even so, at the time both the French language and his readers were so poorly equipped to deal with scientific matters that he had to suppress some chapters, like those on halos and rainbows, because "it's all about geometry and our Romanic language does not have corresponding words."[4]

His most important concerns, however, were of a different order. Mahieu treated with particular care the parts dealing with the eternity of the world, for at least two reasons. The first is that the 1277 Paris condemnations, which shook the Faculty of Arts and which the author may have witnessed in person, still loomed large in the area. The translator was quick to warn that the section in which Aristotle discusses whether the sea ever had a beginning (answering of course in the negative) is "not very profitable," and he will simply skip it: "this truth has been assured enough by the Holy Scripture of the Bible and by the Christian faith."[5] Theologically and philosophically naive readers could take Aristotle's words "too literally" and fail to realize when the philosopher's opinions were at odds with Christian beliefs.[6] The second reason is that his potential readers seem to have been unfamiliar not only with the eternity of the world, but also with the vast geological revolutions of the Aristotelian tradition and with its immense timescales. Le Vilain illustrated the doctrine of the permutation between the ocean and the dry land, so that following the revolutions of the stars "there will be a time when there is dry land, where we have now the sea."[7] However, the possibility that France itself, "the seat of every chivalry, where the Holy Church is honored and served," will one day be erased by an inexorable nature may have appeared disconcerting to his potential readers. The author mentioned that some watery bodies "smaller than the sea" will be destroyed and renewed in different parts of the Earth, since they also age and need to be rejuvenated; were the universe eternal,

little by little this process would embrace the whole of the Earth, as Aristotle
wrote. In the end, though, le Vilain reassured his patron that the world would
not last long enough for the global movement of the ocean to affect large regions of the globe, so that France would never be swallowed by the waters.[8]
"The heavens and the world will not last forever and because of that, God
willing, the Kingdom of France will outlast all the others in its worth and
goodness."[9] Almost one century later, Nicole Oresme still resolved abruptly
that the Earth cannot be very old, or that "of course, there has never been nor
will there be more than one corporeal world" (a widely discussed problem
among philosophers and theologians), after presenting compelling reasoning
that the opposite might be true or at least "probable."[10] Of course, Oresme's
aim could have been to show that the Creation is the result of choice, not necessity, and that alternative Earths and universes were physically possible had
God so wished. Neither faith nor reason, though, compelled him to present
a recent and unique world as definite truths, unless this was a conclusion that
he or his patrons, the king and the court, felt more comfortable with.

Vernacular works composed in the same decades in Italy reveal very few, if
any, of those qualms. While courts were certainly part of the potential public,
the balance of political and cultural power in the central and northern part
of the country had shifted toward the mercantile class and the burgeoning
cities they dominated. Tuscany represents an especially meaningful case. By
the year 1200, most Tuscan cities had tamed the power of lay or ecclesiastic
overlords, established autonomous governments, and built a prosperous
economy based on trade, finance, and textile manufacturing. Bitter internal
conflicts, both within and without the walls of its cities, did not prevent
Tuscany from becoming one of the most densely inhabited regions in Europe.
The growth of Florence later eclipsed its minor centers, but cities like Siena,
Pisa, Lucca, Arezzo, and San Gimignano looked huge and powerful in the
thirteenth century when their commercial networks spawned across much of
Western Europe and the Mediterranean world.[11]

The new political and economic assets emphasized the importance and
dignity of the local vernaculars over the Latin of the Church and of the
universities. Soon enough, the mundane needs of the emerging mercantile
class led to the quick spread of the so-called abacus schools that focused
on commercial arithmetic, accounting, geometry, the measure of volumes,
weights, and areas, and vernacular reading and writing.[12] The son of a Tuscan
merchant would have been familiar with the geometry that troubled le Vilain's
aristocratic readers, but the increased political clout of the mercantile families
also demanded that appropriate tools be available for the governance of

public affairs. Since rhetorical competence was essential to the political life of the Tuscan free cities, around 1260 the celebrated Florentine scholar Brunetto Latini translated into the local dialect a number of Cicero's orations.[13] One century later, a consistent portion of Roman and late ancient literature could be read in Tuscan.[14] Similar considerations apply to the vernacular scientific texts that began to circulate in Italy in the late 1200s. The bulk of most private libraries was made of medical and practical works, often written directly in the vernacular, and sometimes turned into Latin rather than the opposite.[15] Natural knowledge was also incorporated into the Latin encyclopedias that, starting from the late thirteenth century, became the sources of numerous vernacular counterparts. Vincent de Beauvais's *Mirror of Nature* and Bartholomew the Englishman's *On the Properties of Things* provided general overviews of medicine, physics, astronomy, cosmography as well as politics, rhetoric, and natural and universal history.[16] Those works were often meant for quick consultation by laypersons, university students, and even preachers in search of materials for their sermons.[17] Indeed, an oft-stated aim of such encyclopedias was to illustrate unfamiliar phenomena, animals, and places mentioned in the Scripture.[18]

The best known and the most read of the Italian vernacular encyclopedias was Latini's *Book of the Treasure*, written in the mid-1260s in old French but soon translated into Tuscan.[19] As usual in Christian Europe before the mid-thirteenth century, Latini regarded the scriptural account of the Creation as a narrative that needed to be interpreted in light of, and integrated with, natural philosophy rather than strictly separated from it. Yet, Latini's *Treasure* was as secular as was conceivable in his time. Sacred matter was abundantly present, but only insofar as it was inextricably connected in medieval culture to philosophy, ethics, and universal history. Moralizing and sermonizing were conspicuously absent. Latini's account of the creation of the world quickly dispatched the biblical narrative of the six days, focusing instead on the issues that most concerned philosophers and theologians before the advent of Aristotle's cyclical Earth. Among them was the difference between time, in which creatures are immersed, and the all-embracing eternity of God,[20] and the generation of the primordial shapeless matter from which the elements and everything else derived.[21] As Latini pointed out, God created directly the first man but, as for the rest of the physical world, he just "ordered" it to be made. That is, He let the laws of nature He instituted to carry out their operations according to His plan.[22] Matter preceded everything else, not in the sense that it was there "before" but in the same way as "sound comes before singing"; namely, there can be raw sound without singing but no singing without sound, "and yet, they exist together."[23]

What Noah "Reportedly Did"

At a time when private libraries normally consisted of a handful of books, comprehensive encyclopedias like the *Treasure* must have seemed a better investment than more specialized scientific treatises.[24] Still, a limited but significant number of them also became available in Italian or French, particularly on medicine and on the subjects that best helped medieval readers make sense of the natural world, like cosmology, astrology, and meteorology. Le Vilain's *Meteorology* is a case in point. Just like in France, authors would normally be scholars or members of the intellectual professions who attended Latin schools and were often university-trained. Together with Al-Rhazes's and Aldobrandino's medical texts, the Florentine notary Zucchero Bencivenni also vulgarized the standard medieval astronomical textbook, John of Holywood's *Treatise on the Sphere*, in the early fourteenth century.[25] Yet the first scientific treatise in Italian vernacular conceived as an original work, rather than as a translation or adaptation, was written by a tradesman. All available sources suggest that the author of the *Composition of the World* (1282[26]), Restoro of Arezzo, was the same "Restaurus" active in the same years and in the same Tuscan city as a painter, a goldsmith, and maybe even as an illuminator and a bell founder.[27]

Restoro's time coincided with the first emergence of an artisanal elite conscious of its growing social and cultural relevance. Its members began to sign their works, affirm their individuality, and bridge the cultural and social gap that divided them from the masters of the written word. On the one hand, they tried to acquire some of the intellectual and literary abilities that defined the learned, including Latin reading skills[28]; on the other hand, they sought to differentiate themselves from the mass of the rank-and-file manual workers who were incapable or unwilling to follow the same path.[29] Restoro held that the "learned craftsman" did not repeat mechanically what he was taught as an apprentice but, like nature, he introduced in each of his works some changes "because of his nobility and knowledge . . . so as to receive praise."[30] The mastery of the figurative arts, and of the manual practices connected with them, was crucial for the understanding of the natural world. Astronomy (which included astrology) topped every other science, but drawing, engraving, and painting came immediately behind, as long as they were practiced by "learned" artists, turned "almost into gods" by their creative excellence in the

FIG. 3.1 Restoro of Arezzo's *Composition of the World*

The first page of the oldest known copy of Restoro of Arezzo's *Composition of the World*, probably completed no later than 1290. Biblioteca Riccardiana, Florence, Ricc. 2164 c. 1r. Courtesy of the Ministry of Culture.

representation and manipulation of matter.[31] Together with scholars who valued the kind of knowledge inherent in the manual arts, artists like Restoro created small but significant "trading zones" where information, sources, and methods could be exchanged.[32]

In the *Composition of the World*, the author merged his bodily and empirical experience of nature within a robust theoretical framework, mainly derived from Latin translations from the Arabic.[33] From the beginning, Restoro emphasized the nobility of human reason and that God can be known only through the rational investigation of the physical world He created.[34] The "how" and "why" must be assigned not by way of miracles, "which are above reason; but rather by way of reason, and similitudes, and rational examples."[35] Since he did not deal at all with ethics or universal history, Restoro could dispense almost entirely with scriptural sources and, when he mentioned them, they always fitted harmoniously within his conception of the workings of nature.

Restoro's discussion of the emergence of dry land, of the "flood," and of their causes is an impressive document of how events described in the Bible could be absorbed within the structure of a natural philosophy based on Greek and Arabic sources, as well as of how easily the medieval Italian society could accept the ancient Earth it entailed. The Tuscan artist adopted the widespread idea that the magnet-like influence of the northern stars pushed the ocean away from our hemisphere, while at the same time attracting the element of earth so as to form vast mountain ranges that he compared to the backbone in the human body. Smaller heights, however, could originate from earthquakes, or from the inflating of the earth due to the heat of the Sun. Erosion by water then carved valleys, leaving the hardest rocks as mountain peaks.[36] Finally, some mountains were formed by the floods of the sea that periodically submerged parts of the earth, moving dirt "from one location to another" and leaving behind petrified sea creatures like those he had himself extracted on various occasions:[37]

> We once found and excavated, almost at the top of a mighty mountain, many fish bones of the kind we called seashells and others called conches; and they were similar to those used by painters to keep their colors in. . . . And we were once on a large mountain, whose summit was covered with a huge slab made of an extremely hard rock of the color of the iron . . . . And when we dug underneath, we found dirt mixed with sand, tuff and rounded river stones together with fish bones of many kinds and species.[38]

Like Albertus Magnus and countless others, Restoro related diluvial rains to congregations of celestial bodies that produced lesser or greater inundations according to the number and position of the "stars" involved. As was customary in meteorological literature, his use of the singular (*il diluvio*, the deluge) referred to the phenomenon of the periodic inundations of vast regions of the Earth, rather than to the one-time biblical catastrophe.[39]

Likewise, Restoro elaborated on the mountains created by "the earthquake" (*il terremoto*), resorting again to the singular form even though he was obviously discussing a recurrent event, not one that was unique in the history of the Earth. Against Albertus Magnus but in line with many other authorities, Restoro insisted that even the most ruinous flood possible in nature could never be universal, as part of the Earth would always emerge from the sphere of water. A man sufficiently skilled in the science of the stars might then be able to forecast the coming of a deluge in the part of the Earth he inhabits and save himself "like that wise man Noah reportedly did":

> Thanks to the knowledge he was given, he made arrangements in advance, and he saved himself and his family from the danger of the flood in the ark. This major disaster of the flood, which was caused by the greatest possible conjunction that can take place in the world, is said to have happened to purge the Earth from its vices.[40]

The image of a great Flood as a "purge" for the Earth did not refer exclusively to the biblical punishment of humankind's sins. This notion included the renovation of the regions of the Earth that have grown old and corrupted.[41] Since the natural cause of these floods resided in the motions of the stars, they were cyclical in nature and must be counterbalanced by excruciating but equally localized droughts.[42] Restoro saw no reason why this naturalistic account should conflict with the theological one of the Bible. The Noetic catastrophe narrated by Genesis was simply the most recent of the great periodical inundations that scourge parts of the Earth, endowed on that particular occasion with a providential meaning. While Noah received advice of the impending disaster from God, in principle this knowledge was also attainable by human means. Nothing suggests that he was thinking within the narrow temporal limits of the traditional chronology, nor that his approach ever raised religious concerns. Scholars have seen some annotations in later manuscripts of the *Composition* as interventions of preemptive censorship, meant to defuse the impact of his doctrines.[43] In fact, the most significant and widely cited addition had nothing to do with religion and everything to do with Restoro's contention that craftsmen can bring an essential contribution to the knowledge of nature. Invoking the authority of Aristotle, an anonymous interpolator objected that "no one can reach knowledge, unless he is entirely removed from human affairs. . . . Science cannot be learned while plowing, wandering, running and eating; it requires rest and concentration."[44] The author was arguably a traditional scholar, possibly a layman well acquainted

with scientific literature in the vernacular, who found himself confronted
with one of the first manifestations of a different way of conceiving the study
of nature.[45] The chapters explaining that the flood is a recurrent natural phenomenon; that none can ever be universal, including Noah's; and that Noah
himself could have forecast the event through astrology did not elicit any
intervention.

Restoro certainly did not feel bound to a literal reading of Genesis, but
if his book was irreligious so were Albertus Magnus's, even though he was
venerated for centuries as a saint and eventually beatified.[46] Design and
finalism pervaded Restoro's universe: its Builder was the perfect craftsman,
who engineered the world in the best possible way. As he reinstated at every
opportunity, astral influences and natural laws were just ministers of God's
plans. In spite of the overarching role he assigned to the powers of the stars, he
carefully avoided the pitfalls of astrological determinism. Generation and corruption on the Earth, Restoro stated, must follow the motions of the heavens;
but while each of them is cyclical, their combined periods are incommensu-
rable so that they will never reproduce the same exact configuration twice.[47]

A Merchant's Earth

Restoro's book seems to have been more influential than its scant manuscript
tradition suggests.[48] It was still read in Florence in the 1480s, when humanist
and architect Antonio Manetti made the last extant copy. Whether directly
or indirectly, it could have been the source of Leonardo da Vinci's famous
analogy between the circulation of fluids in the human body and throughout
the Earth.[49] Restoro's *Composition of the World* seems to have been valued,
too, as an accessible introduction to astronomy and astrology, and for its
descriptions of constellations and planets and of their influences. Indeed, it
was one of those vernacular scientific works found worthy of a Latin adaptation, rather than the other way around. A Latin *Composition of the World*
written in the early 1400s by Paul of Venice was largely a summary of the
Tuscan artist's book that may have targeted university students or educated
but non-specialist readers and was later printed several times, even in illustrated editions.[50] Physician John of Carrara also borrowed heavily from
Restoro for his Latin *Constitution of the World*, composed between 1483 and
1490 under the patronage of the Marquis of Monferrato in northern Italy.
While John of Carrara knew both the original text and Paul of Venice's
Latin version, his book was mostly based on the former, though he added
quotations from a number of more recent sources.[51] Tellingly, Paul of Venice's

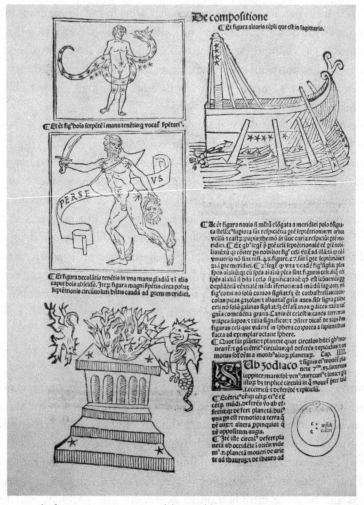

FIG. 3.2 Paul of Venice's *Composition of the World*

Figures of constellations, and a diagram showing the eccentricity of planetary orbits in an illustrated edition of Paul of Venice's *On the Composition of the World* (based on Restoro of Arezzo's work of the same title). Paul of Venice, *Expositio Magistri Pauli Veneti super Libros de Generatione et Corruptione Aristotelis. Eiusdem De Compositione Mundi cum Figuris* (Venetiis: Bonetum Locatellum, 1498), 106v.

Latin adaptation to university standards entailed the careful removal of every religious and scriptural reference, as well as of Restoro's exaltation of the artist and of his special knowledge of nature. A tension between different approaches to the role of faith in works of natural philosophy is sometimes apparent between not only adaptations from one language to the other, but also between vernacular versions of scholarly texts and their readers. In general, the issue was not one of secular versus religious worldviews; nor did anyone question the legitimacy of a rational investigation of nature. Rather, lay audiences found it difficult to grasp the methodological or epistemological curtains erected by scholars between theology and natural philosophy in the course of the thirteenth century.

The reader of a Florentine *Meteorology* (ca. 1310), for example, was not too satisfied with this book, and he let posterity know through his own annotations.[52] In all likelihood, he was neither a scholar nor an ecclesiastic but a merchant, conversant with vernacular scientific literature but also with the routes traded in the early 1300s by Florentine commerce.[53] The anonymous translator presented his work as a version of Thomas Aquinas's commentary on the *Meteorology*, but his reader was too well informed not to realize that the vast majority of the material actually came from Albertus Magnus's less authoritative one.[54] Keen on direct and straightforward information, he did not have much patience either for the bells and whistles (or rather, the *pro* and *contra*) of the commentary format: "Aristotle spoke briefly and to the point, but friar Albertus Magnus of the order of the preachers [Dominicans] . . . exposed it, as you will see hereafter, with much prolixity."[55] Obviously, he mistook the brief synopses given in the vernacular commentary, which functioned as headings to introduce the discussion of a certain subject, for the actual Aristotelian text that he probably never read.

In any case, his most important criticism was not about prolixity: he felt that an account of the physical history of the world should address not only observation, reason, and natural causes, but also the book of Genesis.[56] His point was not that the Creation took place in six days, or that the history of the Earth must be compressed within the limits of the biblical chronology (neither is mentioned), but that the final and supernatural causes of natural phenomena should not be excluded from the narration. Nature comes from God, "Who is above nature"; He ordered that part of the earth be uncovered by water for the generation of plants and animals "like the Bible says in the book of Genesis." The advent of Jesus was a "higher and more marvelous process than the division and ordering of the elements." Likewise, Noah's flood took place in order to punish the sins of men, and the waters withdrew within their present limits in accordance with God's plans.[57]

That said, he readily acknowledged that God operates in the world through natural causes, not miracles, and that the problem of the emergence of dry land must be addressed "naturally and according to astrology." Following Latini (and possibly Restoro), he described how it emerges above the sea "like mountainous islands, and full of caverns . . . by the power and virtue" of the planets.[58] He did not hesitate either to contradict Albertus Magnus's doctrine of the flux of the ocean, based on his own travels "in the sea of England"—a land of conquest for Florentine bankers and wool traders in the early fourteenth century.[59] Experience supported, too, the counter-intuitive but widespread notion that the surface of the ocean is higher than "any high mountains on the earth." As fellow "merchants of Genoa" (or perhaps Brunetto Latini) allegedly told him, from the Atlantic coast of Africa the ocean appears visibly higher than the shore. There was no need to travel, though, to become convinced of the physical reality of this phenomenon: just "take a circular vase, and fill it completely with water; you will see that its level is higher in the middle than along the rim."[60]

Latini's *Treasure*, Restoro's *Composition*, and the Florentine *Meteorology* addressed different levels of the vernacular public, but they all contributed to the diffusion of a lay scientific culture between the late Middle Ages and the early Renaissance.[61] The *Treasure* reflected a time that preceded the massive penetration of Aristotelian philosophy into Christian Europe, and the ensuing need to treat reason and Revelation as fundamentally different cognitive modes. Restoro's later treatise adopted a methodological separation between theology and natural philosophy, but his work was still pervaded with the presence of the divine. Twenty or thirty years later, the Florentine *Meteorology* no longer addressed the providential causes of natural occurrences. Their readers may have criticized the authors' approaches to natural philosophy, but they did not have disturbed or scandalized reactions to the largely secular science they expounded, nor is there any evidence that the diffusion of this literature alarmed the ecclesiastical authorities. Red lines of course existed, and it could be dangerous to cross or even approach them. The world had been created; astral influences did not restrain human freedom; Noah's flood may or may not have been universal, or part of the order of nature, but it did happen and came to punish humankind's sins. Those strongholds could not be touched, but almost everything else was fair game, whether in Latin or in vernacular languages. In spite of the many nuances and differences, there was no such thing as a secular "high culture" in Latin restricted to professional philosophers, where the notion of an ancient world circulated, and a vernacular "low culture" hopelessly tied to a religious worldview and to a literal reading of the scriptural account of the creation.

"Page-like Skins"

Restoro's treatise and the Florentine *Meteorology* were still copied and even printed well into the 1500s.[62] Their longevity was due at least in part to the

affirmation of the culture and values of humanism, which from the late fourteenth century sought a more direct contact with the literary and moral legacy
of antiquity. Humanists created a polished and classicizing Latin that was
adopted as the language of civil administration and international affairs, and
whose mastery became a powerful status symbol.[63] With a renewed emphasis
on the dignity and prestige of Latin as a tool of the governing elites, the production of philosophical and scientific texts in the vernacular almost ground
to a halt, forcing less educated readers to rely on translations or adaptations
composed centuries earlier. Yet those "outdated" works contributed to the
uninterrupted circulation of productive alternatives to the Aristotelian tradition that was by then entrenched in the universities. The directional evolution
of the globe from an initial to a final state, the analogy between the circulation of fluids in the human body and that of water within the Earth, the
physical possibility of universal floods and of the spontaneous generation of
human beings, continued to be accessible to a public with little or no reading
knowledge of Latin.

Mainstream meteorological literature would be a poor indicator, indeed, of the kind of information that was commonly available in central
and northern Italy on the nature of the Earth's surface. In the two hundred
years between Restoro of Arezzo and Leonardo da Vinci, Tuscany turned
into a major mining and metallurgical region whose miners, prospectors,
and builders were tasked with finding minerals for furnaces and guns, construction stones for public and private buildings, and water sources for its
cities. Diverting streams, drying up swamps and mines, and building defenses
against increasingly ruinous inundations was part of the resumes of many
Renaissance artists/engineers. Leonardo da Vinci is just the most celebrated
representative of an extremely rich and variegated tradition.[64] The wealth of
observations provided by those activities would have been out of scope in
learned commentaries to the *Meteorology*, but this does not mean that their
implications for the nature and history of the Earth were lost. The "trading
zones" between craftsmen and the learned, which might have been limited in
Restoro's time, continued to expand in the following centuries. By studying
and finding inspiration in ancient sculptures, architectures, and construction techniques, artists asserted their active participation in the values of
humanism and in the rebirth—that is, the "Renaissance"—of the classical
world.[65]

Earth strata are an outstanding example of a geological feature almost
entirely ignored in medieval and Renaissance meteorological literature, yet
actively studied by miners and engineers for practical purposes. By the early

1400s, Tuscan prospectors had accurately characterized layered rocks and relied on the properties of the various strata to detect the location of water sources that could be channeled into aqueducts, or that had to be avoided when it came to laying the foundations of new buildings. In his landmark Latin treatise *On the Art of Building* (ca. 1440), Florentine humanist Leon Battista Alberti noticed how "Those who have researched [sources of water], have observed that the whole crust of the Earth, and mountains in particular, consists of page-like skins, some denser, some more rarefied, some thicker, some thinner."

> And they have observed that in mountains these skins are heaped and piled up one above the other; so that on the outside their layers and joints run horizontally from right to left, but on the inside the skins slope [downward] toward the center of the mountain, and the entire outer surface is inclined equally, although the lines do not continue uninterrupted. About every hundred feet the skins are cut off abruptly by a diagonal line, to form steps; then by a similar gradation the layers run in continuous steps from either side of the mountain to the very center.[66]

While largely ignored by historians of geology, Alberti's passage is arguably the first detailed description in Western literature of the characteristics and directions of rock strata extending beneath the surface.[67] His account betrays the author's lack of familiarity with actual fieldwork and the use of second-hand information he did not always understand correctly. For instance, he unduly generalized observations that must have been strictly local and particular in nature, like the fixed 100-foot distance between fault lines, or the statement that strata inside a mountain always slope downward.[68] A water prospector would have known that nature accommodates a much greater variety of orientations, inclinations, and spacing between fault lines. In fact, the "master builder of the Italian Renaissance," as Anthony Grafton calls him, gave due credit to the "craftsmen of acknowledged experience and skill" that were his indispensable interlocutors.[69] His sources were neither Vitruvius— his classical model—nor the writings of the natural philosophers, but anonymous "water seekers."[70] Once those "men of sharp wit" figured out the inner structure of the Earth, it was "easy" for them to realize that water collects "in the spaces between the different strata or leaves of rock." Furthermore, they learned how to tell layer from layer based on their thickness and on the nature of the soils, since "the different strata vary in their capacity to hold and yield water."[71] Their understanding of the inner structure of mountains, Alberti added, enabled them to find "hidden water" by boring into the rock at the right locations. Alberti enumerated at least ten of these different layers and ranked them according to their capacity to hold or trap water and to the quality of the water they could provide.[72]

Meteorological works did not normally discuss geological strata, but when they did, they always related their origin to the extreme antiquity of the Earth. In the writings of the "Brethren of Purity" in the tenth century CE, "sands, argil, and pebbles" accumulate at the bottom of the oceans "layer upon layer, in the duration of time and ages, until exposed by the withdrawal of the waters."[73] One century later, Avicenna surmised that they likely formed through successive inundations of the sea in immemorial times, each flood leaving a new layer:

> It is also possible that the sea may have happened to flow little by little over the land consisting of both plain and mountain and then have ebbed away from it. . . . It is possible that each time the land was exposed by the ebbing of the sea a layer was left, since we see that some mountains appear to have been piled up layer by layer, and it is therefore likely that the clay from which they were formed was itself at one time arranged in layers. One layer was formed first, then, at a different period, a further layer was formed and piled (upon the first, and so on).[74]

The Latin version that circulated in Western Europe gave up much of the clarity and richness of Avicenna's original, but it retained the notion that "mountains are made over a great length of time." When "the sea raises [over the dry land] . . . it also brings together and amasses certain soft substances, which following the withdrawal [of the sea] dry up and are turned into mountains."[75] Any medieval reader could have concluded that partial inundations of the sea were the most likely cause for the formation of layered rocks, and that the process must have taken ages. Even alternative explanations implied long timescales. Restoro of Arezzo argued that mountains were the tools used by the heavens to shape the surface of the Earth, in the same way that a blacksmith needs to make his anvil before he can forge metal. The attraction of the northern stars built them one layer upon the other, just like an anvil is made by laying down successive sheets of iron:

> In the appropriate time, the heavens by their virtue can gather mounds of earth, heap them one upon the other and attract this earth towards

the sky, like the magnet by its virtue can attract iron, making as many
mountains and as high as needed, like a blacksmith lays down layers of
iron one upon the other in order to make the anvil he needs.[76]

For Avicenna, the deposit of successive layers by recurring floods of the sea
was only one of the causes that could give origin to a mountain, and not
the most conspicuous. Conversely, Alberti and his sources regarded those
superimposed "pages," "skins," or "leaves" of rock—all hinting at their trivial
thickness when compared to the size of a mountain, let alone the Earth—as
the defining feature of "the whole crust of the Earth, and mountains in particular."[77] Bewildering rocky landscapes, in which cracking and crumbling
mountains exposed their layered interiors to the observer, were ubiquitous
in fifteenth-century Italian paintings. Mantegna, Botticelli, Giovanni Bellini,
and many other masters represented geological features as idealized forms,
rather than as accurate depictions of existing physical objects; yet their visual
language may have embedded commonly held views on the nature of the
Earth's surface. Avicenna's statement that most mountains were in a condition of decay, with new growth and formation being the exception rather
than the norm, would be a good characterization of those sceneries. What
lay before them was a world of ruins, scarred by the unequivocal signs of the
passage of time. Towering rocks made their last stand against erosion, while
a succession of broken and twisted soils attested to the ancient vicissitudes of
the world.[78]

　　Alberti did not have any qualms either about the idea of an ancient Earth.
He reported as a mere "joke" the opinion that the depression occupied by
the ocean was made directly by God, who impressed it in the body of the
Earth when he first created the mountains "as if he used a seal."[79] In the vernacular dialogue *Theogenius*, composed in the same years, he also discussed
the common notion of a universe that went through cyclical renovations,
without a clear beginning in sight. While a given configuration of the stars
repeated itself every 36,000 years, it was impossible to determine "which
one . . . is closer to the end of the world, and further from its beginning."[80]
While no evidence exists that a complete translation of Avicenna's writing on
the subject circulated in Europe before modern times, that theory would have
been a straightforward inference in a culture that regarded the periodical submersion of parts of the Earth as a matter of fact. Indeed, a few decades later
Leonardo da Vinci mentioned this doctrine in his famous discussion of the
origin of fossils and geological strata.

FIG. 3.3 Andrea Mantegna, *Agony in the Garden*

"Geological" landscape in A. Mantegna's *Agony in the Garden*. Isolated peaks like those represented in this painting were usually described in meteorological literature as harder rocks, still standing after erosion washed out to the sea the original plateau. Andrea Mantegna, *Agony in the Garden*. Verona, San Zeno altarpiece (1454).

Leonardo's Deluges

In the two hundred years between Restoro and Leonardo (1452–1519), no other artist left a significant corpus of writings dealing with natural philosophy proper. Even though an increasing number of them took up the pen during the fifteenth century, sometimes filling their pages with learned citations from Aristotle or Vitruvius, they usually restricted themselves to matters directly related to their professional activities, be they painting, architecture, hydraulics, or the construction of increasingly sophisticated machines.[81] Historians have long framed Leonardo's inventive genius within the tradition of fifteenth-century Tuscan engineering, but the paucity of comparable sources and of studies on medieval meteorology have made it much more difficult to contextualize his geological writings.[82] His descriptions of layered rocks, his rejection of Noah's flood as the origin of marine fossils, and his obvious disregard for the chronology of the Bible remain in popular and, to an extent, even academic literature as the hallmarks of a unique genius centuries ahead of his time.

Of course, no one can ever be ahead of their time. Leonardo's notebooks may provide wonderful insights into the workings of a genius's mind, but they represent above all a unique source on the practices, cultural assumptions, and knowledge base that were available to an Italian artist/engineer around 1500. There is no doubt that the interest toward the natural world he began to develop after 1482 far surpassed the technical and scientific notions his employers expected him to master. The tumultuous growth of his library attests to the efforts he made to overcome the limits of a mostly empirical education.[83] Yet, the books he owned in 1495 would have been accessible to any reader of even modest culture.[84] In his fifties, Leonardo was still struggling with the Latin dictionaries and grammars he needed to access more numerous and diversified sources.[85]

His troubles with the international language of learning did not prevent him from appropriating a number of fundamental concepts of medieval meteorology, such as a timeless Earth and an entirely secular approach to natural phenomena. Alongside the oral communications and ad hoc translations he might have used, he was certainly indebted to the medieval vernacularizations that continued to circulate and that provided important alternatives to the Aristotelian Earth of mainstream meteorology. Moreover, the artist squarely placed himself within the artisanal elite that, since Restoro's time, had sought acknowledgment, status, and cultural legitimacy. The study of the Earth represented for him a perfect bridge between the world of the learned and that of the manual workers, ideally suited to illustrate their conception of the role of empirical knowledge and experience in natural philosophy. Inaccessible to the philosopher's mind, as Aristotle himself acknowledged, the subterranean world opened its secrets to the eyes and hands of miners, engineers, and prospectors; yet, origins and causes could not be understood without the theoretical scaffolding provided by learned meteorology. The book on "earth and water" that Leonardo mulled over for a long time was conceived in continuity with medieval attempts to merge artisan expertise and learned meteorology into a more comprehensive discipline, composed of both empirical and theoretical knowledge. The epistemological foundations of the study of the Earth were the same as those of the art of painting:

Those who fall in love with practice without knowledge, are like the helmsman who gets into a ship without rudder or compass, and who can never be certain of where he is going. Practice must always be founded on sound theory.[86]

Leonardo's geological writings include a whole range of different "theories" whose relative chronology is difficult to pin down, owing to the fragmentary state of his notebooks but also to the evolution of his thought. While he never managed to bring the different problems he worked on into an entirely coherent synthesis, the "sound theory" he tried to elaborate between 1506 and 1515 was clearly based upon Albert of Saxony's doctrine of gravity, as expounded in the *Questions on Aristotle's 'On the Heavens'*. The German philosopher's work was never translated, or at least never published into Italian, but it circulated widely in the country. Leonardo himself mentions the book among those he owned in 1508.[87] Drawing upon Jean Buridan's theory of the Earth, Albert of Saxony explained that the offset between the geometrical center of the sphere of the earth and that of the sphere of water allowed the emergence and the continuous renewal of dry land in a self-sustaining and self-regulating cycle.[88] Leonardo's summary of that doctrine is unequivocal:

Because the center of the natural gravity of the earth ought to be in the center of the world, and the earth is always growing lighter in some part, and the part that becomes lighter pushes upwards, and submerges as much of the opposite part as is necessary for it to join the center of its aforesaid gravity to the center of the world.[89]

Albert of Saxony's physics underpinned the whole edifice of Leonardo's Earth, but only as the foundation of a vastly larger building. The philosopher's model remained a broad outline, intended to give general answers to self-evident observations related to large-scale phenomena. Leonardo's problem was more complex, because he needed to provide a common theoretical umbrella to disparate bits of information on stones, geological layers, and fossils that had been circulating for decades or centuries among artists and engineers. Even more crucially, he lived in a world that was expanding not only vertically into the underground, but also horizontally toward other continents. In the late 1400s, new geographical information had begun to undermine the foundations of traditional cosmography and meteorology, revolutionizing the previous knowledge of the Earth's land masses. Countries overlooking the Atlantic Ocean may have led the charge, but they still drew heavily upon the expertise of Italian traders and sailors. Leonardo never mentions recent New World discoveries, but he was well acquainted with the family of Amerigo

Vespucci, the Florentine explorer who gave his name to the Americas, and it is not credible that he was not aware of their implications.[90]

In medieval meteorology, the vastly larger size of the sphere of water, and the offset between its center and that of the sphere of the earth, explained egregiously why dry land was concentrated in a single landmass. This worked well until it became obvious that inhabited islands and continents existed even in the southern hemisphere and at widely different longitudes. America and the antipodes vindicated the once minority view that water must be only a thin layer surrounding the surface of the earth—or, as Leon Battista Alberti put it, a "light sprinkling, like summer dew on an apple."[91] The perplexing images that showed the sphere of the earth immersed in a globe of water almost twice its diameter, barely touching its surface with one extremity and the center of the world with the other, were still ubiquitous but would soon belong to the dustbin of history.

Yet Leonardo did not question the fundamental postulate that water must occupy a much larger volume than earth. If its vast majority can no longer reside outside the earth, then the only place left "is within the body and the veins of the earth."[92] Leonardo's sketches in the *Codex Leicester* show the earth as a crust of variable thickness, surrounding an immense subterranean ocean "in the manner of a container full of water."[93] As Albert of Saxony had already speculated in his commentary on the *Physics*, earth and water may be too intertwined to be treated as separate spheres, each with its own center.[94] Instead of the gradual upward push described by Albert, Leonardo envisioned periodical cataclysms leading to the rise of mountains and dry land—or, alternatively, to their collapse. The veins of water that circulated in the underground constantly eroded the vault of the subterranean ocean; from time to time, vast portions of earth could detach from the dome, falling through the watery abyss toward the center. The displacement of such huge masses perturbs the previous balance, calling for a new one. The whole of the earth then shifts toward the opposite direction in order to restore the coincidence between the center of mass of the terraqueous globe and the center of gravity of the world. As its great body moves, mountains rise on one side of the globe, while on the other, dry land is engulfed by the sea. One of those upheavals may once have formed cavities like those filled by the Dead or Caspian Seas.[95] An even larger event could have caused the very first emergence of dry land or perhaps, the latest occurrence, when its creation outpaced erosion:[96]

> The great elevation of the mountain peaks above the sphere of the
> water may have resulted from the fact that a very large space of the

earth, which was filled with water, that is, the immense cavern, must have fallen for a considerable portion of its vault toward the center of the world.[97]

And thus the rest of the earth, having become lighter in that part where the said weight will have fallen, of necessity will be removed from the centre of the world, and the earth and mountains will have come out of the sphere the water on that lighter part . . . And the sphere of the water in this case does not change site, because its water fills the place left when that part of the earth fell by gravity; and thus the sea remains within itself, without variations of height.[98]

Unlike Albert of Saxony and Aristotelian meteorology in general, Leonardo may not have regarded the existence of dry land as a stable, self-sustaining condition that existed since the Creation and could be maintained indefinitely. Various vernacular sources available to him suggested indeed different scenarios. The Florentine *Meteorology*, a copy of which Leonardo almost certainly owned, described the progressive consummation of the element of water by the heat of the Sun in the context of a directional history, so that "before the end of time, the sea will dry up entirely and only three elements will be left."[99] It is true that the thesis was mentioned only to confute it, in order to reassert the eternally cyclical nature of geological change: the water that evaporates under the Sun's rays would not be lost, but would be fed back into the circuit that supplies rainfalls and rivers to be eventually returned to the sea.[100]

After some wavering, Leonardo appears to have appropriated this criticism, opting instead for the opposite scenario: in the long run (and contrary to Albert of Saxony's steady-state Earth), the creation of new elevations would be unable to keep pace with erosion. Eventually, the whole of the dry land would be washed down to the ocean. The four elements would then find their ideal equilibrium, settling in concentric spheres around their common center according to their weight.[101] The final submersion of dry land, and of terrestrial life with it, would therefore be inescapable.[102] He may have envisioned even that state as provisional: the Sun's heat, and the Moon's influence on the tides, would work together to recreate agglomerations of soil under water, eventually leading to new islands, larger areas of uncovered land, and so on.[103] The history of the terraqueous globe would then be both directional and cyclical, oscillating between eras when the element of water covers entirely the element of earth and other periods when some of it manages to emerge.

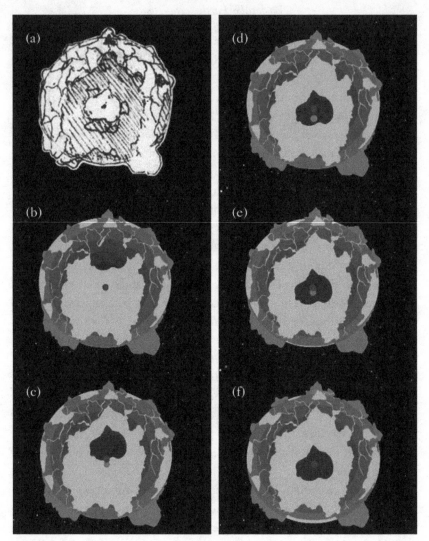

FIG. 3.4 Leonardo's Earth

A. Codex Leicester, f.36r (detail)

B. The element of earth resembles a jar containing a subterranean ocean, which accounts for most of the element of water. The solid crust of the Earth is sprinkled with relatively shallow seas.

C. Subterranean veins of water run within the crust of the Earth. From time to time, erosion can provoke the detachment from the vault of large masses of earth that fall toward the center of the world (or of the Earth's elements) under their own weight.

D. The displacement of such large bodies causes the overall center of mass of the element of earth to shift away from the center of gravity of the world.

E. The detached body is now at rest in the center of the world (or in the center of gravity of the Earth's elements) and can no longer move. In order to restore the previous balance, the crust must move into the opposite direction.

F. The crust is now at rest too, but its movement has caused the raise of mountains and the emergence of new dry land on one side (top), while on the other side (bottom) a new sea has formed.

Vernacular authors like Restoro of Arezzo might also have been the main source of Leonardo's famous analogy between the circulation of blood in the human body and that of water in the body of the earth.[104] After 1508–1510, though, he gave up the most cogent aspects of the similarity. Having dismissed the medieval theory that the surface of the ocean could be higher than mountaintops, Leonardo sought in "the heat of the fire generated within the body of the earth" (meaning right below the surface, not in the center) the engine that could rise water to such heights. In the end, he was unable to find the equivalent of the human heart.[105] Likewise, he renounced the Platonian idea that a vegetative soul in the earth might be responsible for the growth of rocks and mountains.[106]

Leonardo's numerous professional interests provided essential clues to his investigations. The reconstruction of the process of fossilization of sea crustaceans, for example, could be indebted to the studies for the equestrian monument of the Duke of Milan, Francesco Sforza.[107] Some of his hydraulic projects were still underway in the years that saw his strenuous efforts to understand the inner functioning of the Earth. Around 1513 he sketched in pen and ink two diagrams of the "soils of the [banks of the] Adige [river] to be bored", perhaps in anticipation of works to be carried out around the river.[108] His drawing is nothing more than a prospector's note; it shows layers of gravel alternating with more compact soils, and parallel strata that bend and warp in a way that is closely reminiscent of formations actually observable in the region. The second sketch, in particular, could be based on the survey of a specific area, as opposed to Leon Battista Alberti's rather abstract generalization. It is telling, though, that Leonardo's descriptions of layers and folds in the Codex Leicester retained a similar approach, with an almost instinctive emphasis on the relation between geological strata and water sources or infiltrations:

> But with mountains, where the layers in the stratifications of the rocks are placed slantwise or straight up, and little earth is interposed between the layers, the rain waters immediately penetrate that earth and move down through the clefts in the rocks and congregate to fill up the veins and hollow spaces. . . . I shall go on here to discourse a little about finding waters, even though it may appear a subject unconnected with the matter in hand.[109]

In the following lines, he provided his own list of different soils, each characterized in terms of its capability to retain, absorb, or filter water.[110]

While he certainly knew Alberti's book, his description is entirely independent, thus indicating not a common literary source but rather a common base of knowledge.[111] Historians of geology have proposed attributions of paintings, or parts of them, to Leonardo based on his supposedly unique insights into those phenomena. However, it seems clear that this kind of expertise must have been rather common among his contemporary engineers, prospectors, and builders.[112]

Leon Battista Alberti or his sources did not venture to describe the origins of layered rocks, but Leonardo took a further step and elaborated a complex theory of their formation that seems to have no obvious precedent. The examination of the fossils embedded between layers of rock, as well as his expertise in hydraulics, suggested to Leonardo that those "soils" were deposited by the annual overflows of rivers, which in Italy usually take place in the fall. After their withdrawal, they leave on the ground a layer of soft mud; when the silt hardens, organisms both alive and dead, as well as the traces of their activities, get trapped in the new rocks.[113] The sediments brought by rivers also bury oysters, shells, and other creatures that live in coastal waters, and from time to time, large storms can carry this mud "from the seashore . . . into the deep sea" where "they settle and become petrified." Thus, Leonardo concluded, "bed formations of varying thickness are formed their variety corresponding to that of the storms, smaller or greater."[114] In the course of time, the periodical readjustments of the center of mass of the earth lift the sea floor.[115] As rivers cut valleys through the raising plateaus, their flanks expose the old strata. Leonardo acknowledged that layered rocks were extremely common, but since they originated only along riverbeds and in coastal waters, they could not be as ubiquitous as Alberti reported. He remarked indeed that "where the valleys have never been covered by the salt waters of the sea, there the [fossils] shells are never found."[116] A diagram in the Codex Hammer shows a mountain composed of bedded layers, but an annotation warns that "the strata or layers do not continue to any great distance [in the vertical direction] underneath the roots of the mountain," because layered rocks only form in shallow waters slightly below the sea level.[117]

It is hardly a coincidence that Leonardo laid out his most detailed analysis of the subject in the context of his famous discussion of the universal Flood, in which he excluded that fossil sea creatures (and the strata that enclose them) originated from the Noetic catastrophe.[118] The fact that he dared to question its universality, and its role in the formation of marine fossils, has traditionally been touted as outstanding evidence of his "truly prescient character," but meteorological literature had dealt with the issue for centuries.[119]

Aristotelian meteorology admitted, of course, only local or regional floods—but commentaries composed in the previous couple of centuries did not deny that the biblical inundation was universal. They simply ascribed it to a miracle, which did not pertain to natural philosophy and had no place in the history of the Earth. Noah's flood was miraculous in at least two ways: first because it happened at all, and second because such a cataclysmic event did not leave any traces of its passage on the surface of the Earth. By definition, every physical observation concerning the history of the Earth could and must be explained through solely natural causes. The exclusion of Noah's flood from natural philosophy was indeed a cornerstone of its traditional autonomy from theology; yet, at the turn of the sixteenth century, the status quo began to be questioned.

By the late 1400s the unqualified teaching of Aristotelian doctrines like the mortality of the human soul or the eternity of the world was raising serious concerns among the Church hierarchies. In 1513, the bull *Apostolici Regiminis* reinstated that professors of natural philosophy should not teach patently heretical doctrines—or at least, they should not teach them as "absolutely" true. In the same years, the physical plausibility of universal floods was being defended on purely philosophical grounds.[120] By bringing back theological events like the great Flood within the realm of natural philosophy, those developments threatened the disciplinary boundaries that had long guaranteed the philosophers' intellectual freedom. It is impossible to determine to what extent Leonardo was touched by those controversies, but there is ample evidence that an entirely secular approach to natural philosophy had become commonsensical. His writings offer no hint that he ever thought otherwise. His firm denial of the physical impossibility of a universal flood, and of the diluvial origin of marine fossils, should be seen against the backdrop of a growing hostility to the naturalization of Noah's flood among natural philosophers. Leonardo did not intend to question the historical accuracy of the Mosaic story, as it is often argued. Rather, he reiterated—in line with the meteorological tradition—that the biblical Flood was not part of the natural history of the Earth, and that a universal inundation could not explain the phenomena increasingly attributed to it.[121] Leonardo's stance was similar to that of Girolamo Fracastoro's, who in the same years backed the Aristotelian tradition by denying that a general inundation of the globe could have left fossil seashells behind. The only unprecedented thing about these claims was that they needed to be defended.

As Leonardo argued, the "turbid waters" of a catastrophic inundation would have carried various different species "mixed up, and separated one from another, amid the mud, and not in regular rows in layers as we see them in our own times."[122] Moreover, since seashells are heavier than water, they would have remained on the bottom.[123] Given that they can move no faster than snails, it was also unreasonable to believe that they crawled very far in the few months that saw the waters of Noah's Flood cover the earth; yet the remains of oysters and mussels could be found hundreds of miles inland.[124] Finally, a single inundation would have produced a single bed of sediments, so that "you would find the shells at the edge of one layer of rock only, not at the edge of many."[125] What about, then, a great number of such events, like in the hypothesis formulated by Avicenna that arguably continued to circulate?

> And if you should wish to say that there must have been many deluges in order to produce these layers and the shells among them, it would then become necessary for you to affirm that such a deluge took place every year.[126]

Scholars have long assumed that Leonardo rejected this alternative because the idea of other floods comparable to Noah's, and the implied time scales, was too bold even for him.[127] Likewise, it has been argued that he never tried to date geological strata because he chose to ignore the "elephant in the room" constituted by its implications for the belief (held as unquestionable) that the Earth was created only a few thousand years earlier.[128] In fact, Leonardo did not argue that many deluges could not happen, but only that they could not be the cause of layered rocks; otherwise they should take place every year, which is obviously false. Like every meteorological author he could have read, he simply assumed the Earth to be of undetermined and undeterminable age. The absolute dating of strata did not have any relevance in medieval and Renaissance meteorology; when first attempted in the seventeenth century, the aim would be to support, not to disprove, the chronology of the Bible. Leonardo's writings do not provide the slightest hint that the idea of a young Earth ever crossed his mind. The Italian artist failed to bring up the problem not because it was too much of an issue, but because it was not an issue at all. While his world was expanding horizontally toward other continents and vertically in the underground, the one dimension that did not need to be enlarged was time:

> Since things are far more ancient than letters, it is not to be wondered at if in our days there exists no record of how the aforesaid seas extended over so many countries; and if moreover such record ever existed, the wars, the conflagrations, the changes in speech and habits, the deluges

FIG. 3.5 Leonardo's Deluge
Leonardo da Vinci, *A Deluge* (ca. 1517–1518), Windsor: RCIN 912380
"Let there be represented the summit of a rugged mountain with certain of the valleys that surround its base, and on its sides let the surface of the soil be seen slipping down together with the tiny roots of the small shrubs, and leaving bare a great part of the surrounding rocks . . . And the mountains becoming bare should reveal the deep fissures made in them by the ancient earthquakes . . . And let the fragments of some of the mountains have fallen down into the depth of one of the valleys, and there form a barrier to the swollen waters of its river, which having already burst the barrier rushes on with immense waves, the greatest of which are striking and laying in ruin the walls of the cities and farms of the valley."
English translation from Jean Paul Richter and R. C. Bell, *The Notebooks of Leonardo Da Vinci*, vol. 2 (New York: Dover Publications, 1970), 290–91.

of the waters, have destroyed every vestige of the past. But sufficient for us is the testimony of things born in the salt waters and now found again in the high mountains, sometimes at a distance from the seas.[129]

His explanation of why no records exist of the past ages derives from the meteorological tradition, including the evidence brought by marine fossils. Also in line with mainstream meteorology is Leonardo's firm belief that nature cannot produce a universal Deluge. Still, the idea of the periodical destruction and regeneration of parts of the Earth—and of human civilization with it—by large regional floods remained extremely common in his time. Like many contemporaries, the artist developed a profound fascination with the theme of the "deluge," which culminated in the famous series of "deluge drawings" variously dated between 1514 and 1518.[130] Those apocalyptic views, in which the artist fused his "imaginings with his scientific knowledge," did not refer specifically to the biblical event as once thought.[131] He may have found inspiration for his images in the description of Deucalion and Pirrha's flood in the first book of Ovid's *Metamorphoses*, a popular text undoubtedly known to him (even though the only explicit reference to Ovid's poem in his notebooks dated back thirty years). The meteorological culture Leonardo was familiar with represented an even more powerful and arguably closer inspiration. Art historian Geoff Lehman has described the deluge drawings as the pinnacle of Leonardo's concern with "deep order within the apparent chaos of natural dynamism"—the very definition of the meteorological pursuit.[132] Donald Strong saw in the landscapes of *Mona Lisa* and of *The Virgin of the Rocks* the portrait of a world newly emerged from the waters of a flood, after one of its cyclical renovations.[133] In fact, the "deluge drawings" may even prefigure the next renovation of the Earth that many believed to be approaching. The meaning of those collapsing cliffs and sweeping waves lies within medieval and Renaissance scientific traditions as much as their religious or moral allegories and their art.

4

A "Pious" History of the Earth?
1500–1650

*It is unworthy of any Christian author to leave out of his
works God, whose majesty fills and thoroughly embraces the
Heaven and the Earth He created.*

FRANS TITELMANS, *Compendium of Natural Philosophy*, 1547

THE CONVULSIONS OF the sixteenth century affected dramatically
European conceptions of the history of the Earth and its meanings in the
new cultural, political, and material context. The radicalization of religious
tensions, the Protestant and Catholic reformations, and the rise of biblical
literalism shattered the barriers that for centuries guarded the autonomy of
natural philosophy from theology, and vice versa. Restrictions to intellectual
freedom came with unprecedented calls for a Christianized science of nature.
Just like astronomy, meteorology became increasingly pervaded with theological considerations, scriptural sources, and attempts to cram geological phenomena within the short timeframe of the biblical chronology. Meanwhile,
colonial expansion and new geographical and ethnographic knowledge did
more than undermine the basis of the medieval Earth. They raised doubts
on the common origin of humankind, and with it, on the global conversion
of Adam's posterity as the ideological underpinning of political and religious
imperialism. Only a recent universal Flood that drowned but a single family
could safeguard both the historical accuracy of the Revelation and the descent of every human race from that lineage. Once a mainly moral and theological story, which natural philosophy could bypass entirely, the biblical
inundation increasingly solidified into an event that was an integral part of
the physical history of the Earth and was expected to have left material evidence of its passage.

On the Edge of Eternity. Ivano Dal Prete, Oxford University Press. © Oxford University Press 2022.
DOI: 10.1093/oso/9780190678890.003.0005

Nonetheless, around the year 1600 a biblically informed history of the Earth remained an aspiration and an agenda rather than a coherent corpus that could displace the Aristotelian tradition. The latter remained vital, surprisingly adaptable, and largely acceptable as long as it repudiated the premise of an eternal world—but not necessarily its long duration. The development of sixteenth-century Earth history appears far more complex than a crackdown on intellectual freedom by religious zealots, freighted as it was with momentous and far-reaching implications. The cycles of the Aristotelian Earth came to be framed within the directional history that modern geology inherited. Whether to emphasize or to downplay its geological role, the Noetic Flood represented the essential force behind early modern research in fossils and their formation. The absolute dating of geological strata, of little interest for a cyclical Earth of immeasurable age, became a sensible investigation when its existence could be measured in a few thousand years; moreover, it might offer physical evidence for the reality of the biblical inundation. Eventually, though, the pursuit of a more "pious" natural philosophy begat the poisoned fruit that medieval scholars so studiously tried to avoid. As part of the physical history of the Earth, the events of the Mosaic narration could no longer avoid rational investigation, and their truth came to depend on empirical evidence—until such evidence could no longer be found.

"Extremely Pernicious Errors"

In May 1489 the bishop of Padua, Pietro Barozzi, and the local inquisitor resolved to take action against the most troubling teachings imparted at the prestigious university. Under penalty of excommunication and trial, they prohibited public disputations concerning Averroes's doctrine of the "unity of the intellect," namely, that the immortal soul would be common to the whole of humankind rather than individual.[1] The disputations the bishop referred to were held in public spaces, frequently in churches, by a student or university professor who would defend certain theses against whoever might wish to challenge them. Disputations could easily draw a large attendance, especially if—as it was often the case—they involved highly controversial topics. Indeed, the wording of the edict suggests that the bishop intended to insulate a less sophisticated public from heretical interpretations of the Aristotelian theory of the soul rather than to prohibit its teaching.[2] In spite of its apparently circumscribed character, though, Barozzi's intervention was among the early signals of a changing attitude among the Church hierarchies toward university teaching and the dissemination of pagan "errors," in which the "unity of the intellect" was usually lumped together with the "eternity" of the world.[3]

To readers and auditors, the term *eternal* could cover in fact a breadth of different assumptions and interpretations. Most of them would probably take for granted that the universe was created recently and treat its "eternity" as a purely theoretical postulate with no bearing on its true age. Others would understand it as "created but of undetermined (and possibly very great) age"— all the more so since the medieval and Renaissance philosophical vocabulary lacked an equivalent term. This solution could be judged either innocuous or potentially dangerous, but it was hardly heretical. Many would have found both readings possible and valid, rather than mutually exclusive. After all, natural philosophy in general, and meteorology in particular, dealt with the range of what was "probable" (namely, physically possible and not contrary to the faith) rather than pick a single answer with the exclusion of every other. Moreover, theologians did not consider the exact time of the Creation an important issue that needed to be determined univocally. Philosophers like Pietro Pomponazzi (1462–1525) even argued that natural philosophy was mostly an exegetical exercise that consisted of determining and explaining what Aristotle taught about the natural world, regardless of whether he had been factually right.[4] The intentional ambiguity of this terminology had long been a cornerstone of the philosopher's intellectual freedom, but by the late 1400s it was increasingly frowned upon as a smokescreen put up to protect the diffusion of heresy from ecclesiastical interference. Considering how customary it was to speak of the "eternity" of the world without any qualifications, it would be surprising if no one thought that it was indeed the case—or at least that it could be co-eternal with God (or their idea of God). A growing number of clergymen took for granted that this "error" was spreading, either out of ignorance of the Church's true teachings or as a deliberate heresy.

The concerns of the ecclesiastic authorities were hardly unjustified. In theory, the discussion and teaching of philosophical doctrines that contradicted revealed truths should have been presented as a philosopher's opinion and their "absolute" falsity demonstrated. In practice, this had not been the case for a long time. Paul of Venice's compendium of the *Meteorology*, written for university students, assumed the eternity of the world without any remarks or qualifications. The philosopher's main pupil and successor at Padua, Gaetano of Thiene (1387–1465), followed in the footsteps of the master. His own commentary to the *Meteorology* stated plainly how "it is clear

that time does not cease, and everything is eternal."[5] Gaetano of Thiene's chair
was in turn assigned to a former student, Nicoletto Vernia (ca. 1420–1499),
who represented the main target of Pietro Barozzi's edict. Apparently, Vernia's
claim that he upheld Averroe's theory of the soul "for the sake of disputation"
convinced neither inquisitors at the time nor modern scholars.[6] In the same
years, Venetian humanist Ermolao Barbaro (1453–1493), also professor of
philosophy at Padua from 1477, reiterated in his *Compendium of the Natural
Sciences* that "sources and rivers will eternally flow." As it was customary in
meteorological literature, he mentioned Noah's Flood only obliquely, since
that event pertained to theology, not natural philosophy: "nature does not
allow a Deluge to submerge the whole of the Earth and cause a universal ruin,
as its dry and lightest part must always emerge."[7]

Historians used to regard Padua as a stronghold of radical Aristotelianism,
scarcely representative of other geographical or cultural contexts, yet meteorological works published in other parts of Italy or north of the Alps hardly
differed from their Paduan counterparts.[8] In 1512 Jacques Lefèbre d'Etaples
printed in Nuremberg an illustrated edition of the Aristotelian treatise, with
the commentary of the German humanist Johann Dobeneck. While written
in Latin, D'Etaples's version was more accessible, with smaller format and
larger fonts, than typical university commentaries. The commentator did not
make any effort to differentiate between the author's opinions and Aristotle's,
or to show the "absolute" falsity of the Greek philosopher's conclusions. The
Scripture only intervened in his work on occasion of the discussion of the
partial floods mentioned by Aristotle, such as those of Deucalion or Ogyges.
The author added to the list the biblical Flood, but only to set it aside from
natural philosophy—and from his commentary as well: "The one which took
place under Noah does not pertain to nature, but to divine revenge, and did
not involve only one region but the entire Earth was covered with water.
Therefore, we call it universal."[9] To explain the relation between philosophical conclusions and revealed truths was none of his business.

Likewise, Hieronymus Wildenberg's epitome of Aristotle's *Meteorology*,
published in Basel in 1548, reminded the reader of the "perpetuity" of the
sea and that "where now is the sea, there was once dry land and vice versa."[10]
The antiquity if not the eternity of the Earth was a normal assumption, even
in works of a more empirical nature. German physician Georg Bauer, best
known by his Latinized name of Agricola, made of mining and metalworking
a subject worthy of scholarly attention in his lavishly illustrated and widely
read treatise *On Metals* (1556). The book did not deal directly with the age
of the Earth, but its antiquity was implicit in the discussion of the formation

and erosion of mountains. Agricola concluded on an Aristotelian tone that "all of these vast modifications in regions—when, where or how they began—because they happened so long ago, are effaced from human memory, and are not apparent to laymen even though they have occurred to the greatest extent." Contrary to a diffused cliché, Agricola's liberty with respect to any chronological constraint was hardly "striking" or "in advance of its time."[11]

Historical writers, too, felt the lure of the unfathomable past of the Greco-Roman philosophical tradition. In the course of the fifteenth century, the early Christian historian Orosius, who throughout the Middle Ages provided the template for a creationist and providential conception of history, fell progressively out of favor compared to Greek authors such as Thucydides, Herodotus, or Diodorus of Sicily, whose evidence for the first human cultures far predated the biblical chronology of the world.[12] Traditionalist reactions, exemplified by the popularity of universal histories like Sabellicus's *Enneades*, did not stop historians from conjecturing on the human race's antiquity.[13] The possibility of an accurate dating of the time of the Creation gave way to the perception of nebulous origins, with no clear beginning in sight. The 1417 rediscovery of Lucretius's *On the Nature of Things*, which described the Earth as a temporary aggregate of atoms (albeit in the context of an eternal universe), provided a powerful alternative to both Christian providential history and Aristotle's cyclical eternalism.[14] In his *Library of History*, whose first chapters were already circulated in Florence at the beginning of the fifteenth century, Diodorus of Sicily leaned toward the atomistic doctrine of the spontaneous generation of the human species rather than its eternity—although he presented to his readers both theories.[15] Arguably influenced by Lucretius, poet Giovanni Pontano (1426–1503) saw the establishment of civilization—or perhaps, of the current one—not in a first couple already endowed with knowledge and speech, but in "barbarous and savage" ancestors who gradually polished their primitive customs and language.[16] In his poem *On Meteorology*, he contended that the slow but inexorable emersion and submersion of whole countries represented the physical and historical reality of the world. It was not "a fairy tale" that the Mediterranean was once dry land, and that whole ships were sometimes found by miners buried in the deepest galleries:

> In the middle of the sea the tired ox sweated on the furrow; the hull pushed forward by the winds, now swims buried in the earth.... When a new land will emerge from the waves, then elsewhere the sea, overwhelming other countries with its huge mass, will likewise bring into the abyss swallowed cities, castles and fields. No honor will be given to

the tombs of the Kings; and the temples of the Gods will be destroyed
with impunity.[17]

Materialistic accounts of the beginning of the world began to find a place
even in literature that normally adopted the traditional chronology of the
Creation. In 1520, German humanist Johannes Boemus introduced his vastly
popular ethnographic account of the *Customs of Every Nation* with two distinct relations of humankind's origins. While the first followed the conventional biblical story line, he felt it necessary to add a second and completely
different narrative derived directly from Diodorus.[18] Even though the author
specified that the "opinion of the theologians" is the true one, his work effectively exposed a large audience to an entirely secular narration of the origins
of the human species. Besides, not everyone agreed on which "opinion" was
the right one. Unencumbered by religious scruples or scholastic distinctions,
in his *Discourses on Livy* (ca. 1517) Florentine historian and political writer
Niccolò Machiavelli went straight to the point: the world and humankind
were far older than a few thousand years. More ancient records did not exist,
Machiavelli insisted, because of the periodical destruction of the memories of
the past caused by the actions of both men and nature:

> I think the philosophers, who have claimed that the world is eternal,
> could be answered that if it really were so old it would be reasonable
> that there be memory of more than five thousand years, if we did not
> see how the records of the past are obliterated for various reasons,
> some of which come from men, some from heaven [meaning nature,
> or chance].[19]

Tellingly, the arguments of the author of *The Prince* were largely the same
as those of the meteorological tradition: to wars, famines, and plagues, he
added the willingness of "new sects"—including Christianity—to erase the
memory of the previous ones.[20] But the main cause consisted once again in
the great floods that periodically extinguished most of the inhabitants "of
part of the world."[21] The few survivors, mostly "uneducated highlanders," did
not have "any notion of antiquity, and therefore cannot hand it down to their
descendants."[22] Machiavelli considered Diodorus's chronology unreliable not
because humankind could not be so ancient, but because he did not expect
historical documents to survive for such a long time.[23]

A few years earlier, Leonardo da Vinci had described the periodical extinction of civilization in very similar terms, with only one major difference.

More interested in natural productions than in smearing Christianity, the artist did not mention the role of religious "sects," claiming instead that marine fossils were sufficient proof of the past upheavals of the Earth. There is little need to suppose any direct relation between their accounts, as some have speculated, because that kind of knowledge would have been familiar to anyone with minimal learning.[24] Machiavelli is unlikely to have believed that the Earth itself (as opposed to the whole of the universe) was eternal.[25] In 1497, he transcribed a copy of Lucretius's poem, and on occasion he made clear references to the beginning of the world and to the first stages of humankind.[26] "Eternal" arguably meant to him "of great antiquity," just like "world" could designate either the Earth or the whole of the universe. The distinction was rarely made explicit at the time, even when it would have been meaningful.

It is difficult to blame the likes of Bishop Barozzi if they suspected that many of those who wrote or taught on the "eternity" of an uncreated world actually meant it. Moreover, the notion was spilling into literature that could reach a large and philosophically unsophisticated public, potentially unaware of—or uninterested in—the distinctions between "relative" and "absolute" truths that few authors cared to explain. Corrective actions became therefore necessary. The Fifth Lateran Council, summoned by Julius II in 1512 and presided after his death the following year by Leo X, dedicated its eighth session to the delicate issue. In the final document, promulgated in December 1513 as the bull *Apostolici Regiminis*, the conciliar fathers condemned "all those who insist that the intellectual soul is mortal, or that it is only one among all human beings, and those who suggest doubts on this topic," together with those who taught "the eternity of the world, and other similar heresies."[27] Like the Archbishop of Paris in the 1270s, they worried about those who philosophized "without due care," holding truths of faith and conflicting philosophical assumptions as equally true.[28] The Council insisted that heretical conclusions should preferably not be taught and, if taught, their "absolute" falsity had to be clearly shown. Philosophers were mandated to "devote any effort to clarify . . . the truth of the Christian religion."[29]

The conciliar fathers were certainly aware that the humanistic recovery of ancient culture made alternatives to Aristotelian eternalism increasingly attractive. Some of them, like the materialistic atomism propounded by Lucretius, could hardly be more appealing to a Christian.[30] On the contrary, the Renaissance revival of Platonism was probably a major factor behind the promulgation of the bull.[31] The translation into Latin of Plato's works by Florentine scholar Marsilio Ficino (1433–1499) represented a momentous intellectual undertaking, possibly comparable to the translation of the Aristotelian corpus two or three centuries earlier. Ficino's work injected renewed life into a philosophical system that assumed the crafting of the universe from raw matter by an intelligent maker, as well as the immortality of the individual soul. Platonism had underpinned Christian philosophy until the thirteenth century, and it was widely perceived as more befitting a Christian than Aristotelianism. Ficino's influential treatise *On Plato's Theology on the Immortality of the Soul* resolutely attacked contemporary Aristotelians. Florentine Platonists insisted that human reason could demonstrate truths of faith like the immortality of the soul, and they counted numerous admirers and readers in the circles that tried to restrict the influence of Aristotelianism. Among them were Bishop Barozzi and above all Pope Leo X, who promulgated the bull.[32]

"Creationist" accounts of the formation of the Earth, which never ceased to circulate but had been marginalized for more than two centuries, acquired new intellectual legitimacy. Discussing with Pontano the origin of fossil wood and seashells, Neapolitan lawyer Alessandro Alessandri (1461–1523) sketched the beginning of the world as a natural process overseen by the Creator, in which the globe either accreted out of smaller bodies of the same nature or differentiated according to the specific weights of its constituents.[33] The first option nodded to a Christianized atomism, while the second looked at the creation myth in Plato's *Timaeus* and recovered earlier medieval theories of the universe's origins. Further dismissing the Aristotelian tradition, Alessandri attributed petrified animals and plants to the deluge that, as "some related," once flooded the Earth, mixing all of its constituents.[34] Noting that Alessandri never mentioned Scripture or the chronology of Genesis, historian of geology François Ellenberger argued that his Creator might not have been the Christian God.[35] More plausibly, a "philosophical" account of the beginning of the world along Platonian lines was seen as largely compatible with the Christian Revelation, which did not need to be interpreted literally.

The *Apostolici Regiminis* required nothing of the sort. The notion that theology and natural philosophy belonged to entirely different spheres remained so ingrained that even such a moderate document met with internal dissent. The General of the Dominican order, Tommaso da Vio Cajetan, who took part in the drafting of the bull, objected that philosophers should not be asked to explain Christian doctrine.[36] Indeed, the bull remained largely unenforced and had no significant impact on university teaching or meteorological literature. The polemic on the human soul continued to rage in the following years, but even its most controversial protagonist, Pietro Pomponazzi, managed to avoid any personal consequences—even though he narrowly escaped an inquisitorial trial in 1518.[37] As for the "eternity" of the world, the issue remained largely under the radar.

Meteorology for the Layman

The deliberations of the Lateran Council continued to be ignored not only in university teaching and learned commentaries, but even in the popularizations of Aristotelian natural philosophy that became increasingly common in the course of the century. After the wave of translations and adaptations composed in the thirteenth and fourteenth centuries, the use of the vernacular had been strongly devalued by fifteenth-century humanists that emphasized a refined and classicizing Latin. By the early 1500s, however, the stage was set for a spectacular return of interest in scientific and philosophical texts in the vernacular. Once again, Italy offers an exceptionally rich landscape despite the fact that the eclipse of the vernacular as a language for natural philosophy was deeper and more prolonged than elsewhere. Astonishingly, half a century separated the first printed edition of an Aristotelian treatise in French (1488) from the first one in Italian (1538).[38] Nonetheless, the country enjoyed a still unparalleled combination of economic vitality, availability of capital, technical prowess, and cultural infrastructure. Once the trend took off, Italy became the only European country to elaborate a systematic program for the popularization of Aristotle's natural philosophy.[39]

The growing economic influence of the publishing industry was certainly a crucial factor. Its tumultuous expansion would have been impossible without tapping into the vast market of vernacular readers created by the plummeting costs of the printed book.[40] Owing to its economic dynamism, extensive commercial networks, and a large and relatively educated population, Venice quickly became the foremost printing center in Italy and one of the largest in Europe.[41] Its publishers were constantly striving to find new markets and new audiences, while the industry attracted editors and collaborators from throughout Italy. In the late 1530s they identified a public for printed scientific and technical texts in the vernacular and promptly set out to exploit it. The targeted audience included aristocratic as well as more popular readerships. Italian, not Latin, was the language spoken in the brilliant courts spread across central and northern Italy. Models of gentlemanly behavior and polite conversation (like those defined in Baldassarre Castiglione's immensely popular *Book of the Courtier*, first published in 1528) insisted that pedantry and tedium should be banished from those places. The ban certainly included scholastic commentaries in Latin, which epitomized those undesirable features. Women played an important role in the intellectual life of courts and affluent households, yet they rarely received a formal education. If natural philosophy was to gain access to those prestigious spaces, it had to conform to their needs and embrace not only a different idiom but even different genres, which came to include dialogues, summaries, and paraphrases.[42] Forewords and dedicatory letters often addressed aristocratic or female readers, highlighting the fact that vernacular literature was not necessarily intended for the lower classes. In many cases, it may simply have targeted a certain elite instead of another, or added one more idiom to the multilingualism (which already included Latin, Greek, Hebrew, Arabic, and Syriac) of Renaissance philosophy.[43]

Yet, a number of authors did conceive of the vernacular as a means to disseminate knowledge hitherto restricted to the learned. While many texts were as complex as their Latin counterparts, others aimed to adopt a language and style that could be accessible to a less sophisticated public. Crucial figures on the Italian literary scene such as Sperone Speroni (1500–1588) and Alessandro Piccolomini (1508–1579) denounced the imposition upon the Italian vernacular of artificial canons, which turned it into another elite language. Tellingly, their linguistic theories drew directly upon a cyclical conception of the history of the Earth, which seems to have been plain common sense: as countries and civilizations rose from the ruins of the latest deluge, only to be annihilated by the next, languages followed their fate. If so, it made little sense to fixate on past languages and inalterable rules. Authors should use the living idiom of their country and not be afraid of modifying it as needed, with the primary aim to reach as many readers as possible.[44] In some cases, like that of Giovan Battista Gelli (1498–1563), historians have even talked of a "rebellion against Latin."[45] The son of a wine trader, Gelli learned Latin for the practical need of acquiring knowledge that was not otherwise available. The "Florentine Academy" he co-founded in 1540 was explicitly dedicated to the advancement of the vernacular, justified once again by the cyclical nature of human and Earth history:

When, in a certain part of the world, arts and sciences have attained their perfection, they soon decline because of deluges, wars and uprisings, or the intermixing with barbaric and unlearned nations, or the mortality brought by plagues, so that they almost get lost and have to start anew to regain their perfection.[46]

The scientific literature in Italian that began to be published during those years made no efforts to assuage the secular tone prevailing in universities or in Latin literature. Most popularizers were in fact educated humanists, who presented themselves as mediators between Greek and Latin learning and the vernacular world of courtiers, women, merchants, and tradesmen they regarded as their typical audience.[47] In most cases, they limited themselves to conflate, modify, and simplify the standard Latin compendia and summaries that represented their main sources. The same approach was retained in manuscripts or academic lectures held in the vernacular, regardless of whether they were conceived for a select or for a broader audience.[48] Publishing for a philosophically illiterate public in no way implied replacing the "eternity" of the world with the short chronology of the Bible.

The new genre was inaugurated in earnest by Alessandro Piccolomini's *On the Fixed Stars* and *On the Sphere of the World*, published in Venice in 1540 and reprinted many times until the end of the century.[49] A prominent figure in Paduan intellectual life, Piccolomini engaged with matters as varied as poetry, playwriting, and the condition and dignity of women; astronomy and natural philosophy were also a cornerstone of his activity.[50] Like many medieval vulgarizations, Piccolomini's work was conceived not as a direct translation from Aristotle but as a collection "in part from the best authors and in part newly produced."[51] The aim was to popularize not Aristotle's thought per se but natural philosophy, which happened to be largely framed in Aristotelian terms. The subject of *On the Sphere of the World* overlapped only in part with meteorological topics, but the treatise featured a thorough explanation of the reasons for the emergence of the sphere of the earth from that of water. Piccolomini's discussion, based on Pietro d'Abano's *Conciliator*, always remained on entirely philosophical grounds, so he rebuffed the teleological argument that the Earth might emerge from the water in order to make terrestrial life possible as "rather theological, than physical."[52] The author declared his intention to work on a translation of the *Meteorology*, but in the competitive environment of the Venetian publishing industry one had to be quick to catch opportunities. Sebastiano Fausto of Longiano (1502–1565), perhaps inspired by the resounding success of Piccolomini's work, beat him to it and composed a short vernacular treatise *On Meteorology* in 1542.

Longiano had no more background in this field than could be expected of any educated person of the time, but he was an accomplished example of the new breed of polygraphs that proved instrumental to the expansion of the Venetian press industry after the 1530s.[53] Apart from his *Meteorology*, in

1542 alone he published treatises on education, on gentlemanly conduct, and on the divinatory arts of the ancients as well as a translation of Dioscorides's pharmacopeia.[54] The first and last subjects were among the most likely parts of natural philosophy and medicine to find a market among general readers. As Francesco del Garbo wrote to Benedetto Varchi during these years, a popularizer had better begin "with some of those Aristotelian books that can be useful and enjoyable to that sort of men, who read vernacular books more than the others."[55] It may not have been a coincidence that the same printer (Curzio Navò) had just published Vannoccio Biringuccio's vernacular treatise on mining and metalworking.[56] Composed in the aftermath of Piccolomini's highly successful vernacular treatise, Longiano's book might have been an attempt to forestall a competitor, to exploit the expectations Piccolomini had created, and to obtain a share of the new market of scientific popularizations. The author's unpretentious style and the avoidance of uncommon terms followed a precise program, and his emphasis on concepts, rather than words, echoed common sentiments among the authors of vernacular treatises:

> It may happen that others will follow this same path, so that soon every philosophical concept can be learned without the Greek and the Latin language; and in a short time the sweet fruits of the philosophical garden would be enjoyed, for the labor would only be about the meanings, given that words in the mother tongue—a few terms excepted—would be easily understood.[57]

His chapter on the origin of mountains was sourced directly from Paul of Venice's Latin *Exposition of Aristotle's Natural Books*, composed in the early 1400s as a compendium for university students.[58] Longiano attributed the origin of mountains to the same causes listed by Paul of Venice, translating his source almost literally when he added that "each one of these causes is enough to create a small mountain, and all together they can make a very large one."[59] The author followed Paul of Venice closely even in the description of the permutation between seas and dry land, whose cycle of 36,000 years represented, of course, only the period of the cyclical renovation of the surface of the Earth—not its age, which could be considered "eternal" for all practical and philosophical purposes:

> From what has been said, it is possible to prove—and indeed it follows as a plain fact—that since [or, "in the assumption that"] the world is

eternal, the sea will necessarily move to a different place, and where now there is the sea, there will be land, as it was in other times; conversely, where now there is dry land, there was, and will again be, the sea. It is true that this process will take place in, so as to speak, an almost infinite amount of time.[60]

Even though those transformations may escape human senses, every extant mountain will eventually be flattened, and the earth they are made of will be washed out to the sea. Filled with the sediments carried by rivers and rainfalls, the sea will in turn recede from the current coastline, "where buildings and cities will be constructed, as we can see today in Venice."[61] Longiano concluded the chapter by explaining the medieval doctrine of the offset between the sphere of water and that of the earth, and why a universal flood was impossible "according to natural principles."[62] The only scriptural reference occurred in the very last sentence of the book, following the description of the natural causes of the rainbow:

> Although we freely acknowledge that the rainbow is a natural phenomenon, yet it could be said according to the Christian theology that it is also plausible that God made the celestial arch for mortals as a sign of the covenant after the flood; and that in this image he wanted to do things which the human mind cannot understand.[63]

The only purpose of this sentence was to provide an edifying conclusion to the small treatise and to remind the reader of the limits of human reason. While natural philosophy and theology may have different explanations for the same phenomenon, the discrepancy did not entail any explosive contradictions. As in the case of Noah's Flood, God could have used a particular occurrence of a natural phenomenon for a higher aim and invested it with a salvific meaning. Even this reading is presented simply as a "plausible" alternative to a purely metaphorical or allegorical interpretation of Genesis. Thirty years after the promulgation of the *Apostolici Regiminis*, Longiano's readers would still have learned of the "eternity" of the Earth as a matter of fact. Such knowledge was never secretive, but in Leonardo's time vernacular readers had to scrounge around for the rare manuscripts of outdated works, or seek help in translating relevant passages from a language they did not understand. A few decades later, that information could be easily purchased, for the equivalent of a few dollars, at any bookseller in Venice and elsewhere.

Meteorology Reformed, and Counter-Reformed

The popularization of a secular meteorology raised n o more alarms within the Church's hierarchies than it had a couple of centuries earlier. The field had been clearly defined, normalized as part of the education of the elite, and hardly seen as a threat to the social and political order. The parallel tradition that looked at celestial and Earthly phenomena as signs of divine will or as omens of portents to come was a different matter. While learned meteorology dealt with natural phenomena exclusively as natural events, this was only one of many possible ways to consider them. Poet Giovanni Pontano found it appropriate to discuss meteorological occurrences in relation to wars, famines, plagues, and political upheavals.[64] In the same years, Burgundian chroniclers wondered about the place that earthquakes and other meteorological "prodigies" should have in their historical works, debating their divinatory value.[65] The ominous nature of heavenly phenomena was of course common currency in popular culture.[66]

Around 1500, astrology and religious prophecy merged with learned and popular meteorology into a new kind of apocalyptic literature that forewarned the destruction or reformation of the moral, social, and natural status quo. The end of the world or its renewal would be announced by plagues, rains of fire, earthquakes, and, above all, new exterminating floods.[67] Resounding events seemed to confirm the crisis of the present order and the inevitability of the natural catastrophes that were expected to bring forth its collapse. The seemingly unstoppable advance of the Turkish Empire in the eastern Mediterranean and in the Balkans loomed as a deadly threat to Christianity, as well as a well-deserved punishment for its sins. Still, the expansion of Christian powers into previously unknown countries could signal that the final conversion of the infidels and, with it, the end of times were at hand. Ominous monsters and extraordinary generations, once mentioned in bestiaries and universal chronicles as the marvels of distant countries, were being reported with increasing frequency within the borders of Europe.[68] The coming of a great reformer, or of the Antichrist itself, appeared imminent. On the other hand, Rome seemed quite comfortable with the current state of affairs and showed very little appetite for millenarian apocalypses. In 1516, the still ongoing Fifth Lateran Council imposed restrictions on popular preaching and prophesizing, condemning those "preachers in our times" who threatened "various terrors, menaces and many other evils, which they say are about to arrive and already growing."[69] Yet the abysmal corruption of the papal court offered powerful arguments to those who anxiously scrutinized the Earth, the heavens, and the Holy Writ for signs of the regeneration of the world. Mere months after the conclusion of the Council, Martin Luther affixed his 95 Theses to the doors of Wittenberg Castle's chapel. The Reformation had begun.

Luther did not spare criticism of the medieval philosophical tradition, based on a pagan culture that in his view a tarnished Church had too long tolerated and even assimilated. Voicing diffused preoccupations, he held that notions like the mortality of the human soul and the eternity of the world could not be sanitized into a separate sphere of learning. In fact, they were spreading like a cancer within the body of Christianity, undermining its core beliefs. Luther especially targeted Aristotle's *Meteorology* because of its assumption that every physical phenomenon had to be explained through natural causes, therefore denying the possibility of their interpretation as supernatural omens.[70] After 1520, Lutheran philosophers and theologians began to emphasize final causes in meteorology, looking at unusual meteorological occurrences as proof of God's providence and of his direct intervention in the universe. The Lutherans' preoccupation with preserving a science of divine signs was part of the larger aim of merging natural philosophy and Christian theology. Meteorology lent itself particularly well to the implementation of this program, thus gaining a conspicuous position at the Lutheran university of Wittenberg.[71] Eventually, Philip Melanchton and other Lutheran scholars were unable or unwilling to overcome and replace the Aristotelian meteorological tradition. Instead, they bent it to their needs by making room for divine causation in some cases, allowing for natural explanations in others, and adopting textbooks that focused on the description of physical events rather than on their causes.[72]

The reforming wing of the Catholic Church entrusted its hopes for reconciliation to the council summoned in Trent in 1545. It did not go too well for them. As its first historian, Venetian Paolo Sarpi (1552–1623), famously wrote: "desired and procured by godly men to reunite the Church," the council had "so established the Schism . . . that the discords have become irreconcilable." Hoped for by the bishops to regain their Episcopal authority, "usurped for the most part by the Pope, hath made them lose it altogether." Feared by the See of Rome as a means to tame its power, "it hath so established and confirmed the same, over that part that remained subject onto it, that it was never so great nor so soundly rooted."[73] The assembly was soon taken over by the uncompromising wing of the Roman hierarchies, mostly composed of Italian prelates loyal to the papacy. While important steps were taken to reorganize the Church and improve the quality of the clergy, the primary item in the assembly's agenda became to demarcate and destroy heresy.[74] To serve this purpose, in the eighteen years in which the council dragged on—and even before it opened—the Church developed new organs of surveillance and repression like the Roman Inquisition (1542) and the Index of the forbidden books (1559). Even though the conciliar fathers did not directly address matters concerning natural philosophy, influential Cardinal Girolamo Seripando singled out the eternity of the world as one of the "errors" that must be extirpated:

> Some heretics said, that it is true that God created the world; but in the same way as a body casts a shadow, and a luminous source spreads light. From these false examples two errors were born: that the world was not produced through God's free will, but naturally, and that it was eternal.[75]

Seripando was asking for nothing more than the application of the *Apostolici Regiminis*, which by that time had become a rather minimalistic program for much of the Catholic world.[76] The tolerance toward the most dangerous tenets of Aristotelian natural philosophy, and its relationship to Christian theology as it was established in the Middle Ages, began to be questioned well before Trent. The discomfort with a natural philosophy that excluded the divine from the physical world had been growing for decades; while particularly visible in the Lutheran camp, the propensity to consider exceptional natural phenomena as messages from God was in no way unique to it. In the wake of the Reformation, calls for a Christianized science of nature became louder even among those who remained loyal to Rome.

When Gasparo Contarini (1483–1542), a man of "profound and personal" piety and a leader of the reformers, arrived at Padua University as a student of Pomponazzi, he was appalled at the diffusion of Averroism among professors and fellow students.[77] Unlike his teacher, he thought that natural philosophy and theology should not be kept separate, "since reason acted as a check to speculation while faith gave answers unattainable through reason."[78] In his otherwise conventional treatise *On the Elements and their mixing* (first published in 1548, but composed around 1530–1535[79]), he refused to adopt the hypothesis of the eternity of the universe without any qualifications, stating explicitly that the world is "perpetual" only "according to Aristotelian philosophers."[80] The Flemish Franciscan Frans Titelmans went much further along that road. In the early 1530s, he set out to breach the monopoly that the Aristotelian tradition still enjoyed in the universities by offering students an alternative to the secular abridgements (such as Paul of Venice's) that were available. In a radical break with the past, he presented his *Compendium of Natural Philosophy* (1530) "not as a treatise of pure philosophy, but as a mixture of theology and philosophy at the same time." It was unworthy of a Christian writer, he added, "to leave out of his works God, whose majesty fills and thoroughly embraces the heaven and the Earth He created."[81] The conjectural nature of meteorology, which had traditionally favored criticism of and elaborations on the Aristotelian text, began to be exploited to foster a more certain source:

> Some say . . . that even if God had not miraculously gathered the waters [in a lower place] in the third day, this would have happened over time for natural reasons. . . . But since such predictions are uncertain, we will content ourselves with drawing from the infallible doctrine of the Holy Scriptures, that dry land emerged by the Word and Infusion of the Almighty God. . . . Since the Scriptures entirely assign this miracle to God in this way, it seems unnecessary to further inquire into other causes.[82]

Titelmans's approach received harsh criticism from leading intellectuals such as Erasmus of Rotterdam, but his *Compendium* proved attuned to the new times. With at least thirty-six editions before 1596, it became the best-selling handbook of natural philosophy of the century, and its success continued well into the 1600s.[83] The Franciscan order seems to have spearheaded the subordination of natural philosophy to a more physical reading of Scripture, while the Dominicans appeared divided. In 1551, though, their General Chapter demanded that its members rally behind the decrees of the *Apostolici Regiminis*. Three years later, the election to Saint Peter's throne of Paul IV Carafa (the actual mastermind of the Roman Inquisition) sanctioned their implementation, underscoring the increasing difficulty to keep theology out of works dealing with natural philosophy.[84]

When acclaimed popularizer Alessandro Piccolomini published his new treatise on *Natural Philosophy* in 1551, he had to take into account the new circumstances. In spite of his premise not to depart from the doctrine of Aristotle "unless the senses or a most convincing demonstration dissuaded me," he added to the discussion of the eternity of the world a chapter that removed the ambiguities usually embedded in this kind of literature and clarified the limits of natural knowledge.[85] The world was only eternal "philosophically speaking"; "absolute and infallible necessity is reserved to the judgement of the Holy Church." He held "as a certain truth" that the world had a beginning, and can have an end at "any time this will please his great Architect."[86] He also found it problematic to retain the common notion of the cyclical extermination of humankind by floods similar to that of Noah's. While his description of the origins of civilization owed more to Lucretius than to the Bible, he had to concede that human beings lived without memory of a deep past either because "the Earth was newly created into the world (as we must believe); or because it was like reborn and renovated when it re-emerged and was freed from a vast deluge of waters (as many judged)."[87]

Still, those disclaimers did not have any organic connection to the rest of the treatise, which was conducted along traditional lines. While Titelmans assigned an important role to Scripture in the explanation of physical phenomena, Piccolomini continued to treat natural philosophy as an autonomous field, and he was allowed to do so in a work published in Rome itself. Their works are indicative of two major trends that developed in the course of the late sixteenth century: on the one hand, a mounting cultural pressure to construct a Christian science of Earthly phenomena, sometimes entailing the explicit compression of their history within the limits of the biblical chronology; on the other, the absence of any definite imposition to do so. Broadly speaking, the ecclesiastic censorship continued to require nothing more than that authors clarify the Christian truth every time they enounced a clearly heretical proposition. When it came to meteorology, this boiled down to explaining that the world was created. The semantic ambiguity that long enshrouded the term "eternal"—and the authors' own beliefs—was no longer allowed to exist. For the rest, a literalist interpretation of Genesis assigning a geological role to Noah's Flood, or squeezing the history of the Earth to within a few thousand years, were not needed at all to obtain the ecclesiastical imprimatur. As long as authors refrained from explicitly mentioning the eternity of the world, they might even continue to forego any religious references in their works.

The Inquisitor on the Lagoons

The meteorological literature produced in Venice after 1550 represents an especially instructive case study given its sheer size and variety, the economic and cultural weight of the metropolis on the lagoons, and the city's traditional independence from Rome. In the most important printing center in Italy and possibly in Europe, no ecclesiastical control over the press or the circulation of culture existed before the reorganization of the local Inquisition in 1547.[88] Until then, Lutheran books were freely imported, and Venetian patricians collected them, to the dismay of the Vatican.[89] Gasparo Contarini, a Venetian nobleman, held justification by faith—one of the main tenets of Lutheran theology—as a cornerstone of his spiritual life in spite of his loyalty to Rome.[90] In the 1540s the government itself leaned toward schismatic England and the Schmalkaldic league of the German protestant princes, seen as a counterweight to Spain and to the Catholic emperor, Charles V. The passage of Venice to the Reformation, however, was never a realistic possibility. After the imperial victory at Mühlberg (1547), political and religious considerations persuaded the Venetian government to rally behind the Catholic field and comply—in part, at least—with the requests of Rome.[91] Amidst widespread worries for the fate of the flourishing press industry, in the 1550s the Venetian Inquisition quickly increased its grasp and widened its influence over the cultural life of the city. Counter-reformation policies were accompanied by a genuine religious revival, attested to by a sharp rise in the consumption of religious and devotional literature; in the second half of the sixteenth century, such went from 13% to 35% of the total output of the Venetian presses.[92]

The problem of how to bring together traditional methods, recent geographical discoveries, counter-reformation political issues, and new religious sensibilities was already apparent in the Latin dialogue *On the Nile*, printed in 1552 by the Veronese Count Ludovico Nogarola.[93] His work was occasioned by the publication of the first volume of Giovanni Battista Ramusio's *Navigations and Travels* (1550), the first general synthesis of the ethnographic and geographical knowledge that was transforming the European perception of the Earth.[94] Ramusio's choice of publishing in Italian was warranted by the popularity of travel literature, which raised not only interest but also anxiety in a city whose economic prosperity was threatened by the new Atlantic routes.[95] Dedicated to Girolamo Fracastoro, the book included part of their epistolary exchange on the annual inundation of the Nile that was traditionally regarded as one of the most obscure phenomena in meteorology.[96] The topic provided the author with the opportunity to review the state of the physical knowledge of the terraqueous globe, but as Fracastoro reminded him, a further element bestowed intellectual dignity upon the subject: the unfathomable temporal depth of the processes involved. "While these things concerning rivers, mountains and the Earth are not eternal," Fracastoro noted, "nonetheless they come close to the eternal ones."[97]

As one of the characters in Nogarola's dialogue, Fracastoro (FRA.) quickly dismissed the ancient doctrine of the perpetual heat and drought of the tropics. Instead, he pointed at the newly discovered equatorial rains as the likely cause of the annual overflow of the river.[98] FRA. "I confess that I spoke against those great men [of the antiquity]; but not recklessly, nor without a good reason."[99] As soon as the topic seemed exhausted, however, one of the interlocutors (Adamus, ADA.) abruptly hinted at a major omission in Ramusio's work, much to Fracastoro's surprise:

FRA. "What has he left out then?"

ADA. "I think it is very serious that, in the discussion on the origin of the Nile, he seemingly neglected the mysteries of the Holy Scriptures. It is written indeed in the book of Genesis, that a river flowed out of that pleasure place in order to irrigate [the terrestrial] Paradise, and then it did split into four branches whose names are Pishon . . . Gihon . . . Tigris . . . and Euphrates. Instead of Pishon and Gihon, Augustine reads Ganges and Nile, and almost all who deal with geography wonder in what sense Moses's words are true. Ramusio should have explained that."[100]

The geography of Eden and the identification of its rivers were common topics in medieval and Renaissance cosmography, but seemed entirely out of place in a meteorological dialogue.[101] After a long debate on how to square natural philosophy and modern geography with Genesis, the interlocutors agreed that a figurative reading of Scripture was in order: in line with a widespread tradition, the terrestrial paradise was not a particular place on Earth but rather the primeval condition of the world, which changed after Adam's fall. Moses was therefore correct when he wrote that the Nile flowed from Eden, because Eden used to coincide with the whole of the Earth.[102] In Nogarola's book the Bible was interpreted in the light of natural philosophy and geography, rather than the other way around, but some interlocutors must have found it troubling that the issue was raised at all. It may not be a coincidence that Fracastoro (who was well and alive at the time, and an acquaintance of the author) did not utter a single word in the long exegetical debate, reappearing only when the interlocutors propose to close the theological interlude: "FRA. I would really like that."[103]

The awkward integration of scriptural sources into what had been a purely secular field became a characteristic of works published in the 1550s and 1560s. Many ecclesiastics, caught in the middle of epochal changes that often collided with the education they had received, found themselves in particularly difficult predicaments as they tried to chart their way across unexplored waters. The Dominican friar Valerio Faenzi, who belonged to an order in which the controversy had been particularly fierce, might have been one of them. A member of the Venetian Academy of Fame, in 1561 he published under its aegis a small Latin dialogue, *On the Origin of Mountains*, that was arguably the first book devoted entirely to the topic.[104] In the prologue, a party of gentlemen meet for a fishing party on the idyllic shores of Lake Garda, whose northern branch nestles deeply into the Alps. More impressed with those towering mountains than with the catch, they decide to discuss their origin and nature according—as usual—"to human reason only."[105] The dialogue then unfolds along traditional lines, as a critical review of the opinions of ancient and medieval authors such as Aristotle, Herodotus, Pliny, Seneca, Isidore of Seville, and Albertus Magnus, supplemented by the author's geological observations of the region between Venice and the Alps. Yet the second part of the book introduces a sudden fracture: troubled by the extreme slowness of most geological processes, which could suggest an eternal or even immensely old Earth, Faenzi (disguised in the dialogue as "Camillus"[106]) finally hints that some mountains could be there simply because God created them "when he made the Universe and the elements."[107] One of his companions promptly declares his relief: "I am glad to hear that you call God the creator of everything. . . . I suspected that you were in the same error as those philosophers, who deny that God is the author of the world, and reject the same name of creation."[108] In spite of the initial declaration to stick to human reason only, the Mosaic story emerges in the last pages of the dialogue as a factual account of the history of the Earth that could not be ignored. In the end, the dialogue fell short of a full endorsement of a Mosaic Earth history, as the author confessed that none of the proposed opinions satisfied him entirely.[109]

It is not inconceivable that the last part of the dialogue was added at a later time, in order to correct the original plan. Its hesitations and contradictions were typical of years when the authority of the biblical text in natural matters began to be asserted, yet it could not be backed by any well-established philosophical tradition. Nonetheless, its publication by the ambitious, albeit short-lived Academy of Fame (active between 1557–1561[110]) signaled a significant turn in the cultural life of the city, and Faenzi's superiors may have appreciated the effort. At odds with the uncompromising Franciscans who held the office of ecclesiastical inquisitor, the Venetian government asked that they be replaced with Dominicans. In April 1566, Pope Pious V obliged, appointing Valerio Faenzi.[111] His moderate profile was precisely what made of him a good choice. In order to implement its policies, Rome needed the collaboration of the secular authorities, which in Venice was never granted. The city remained fiercely independent, concerned with the fate of the press industry and determined to uphold the prerogatives of the government. An effective inquisitor had to be considerate and tactful, lest he provoke strong and counterproductive reactions. Assisted by the War of Cyprus against the Ottoman Empire (1570–1573), which required the mobilization of every material and spiritual resource, his unusually long tenure coincided with the zenith of the inquisitorial influence in the Republic.[112] In June 1570, Faenzi reported to the papal ambassador that "in no other time, and in no other occasion there has been a similar purge of certain kinds of prohibited books." In a city gripped with fear at the possibility of a Turkish victory, the books' owners spontaneously brought them to the inquisitor, atoning and asking for forgiveness.[113]

"I Wish You Not to Believe that the World . . . Is Eternal"

In spite of the tight control the Inquisition managed to exert on the Venetian press until the early 1590s, and the preference for a Christianized history of the Earth, there is no evidence that the authors of meteorological works were required to endorse it. The Italian liberal historiographical tradition has long maintained that inquisitorial trials and condemnations stifled the intellectual development of the country, thus precipitating the cultural decadence of the following centuries. An exhaustive analysis of contemporary meteorological literature, however, supports the revisionist approach of more recent scholarship.[114] In many ways, the new tendencies that emerged in the second half of the century enriched and expanded the field rather than strangled it. Meteorological publications both in vernacular and in Latin continued to thrive. Exploration travels and ever-increasing geographical information placed even more strain on the edifice of Aristotelian meteorology, stimulating new or revised debates on the origin of rivers and sources, on tides, and on the shape and relative sizes of the spheres of the earth and of water. Together with general introductions to meteorology or natural philosophy, treatises and dialogues dealing with specific topics maintained a significant market.[115] The increasingly common reading of meteorological events as prophetic signs made of them a matter of political confrontation, especially as the Church used its authority as sole mediator between the human and the divine to appropriate their interpretation. The swarm of earthquakes that struck Ferrara around 1570 prompted a number of writings and a charged controversy over their nature: natural phenomena, as court physicians and literati argued, or divine punishment for the Duke of Ferrara's tolerance toward the local Jews, as the Vatican thundered?[116]

Late sixteenth-century meteorological literature escapes neat classifications, but for the sake of clarity it can be divided into three categories. The first includes authors who continued the Aristotelian tradition, only bothering to meet the most basic requirements for publication. In some cases, they simply ignored the issue of the age of the world: extremely slow processes leading to the saltiness of the sea or the permutation between oceans and dry land were thus presented without any references to the temporal scales they entailed. Explicit figures on the length of Earth's geological cycles became less frequent. Another strategy consisted of the inclusion of short disclaimers affirming the relative value of philosophy and of its inferences, or differentiating between Aristotle's teachings and the author's own beliefs as a Christian. In his *Short Treatise of the World and of Its Parts* (1571), Diego de Nores simply warned the reader that Aristotle's opinion on the age of the world "is contrary to the word of He, who for his great goodness created [the world]. . . . In this discussion, we will have to trust [his word] rather than the sophistries of learned mortals."[117] In his *Natural Ladder* (1564), Giovanni Camillo Maffei recovered instead the medieval thesis (not seriously defended for more than two centuries) that Aristotle did not intend to teach the eternity of the world, but only point out that natural laws cannot explain its origin thus implying a supernatural intervention: "I wish you not to believe that the world according to Aristotle is eternal, as most learned men today affirm. On the contrary, his real opinion was that the world had a beginning, and must have an end too."[118]

A second group of publications made room for limited supernatural intervention, accepting for example the physical implications of a universal Flood of supernatural origin, and/or embedding the Earth's cycles within a linear history. For instance, the vernacular *Discourses on Aristotle's Meteors* by the Ragusan philosopher Nicolò Vito di Gozze (1584) amended Aristotelian meteorology with the frequent recourse to medieval sources, and reminders to what Catholics must believe.[119] Still, the world might have been much older than that of the Bible and of contemporary chronologists: humankind was composed of almost innumerable generations of men, but we have no knowledge of those ages because "many times the world ran out men either because of the Flood, or unceasing wars, or plague . . . so that books, writing and eventually sciences disappeared.[120] The same goes for the permutation between seas and dry land, which takes place "little by little, over an extreme length of time."[121] What about, then, Noah's Flood? We have sure knowledge—one of the characters insisted—of the year, month, and day when it happened. The answer implied that unlike earlier events, the biblical catastrophe took place close enough to our age that we have accounts of it even in the writings of profane authors:

> We have memory of it, because many who wrote on the antiquity of the world mentioned it, such as Berossus, Archilocus, Fabius Pictor; but Aristotle, who supposes that the world exists from the eternity, and then Averroes says that the time of written records is almost nothing compared to the eternity of time; so that they disappear in the great intervals of time.[122]

In 1562, the Venetian philosopher Francesco Patrizi boldly speculated on the vicissitudes of the Earth, of humankind, and of their fall from a blessed state in his vernacular *Ten Letters on Rhetoric*.[123] A member of the Academy of Fame and a leading critic of Aristotelianism, Patrizi was also a firm believer in the wisdom of the ancients and in the immense length of their history.[124] In a Platonian vein, he feigned that ancient Ethiopian documents had preserved the memory of the first ages after "the latest renovation of the world," when the Earth had a smooth surface and occupied a much larger volume.[125] Humans and animals lived in perfect happiness not only on its surface but even underneath, where a central fire warmed and illuminated immense subterranean spaces. Like Leonardo da Vinci before him, Patrizi believed that major natural catastrophes could be provoked by the collapse of the vault or parts of it. The latest one was a global cataclysm that accompanied the moral decadence of the human race, leaving only a sparse number of shocked and terrified survivors. Among the ruins of a smaller and inhospitable Earth, modern "investigators of metals and marbles" could thus find, in the form of fossils, the petrified remains "of the previous life."[126]

With due caution, it was still possible to hint at the idea that the creation of the world was not incompatible with its eternity and to do so in places that were. in every sense. much closer to Rome than Venice. Philosopher Girolamo Borri (or Borro) argued that the world could be created yet without a beginning, in his 1561 vernacular *Dialogue* on the nature of tides, and again in the revised versions he published in Florence in 1577 and 1583.[127] Like Nogarola and Patrizi, Borri assumed that present-day Earth must be very different from the primeval one, advocating an active geological role for Noah's Flood that "changed the whole face of our lower world."[128] Borri had received frequent attention by the Inquisition since the early 1550s, owing to his alleged Lutheran sympathies.[129] In 1582 he was once again arrested for heresy, but according to a contemporary source his troubles had little to do with his meteorological writings and everything to do with the "foolish opinion that the soul is mortal."[130]

Finally, in spite of the resilience of the secular tradition, a third group of publications followed in the footsteps of Titelmans. His *Compendium of Natural Philosophy* was not printed in Venice until 1571, but an Italian translation of the Spanish physician Juan de Jarava's *Natural Philosophy* appeared in 1557.[131] Quoting widely from Genesis and Ecclesiasticus, Jarava not only refuted the eternity of the world but also expanded on the sphere of the Empyrean sky as the physical dwelling of God and of the saints. In the preface, the translator compared that book to the devotional literature that especially addressed the female public, since it was full "of divine things pertaining to nature."[132] Italian authors still tended to resist the penetration of theology in philosophical works. Some, like Bartolomeo Arnigio, chose to present peripatetic or Platonic teachings alongside the Mosaic history, without decidedly endorsing either,[133] but around 1580 a biblically informed history of the Earth began to emerge as an explicit program in Venice and elsewhere.

The first Italian meteorological text to adopt in earnest the chronology of the Bible was probably Cesare Rao's vernacular booklet *On the Origin of Mountains*, published in Naples in 1577.[134] Following Isidore of Seville, he placed Noah's Flood in the year 1656 from the Creation; he then cited Diodorus of Sicily for the great Egyptian inundation of the year 2165, shifting back to a Christian author—Eusebius of Cesarea—to locate that of Deucalion's in the year 2438, almost three centuries before the war of Troy. Showing none of the hesitations that fifteen years earlier tormented Valerio Faenzi, Rao discussed only those geological agents that allowed for a young Earth.[135] The ship allegedly excavated by Swiss miners in 1460 (complete with sails and a crew of forty) may have been sunk into the Earth by Noah's Flood or one of the later and lesser ones, rather than by slow revolutions of the globe or colossal earthquakes.[136] Against the diffuse opinion that "there were no mountains before the Great Flood," the author cited both Scripture and reason: "some mountains were born in the beginning of the world by divine command, others were caused in different times, and will in the future, by Earthquakes, Floods, the virtue of the stars, of the Sun and of winds, helped by the mineral 'virtue.'"[137] In the vernacular *Meteorology* he published in Venice in 1582, the amalgamation of philosophy and Scripture was a fait accompli.[138] The Aristotelian opinion that water never entirely submerged the Earth was deemed "false" as it presupposed too many things that were "contrary to our faith." Rao's main source, Albertus Magnus, spoke otherwise either because he wrote "as a philosopher, who presupposes the eternity of the world," or because he generalized inadequate observations rather than stick to universal reasons.[139]

In the same years, Agostino Michele brought scriptural evidence even into the centuries-long controversy on the relation between the spheres of earth and water. This debate had been revived by Alessandro Piccolomini in 1557. The popular author denounced the inertia of contemporary meteorological literature, which continued to elaborate on medieval models and sources even when they were clearly refuted by new geographical information. Piccolomini argued that the theory that the sphere of water is much larger than that of the earth, still maintained by most authors, was demonstrably false because dry land had been found even at the antipodes; therefore, water can only cover the earth as a relatively thin layer.[140] In his vernacular treatise *On the Magnitude of the Water and of the Earth* (1583), Michele objected not just to Piccolomini's theory, but first and foremost to his secular approach. Echoing Titelmans's stance, Michele affirmed that, since theology is certain and infallible, it should be used to grant certainty to other sciences.[141] The discussion must then include Scripture, which showed the existence of waters not only in the bowels of the Earth but even above the sky, in spite of the many "philosophers" denying it.[142] Michele explicitly questioned the piety of those who denied its universality: who would be so irreligious or so obstinate as to "deny this truth of the universal Deluge, which has been described by Historians, proved by Philosophers, and authenticated by Theologians?"[143] The Camaldolese monk Vitale Zuccolo argued in his *Meteorological Dialogues* (1590) that Noah's Flood, as well as local ones of the kind mentioned by Aristotle, must depend directly not on the revolutions of the stars but "on the power of the Almighty God. . . . And I think that the reason is the congregation of all human evil, which induces God to such devastating effects."[144]

The sheer diversity of the meteorological works published in those decades shows that censorship (or self-censorship) cannot have been an important factor in the elaboration of a "pious" history of the Earth. Such books appeared in years when the ecclesiastic Inquisition closely oversaw the Venetian publishing industry. Between the 1560s and the early 1590s, Venetian inquisitors were effectively able to suppress the publication (though not the importation) of prohibited or even just immoral or inappropriate books. Their pervasive competences covered matters that had little to do with heresy proper, and aimed in general to impose a stricter moral and doctrinal discipline upon the unruly Italian society.[145] Had they believed that it would be wrong or dangerous to propagate the possibility of an older Earth than that of Genesis, especially in vernacular and didactic literature, they had every opportunity to intervene. When local secular authorities were strong and independent enough, even the minimum program of banishing eternalism could be jeopardized. As soon as the Turkish pressure in the eastern Mediterranean eased, the anti-Roman wing of the Venetian aristocracy took control of the government and immediately set out to restrict the interferences and privileges of the Vatican. Their aim was not to undermine the Church and its institutions (which represented an integral part of every early modern state), but rather to push back against the overreach of Rome in secular matters. In the 1590s, the ecclesiastical Inquisition began to lose its grip on the circulation of books and ideas that it had maintained for about three decades. In 1606–1607, a dispute with Rome over jurisdiction on ecclesiastics and ecclesiastic properties led to the expulsion of the Jesuits from the whole territory of the Republic for over fifty years. Their efforts to open their own university, which could have offered an alternative to the secular tradition of Padua, ended in failure.[146]

Young aristocrats from the capital and from the rest of the state heard instead the courses taught by Cesare Cremonini (1550–1631), the prestigious professor of Aristotelian philosophy.[147] In 1615, the renewal of his appointment was contested even by parts of the Venetian aristocracy. Galileo's close friend Gianfrancesco Sagredo informed the astronomer that Cremonini "with his doctrine of the soul impressed atheism in much of our youth . . . so that many judge him a scandalous man, unwise, and unworthy of being confirmed at the University of Padua."[148] Sagredo's powerful father shared this opinion, but his faction was a minority in the senate. The philosopher retained his chair until his death. The influential academies of the Recovrati in Padua (which he co-founded in 1600), and that of the Unknown in Venice, which numbered patricians among the most influential, helped preserve and spread his legacy.[149] An anonymous informant of the Inquisition complained in 1652 that the whole state was infested with the doctrines of that "damned Cremonini, who in accordance with Aristotle's beliefs taught in Padua that the soul is mortal, that the world is eternal," and many more impieties.[150]

Jesuitical Chronologies

Cracks and fissures showed up even within religious orders that at a first glance stood solidly behind a young Creation. Among them, the newly founded Society of Jesus deserves special attention given the immense cultural influence it exerted as the intellectual vanguard of the counter-reformation Church and in the education of the Catholic elites. The position of the Society on the age of the Earth stemmed from the theological tendencies prevalent after the Council of Trent. The assembly reaffirmed against the Protestants that the Catholic Church was the sole authorized interpreter of Scripture. Its doctrine rested not only on the Revelation but also on the tradition of the Church, defined as the "unanimous agreement of the Fathers on matters concerning faith and morals."[151] In subsequent years, those principles (which had to do with authority rather than exegesis) were often interpreted in the sense of conferring primacy to a literal reading of the Bible.[152] Furthermore, the criteria for what counted as "matters of faith and morals" continuously expanded, to the point that many theologians included among them even the stability of the Earth and the motion of the Sun.[153]

The main rules for the interpretation of the book of Genesis, which was even more relevant to Earth history than to astronomy, were laid out in 1590 by the Spanish Jesuit Benito Pereira (or Peyrera) in his commentary on the first book of the Bible.[154] Arguably the most influential among early modern commentaries on Genesis, Pereira's tomes were commonly used and grudgingly praised even by the Protestants.[155] The Jesuit theologian's "first rule" clarified that Moses's work was "historical" as opposed to "philosophical." That is, it was an account of the main events of the Creation in narrative form and did not require the exposition of "[scientific] reasons and arguments," but it still exposed facts in the same order and length of time in which they unfolded.[156] Yet Pereira, who also taught and wrote on natural philosophy, was acutely aware of the risks of literalism.[157] His second rule advised against the recourse to unnecessary miracles.[158] Even more importantly, the fourth rule stated that the interpreter of Moses must "carefully guard against, and absolutely refrain from" saying anything contrary to "patent experiences and reasons of philosophy, or of other disciplines." Since there was only one truth, that of the Scripture could not oppose the "true arguments and evidence of the human sciences."[159] The meteorological field, however, was considered tentative and conjectural by nature, so it could hardly be expected to provide the unquestionable evidence the fourth rule called for. Meteorology could not demonstrate conclusively the antiquity of the Earth, any more than astronomy could demonstrate its motion. In the absence of such proofs, the letter of Genesis took precedence. Pereira attacked resolutely not only the doctrine of the eternity of the world, but also the exaggerated antiquity attributed to the Egyptians, the "nonsenses of the Chaldeans," and Plato's "fictions" on the history of Athens:

I thus wanted to argue briefly . . . against the two errors of the Gentiles: the eternity of the world, and its excessive antiquity; so that it would be clear to everyone not only that the world is not eternal, but also that from its inception to the present day no more than 5,600 years have elapsed.[160]

Sure enough, some passages of the Bible gave Pereira a difficult time. The book of the Ecclesiastes, for example, rhetorically asks: "Who will count the sand of the sea, the water drops in the rain, and the days of time?"[161] The straight and literal meaning being that past time is too large to be embraced or that its length cannot be reckoned with any certainty. Pereira proposed instead that, while the time of the Creation could not be reckoned with human means, it was known through divine Revelation. Alternatively, it may not have been certain but it could at least be "probable." Or the Holy Writer may have referred not to the time elapsed since the Creation but to future time until the end of the world.[162] Yet, the increasing sophistication of chronological studies seemed to undermine the same certainties they tried to shore up. French Calvinist Joseph Scaliger (1540–1609), who contributed crucially to the reformation of the field, published lists of Egyptian dynasties that went back to a time preceding not only Noah's Flood, but even the earliest possible dates for the Creation.[163] Those documents might have been genuine, according to Scaliger's own stringent criteria for historical authenticity.[164]

The leadership of the Society of Jesus constantly emphasized uniformity of thought and teaching among its members, whether they operated in Poland or in Peru. The definitive version of the *Ratio Studiorum*, or "Plan of Study," was released in 1598 with the aim to standardize the Society's educational system. In subsequent years, General Aquaviva recalled its members to the observance of "solid and uniform doctrine," namely the teaching of Thomas Aquinas in theology and Aristotle in philosophy—unless the latter explicitly contradicted Christian truths.[165] The fact that Aristotle provided the most complete and coherent philosophical system was deemed more important than problems with specific points of natural philosophy. This program proved difficult to enforce over an intellectual elite at a time of tumultuous scientific developments. Behind the façade, debate was intense within the Society, and concerns for uniformity always had to reckon with a plurality of views and internal dissent. Only their vows of obedience bound a number of the most prominent Jesuit astronomers to the defense of geocentricism and of other doctrines they deemed surpassed.[166] While Jesuits were apparently expected to abide by the chronology of the Hebrew Bible, Pereira's admonishments might have targeted some of his brothers, too.

Mathematician Christoph Grienberger (1561–1636) observed, against Aristotle's thesis that no change can happen in the heavens because none had ever been detected, that this argument was of little import "in a world that has not lasted 8,000 years yet."[167] In other words, Grienberger seems to have thought that the Earth was even older than the Septuagint version of the Bible, which placed its creation about sixty-eight centuries earlier.[168] The most influential Jesuit authors dealing with the history of the Earth were certainly determined to fight eternalism. In many cases, they abandoned Aristotle's cyclical Earth in favor of a linear history, albeit interspersed with the recurrent floods of the meteorological tradition.[169] Yet they also tended not to provide a definite timeline for its creation. Like other authors of the time, they often recovered medieval conceptions that had been shunned for more than three centuries in favor of Aristotle's steady state Earth, blending them with updated geographical knowledge and more modern Aristotelianism.[170] In his commentary on Aristotle's *Loca Mathematica* and in his widely read *Cosmography*, Giuseppe Biancani (1566–1624) envisioned the newly created world as a perfectly spherical body, entirely covered with water. Since this condition corresponded to a state of balance of the elements, it took a direct command of the Almighty to raise part of the earth in order to make it inhabitable.[171] Being unnatural, this situation could not be maintained forever: in a distant future, erosion would smoothen every elevation allowing the ocean to cover it again as it did in the beginning—if not preceded by "the fire mentioned in the Holy Scripture."[172]

Niccolò Cabeo's ambitious commentary to the *Meteorology*, published in 1646, expanded considerably on Biancani's Earth history, but it retained its approach and its main tenets.[173] Cabeo acknowledged that the problem of the age of the Earth divided even Catholic authors, as some of them endorsed the possibility of a beginningless Creation.[174] He maintained, though, that the existence of mountains in the face of unrelenting erosion represented impregnable evidence against a timeless Earth or even an extremely ancient one. Accordingly, he disposed in short order of the most cited causes for the formation of new mountains, like earthquakes or Aristotle's account of the displacement of the ocean.[175] As for the Deluge, Cabeo could only think of supernatural causes. In the first scenario, the element of water was temporarily made less dense, thus occupying a larger volume. In the second, the surface of the sphere of water, which was slightly eccentric to the center of gravity of the universe, miraculously became concentric so as to submerge dry land.[176]

Finally, the waters that, according to Genesis existed "beyond the firmament," may have been allowed to fall down to Earth "not as single raindrops, but as gushes of water pouring down from the sky like rivers."[177] Albrecht Dürer's watercolor of a deluge he dreamed in 1525, with mighty pillars of water crashing down to Earth from the sky, immediately springs to mind. The miraculous nature of Noah's Flood had a different meaning than it had in the Middle Ages. Once an expedient way to leave theology out of natural philosophy, later it served to explain historical events that appeared incomprehensible to reason. On the other hand, the accuracy of the biblical story could and should be supported with physical observations in order to fight the swelling ranks of skeptics and unbelievers.[178]

German polymath Athanasius Kircher built on the foundations laid by his Jesuit predecessors in his *Subterranean World* (1665) and in the description of the Flood and of its aftermath that he gave in *Noah's Ark* (1675).[179] Placed at the center of the information web that the Society cast over much of the world, he strove to accumulate and disseminate the largest possible body of "probable knowledge" rather than try to discriminate between "correct" and "wrong" information.[180] This approach, common in Jesuit culture, was especially suited to the meteorological field. The spectacularly illustrated *Subterranean World* represented one of the last works in which this tradition was still recognizable in at least three key aspects.[181] First, the author assumed a geocentric cosmos; second, he spoke in conjectural and probabilistic terms; and third, he described a steady state Earth rather than an evolving one. Kircher adopted the by-then common idea of a central fire, which provided the source of heat for the circulation of subterranean waters that Leonardo da Vinci and many others had sought in vain.[182] His chapter on the origin of terrestrial life reveals the profound influence that Renaissance chemical philosophy exerted on Jesuit science.[183] Against "official" Jesuit exegesis, Kircher explained the process of the Creation in chemical terms, which also happened to be closer to the Augustinian view of the instantaneous origin of the universe. God, Kircher confidently asserted, created "everything at the same time;" His power produced only a certain "chaotic matter," endowed with the "seeds" of the living things whose development would take place at the appropriate time according to natural laws.[184]

Modern-day paleontologist and historian of science Stephen Jay Gould wrote that he developed great respect not so much for the quality of Kircher's insights as for the quality of his doubts.[185] Kircher's metaphorical reading of Genesis resonated with the suspicion that fifty-six centuries might not be able to accommodate the whole history of the world. Even if no historical

FIG. 4.1 Albrecht Dürer's Dream of a Deluge

On the morning of June 7th, 1525 Albrecht Dürer recorded in watercolors and words the still-vivid nightmare of a Deluge he had just experienced:

I had this vision in my sleep, and saw how many great waters fell from heaven. The first struck the ground about four miles away from me with such a terrible force, enormous noise and splashing that it drowned the entire countryside. I was so greatly shocked at this that I awoke before the cloudburst. And the ensuing downpour was huge. Some of the waters fell some distance away and some close by. And they came from such a height that they seemed to fall at an equally slow pace. But the very first water that hit the ground so suddenly had fallen at such velocity, and was accompanied by wind and roaring so frightening, that when I awoke my whole body trembled and I could not recover for a long time.

Watercolour on paper, inscribed and signed in the Albrecht Dürer *Kunstbuch*.

Vienna, Kunsthistorisches Museum, Kunstkammer.

documents survived the Flood, the materials that fellow Jesuits continued to provide from Arabic, Chinese, and American sources seemed to reflect the memory of the pre-diluvial world, transmitted through Noah's family and their descendants. Given enough time, it was conceivable that a number of different kingdoms and cultures could have been established even before the cataclysm. Like Grienberger, Kircher envisaged a history that encompassed at least 8,000 years. Whether he considered the possibility of even deeper antiquity is difficult to determine due, as Anthony Grafton argued, to the "deliberate ambiguity and complex wit" of his language.[186]

Kircher's opacity may easily have been an answer to the restrictions imposed on Jesuit authors by the internal censorship of the Society. In addition to theological orthodoxy, the criteria included scientific excellence in the sense that authors were held to high standards of scholarship, not that they were expected to be original. Innovation was actually discouraged, and censors routinely quashed harmless theories simply because they were not mainstream. The publication of Biancani's *Cosmography* was delayed for years on account of the author's praise of Lutheran mathematicians like Tycho Brahe and Kepler, his criticism of Aristotle on the problem of the floating bodies, and his indulgence toward Copernicans. Some parts of the published version would reflect poorly on the author, if it were not known that they were imposed against his strong objections and that he supported them with weak arguments on purpose.[187] It was not uncommon for Jesuit authors to adopt oblique strategies that were often signaled by such awkward or inconsistent passages. Problems related to the chronology or age of the world surface only rarely, though, in the papers of the Jesuit censors.[188] There is no evidence that they raised objections against the chronological vagueness of Biancani's Earth history, or Kircher's description of the Creation in the *Subterranean World*.[189] In a century that theorized "honest dissimulation," the German polymath mastered instead the art of "pragmatic disregard" of the censors of his order.[190] Yet his last works, dedicated respectively to *Noah's Ark* (1675) and to *The Tower of Babel* (1679), seemed to disavow the bold enquiries of the previous decades, as if to remind the historian of the difficulty of penetrating the mind, beliefs, and motivations of a seventeenth-century Jesuit. Whether as a gesture of humility, reconciliation, or faith, Kircher concluded his life with a profession of confidence in the timeline of the Creation of Pereira and of other Jesuit chronologists, supported, as he claimed, by the "irrefutable testimony of Sacred Scripture."[191]

The Rise of Diluvialism, 1650–1720

> *I do not know whether [the formation of fossils] happened*
> *during the universal Flood, or in other particular*
> *inundations. Neither know I . . . if in a certain time the*
> *world, tired of lying on one side, turned on the other and*
> *exposed to the rays of the Sun the part that was underwater,*
> *packed with the leftovers of the sea.*
>
> AGOSTINO SCILLA, *The Vain Speculation, Undeceived by Sense*
> (1672)

THE POSSIBILITY OF an ancient Earth remained legitimate in late sixteenth-century Europe, and the notion continued to circulate in easily accessible literature. Even Aristotelian or Lucretian eternalism, considered plainly
heretical and treated as such, proved difficult to contain and impossible to
root out. In fact, it represented the main intellectual foundation of countless forms of unorthodoxy, libertinism, and freethinking that never ceased
to challenge doctrinal conformism and religious intolerance.[1] This does not
mean that the idea of an old Creation was not fiercely combatted. On the
contrary, powerful forces across the Catholic and the Protestant world alike
converged to promote a biblically informed natural philosophy, upheld a recent Creation compatible with a more literal reading of Genesis, and touted
the Noetic Flood as the central event in the geological history of the Earth,
thus making its antiquity unnecessary.

The idea of a "Mosaic" natural philosophy met with considerable success, and its influence was profound. In the works of seventeenth-century
naturalists, the concept of a linear history largely displaced Aristotle's steady-
state Earth. The Noetic Flood was increasingly integrated into a natural philosophy that once excluded it explicitly. Even those who declined to assign a
unique role to the biblical inundation, of whom there was no shortage, often
acknowledged that it might have contributed to the present conditions of

On the Edge of Eternity. Ivano Dal Prete, Oxford University Press. © Oxford University Press 2022.
DOI: 10.1093/oso/9780190678890.003.0006

the Earth's surface together with more ancient floods and upheavals. The authors who pursued a Mosaic history of the Earth certainly did not do so out of constriction or censorship. Many sincerely believed that too much had been conceded, and for too long, to pagan philosophies that could not be reconciled with the Christian Revelation or would reduce it to simple metaphor. In their place, they envisioned a purposeful nature whose manifestations revealed the plans and warnings of a caring Creator.

Unlike Aristotle's inexorable and impersonal juggernaut, that nature responded to human actions and conduct. Pointing to the centrality of geology in early modern debates about human origins, Lydia Barnett has argued that the Flood epitomized humankind's agency over its own environment— through sin or good behavior, if not yet technology. Moreover, at a time when Christianity attained a worldwide reach, the global impact of the biblical catastrophe guaranteed the universality of its message and the common descent of every human race from Noah's family.[2] The religious institutions that mediated between the human and the divine, and the European powers that justified colonial expansion with the conversion of the infidels, found that universe appealing. Yet, around 1600 there seems to have been little need to back the Mosaic account with material evidence or natural reasons. In medieval and early modern Europe, it was much more common to attribute "marine petrifications" to vast revolutions of the Earth in a distant past (or incomplete spontaneous generations), rather than to the Noachian Flood. In the course of the seventeenth century, however, a young Earth and the reality of a global inundation in historical times relied increasingly on rational arguments, often based on the study of natural documents such as fossils and stratified rocks in addition to historical and exegetical sources. That approach came to be known as "diluvialism." Twentieth-century fundamentalists called it "Flood Geology."

Making the Flood Real

The discovery of inhabited countries well south of the equator, and of America, a previously unknown continent in the middle of the ocean, shook many of the most trusted pillars of medieval cosmography and meteorology. The notion that dry land existed because of the offset between a sphere of earth and a much larger sphere of water was common knowledge, but it entailed a single land mass in the so-called "inhabitable quarter" of the world. Everywhere else, the ocean would be too deep for any islands or continents to surface. As Jean Buridan argued in the fourteenth-century:

Some have tried to cross the sea to arrive at other quarters, but they have never been able to reach inhabitable land; and so they say that Hercules placed pillars at the limits of the quarter that we inhabit [Gibraltar strait] to signal that beyond them there is no inhabitable land, or seas that can be crossed.[3]

His sophisticated model for the continuous generation of new mountains, which in turn supported the plausibility of an ancient Earth, rested on that assumption. Since islands and continents were being found everywhere on the globe, this could not be right: the ocean must cover the crust with a thin layer—a minority view in the previous centuries. Leonardo da Vinci tried to adapt Albert of Saxony's model of the Earth to the emerging geographical realities, but he was probably one of the last to do so.[4] In the following decades, deniers of an ancient Earth could claim again that no credible mechanism existed to counterbalance the continuous erosion of hills and mountains.

The real stakes though—economic, political, and religious – revolved around the bodies and souls of the Americans, and their condition within the newly founded Christian empires. In the eyes of sixteenth-century Europeans, it was far from obvious that those races shared their same nature. The definition of "humanity" was as theological as it was biological: humankind consisted of the descendants of the first couple, created directly by God and endowed with immortal souls but also marred by the original sin. The medieval Church had usually adopted Augustine of Hippo's doctrine that every rational, yet mortal human-like being must come from Adam "regardless of the shape of their body, or the color of their skin."[5] Nonetheless, the existence in remote parts of the world of more or less human creatures, who may not have descended from Adam and were perhaps excluded from salvation, was a given. Drawing mainly from the authority of ancient sources like Pliny the Elder's *Natural History*, books like the magnificent *Chronicle of Nuremburg* (1493) delighted their readers with images of crane-men, dog-headed men, and one-eyed cyclops. As the text explained, God made such races and the biblical Giants because of His love of variety in the Creation.[6] The tritons, mermaids, and other "monsters" that populated Renaissance natural history may have belonged to the same category.[7]

Encounters with similar creatures in the farthest corners of the Earth would not have been unexpected. The existence of human beings in isolated lands like the Americas was a different matter. The old world could easily have been populated by Adam's offspring, but no evidence existed of trans-Atlantic contacts before 1492. Human-like physical traits and intellectual capabilities alone were not enough to guarantee the possession of an immortal soul, let alone a common ancestry. By and large, three possibilities were available. Human-like beings may have formed in other parts of the world by spontaneous generation after a partial or universal Deluge, as argued by Avicenna, discussed by Albertus Magnus, and taught in the sixteenth century by leading academics like Pietro Pomponazzi, Andrea Cesalpino, and others.[8] Alternatively, Americans were the result of a distinct divine Creation that may have preceded that of Adam's (a doctrine known as pre-Adamitism) or followed the Noetic Flood.[9] Finally, Noah's descendants could somehow have managed to reach America, in spite of the fact that their offspring seemed entirely ignorant of ocean-going navigation and that they had no memory of that migration.

The Spanish settlers, who sought a justification for the mass enslavement of the natives, favored the non-humanity of the Americans. The crown, which in theory took possession of those lands in order to convert the infidels, held the opposite position because, in the absence of souls to save, its claims would have been pointless.[10] In a series of bulls issued in the early 1500s, Rome resolved the controversy in favor of the Spanish monarchy by declaring that the Americans were indeed descendants of Adam and capable of receiving the Catholic faith.[11] The rights of the King to the lands and bodies of the Americans were thus confirmed, together with those of the pope on their souls. Both pre-Adamitism and spontaneous generation threatened the unity of the human race, on which global imperialism and Christian universalism depended. On the other hand, the reality and universal nature of Noah's Flood had the double benefit of saving the literal meaning of the Bible while guaranteeing that through that patriarch, every human being descended from Adam—and inherited his sin.

In the course of the sixteenth and seventeenth centuries, the Deluge grew larger in the European conscience as the lynchpin of human and Earth history. Medieval meteorology had essentially excluded it from the history of the Earth, or reduced it to the latest of the great inundations (either universal or local) periodically unleashed by recurrent astral conjunctions. In this form, the Flood easily resonated with popular astrology and with the familiar biblical story of Noah. No meteorological doctrine was more widely known than that of the periodical exterminating deluges brought or announced by the stars. In Geoffrey Chaucer's *Canterbury Tales* (ca. 1390), an Oxford student can easily convince a gullible carpenter of the imminence of a new flood similar to that of Noah's, whose coming he determined precisely thanks to his astrological science. The idea does not seem implausible to the victim, who spends the fated night on the roof of his house in a makeshift boat.[12] In the late 1400s, the "Deluge" became in various parts of Europe an increasingly real, alarmingly common, and painfully tangible experience. By that time, northern and central Italy were bearing the full brunt of the deforestations of the previous centuries that left a hydrologically devastated landscape.[13] Frequently employed as a hydraulic engineer, Leonardo da Vinci was an impotent witness to the ruins and tears brought by the frequent mudslides or overflows of rivers:

> Inundations caused by impetuous rivers ought to be set before every other awful and terrifying source of injury. . . . In what tongue or with what words am I to express or describe the awful ruin, the inconceivable and pitiless havoc, wrought by the deluges of ravening rivers, against which no human resource can avail?[14]

The term *deluge* was itself ambiguous. In Flood literature, common parlance, and even Leonardo's notes, its meaning ran the whole gamut between the drowning of every earthly creature, one of the vast but circumscribed inundations of the meteorological tradition, and yet another breaking of the embankments. In the spiritual climate of the early sixteenth century, those local tragedies were often perceived as mere warnings of a vaster catastrophe. Announcements of a forthcoming "Deluge" of water occupied a central place in early sixteenth-century apocalyptic literature. Predictions of a rare conjunction of all the planets in the watery sign of the Fish, expected to take place in February 1524, caused widespread fears and even panic in the preceding months and years. The "great fear" of 1524 has been studied as a phenomenon at the intersection of popular astrology and prophecy, millenarianism, and modern propaganda due to its pervasive presence in mass-printed pamphlets and broadsheets.[15] Yet its wide reception across the entire social spectrum would be impossible to explain without the capillary diffusion of the meteorological culture, which made the occurrence of such floods entirely credible in the first place.[16] When Machiavelli ridiculed the general hysteria, he did not mean to disparage the possibility of these events (cyclical deluges were in fact a cornerstone of his conception of human history), but rather the excesses of popular credulity and the misplaced trust in astrology's predictive capabilities.[17]

While there was a certain consensus that a flood of some sort should be expected, none existed on its form, magnitude, and timing.[18] When the

FIG. 5.1 The 1524 Deluge
The 1524 planetary conjunction in the Fish (above) and the physical and social Deluges it was supposed to bring. Leonhard Reymann, *Practika* for the year 1524 (1523).

dreaded catastrophe failed to materialize, the backlash against astrology and popular prophecy was limited.[19] After all, the conjunction might have set in motion a chain of events that would take a long time to unfold, or perhaps it forewarned of a number of geographically circumscribed natural disasters instead of a single, larger one. In the following decades, the "Flood" remained a close and tangible presence. The overflows of the Tiber in Rome, or the chronic surges of the North Sea in the Low Countries and East Anglia, were lived and visually represented as local versions of the biblical tragedy.[20] In line with the new religious and cultural sensibilities, divine wrath or humankind's misguided behavior eroded the position of astral agency as the direct cause of both local and global environmental crises.

The doctrinal and professional barriers that led to the omission of the biblical Flood from learned meteorology had always been difficult to grasp for the public of the scientific popularizers. Their readers were quick indeed to embrace its physical reality and universality, and eager to learn about the possible natural causes of the phenomenon. Camilla Erculiani, the wife of a well-to-do Paduan apothecary, attributed the Flood neither to nature's inexorability nor to God's supernatural intervention, but to the excessive numerical growth of humankind. In her *Letters on Natural Philosophy* (1584), she reasoned that their bodies absorbed so much of the element of earth that not enough was left to emerge from the water.[21] The inundation ceased only when their corpses dissolved, returning to the world their constituent elements.[22] Involved in the family business, Erculiani could nurture her intellectual ambitions in the stimulating environment of the early modern apothecary shops. Their services included the sale of books, the delivery of mail, and the availability of spaces in which customers could play cards, chess, or engage in discussion.[23] Her conversations, epistolary correspondences, and readings (she mentioned explicitly Alessandro Piccolomini) dragged her into risky theological arguments.[24] She suggested, for example, that Adam was mortal even before sinning, because his body was composed of common elements subject to corruption and decay. By claiming that Noah's inundation was unrelated to divine anger, she overstepped a centuries-old boundary. Grilled by the local inquisitor, she acknowledged that universal floods were unleashed by natural causes, but God did employ toward providential ends the particular occurrence described in the Bible.[25] The high-profile lawyer Erculiani's family could afford to hire had every reason to point out that those opinions had never been heretical and that God could employ natural phenomena toward His aims. It would only be heretical to state that there was no relation between the particular Flood of the Bible and the punishment of humankind's sins.[26] The inquisitor did not raise any objections to "the other universal floods" whose existence was a given to Erculiani. In the end, the book was not prohibited and it seems unlikely that the author suffered any serious consequences.[27]

Meanwhile, the physical reality and universality of the Flood was pressed into service to uphold the legitimacy of Christian imperialism. Cosmographers, naturalists, and missionaries confronted the conflicting priorities of assuring the common descent of every human being from Adam, while at the same time emphasizing the diversity and inferiority of the colonized. The universal Flood could offer the former, but in order to present it as a plausible solution they needed to solve the problem of the migration of Noah's descendants to America. Travel by sea appeared unlikely, and such travels impossible by land. However, this might have been true only for the present era, since the global inundation could have altered the geography of the globe.[28] The idea of vast transformations between the pre-diluvial and the post-diluvial world was common in the Renaissance and often mentioned to explain the incongruence of the Edenic geography. This might have been the reason, too, why modern explorers failed to discover any hints of the terrestrial paradise, whose location might no longer exist.[29] If so, the post-diluvial world could have featured landmasses between the continents that subsisted for centuries before earthquakes or particular floods plunged them into the sea. It was those "land bridges" that allowed Noah's posterity to spread to the new worlds of America and Oceania.

Scripture itself was silent on the subject of ancient land bridges, but their supporters could present a range of plausible sources. The most convenient was Plato's description of Atlantis, the lost continent beyond the Strait of Gibraltar that an earthquake sank into the ocean thousands of years earlier. Perhaps attached to America, or not far from it, Atlantis would have required only a short navigation from the western coasts of Europe. As Pedro Sarmiento de Gamboa suggested in 1572, maybe America was Atlantis, before a mighty upheaval caused it to drift away from the shores of the Old World.[30] One century later, Athanasius Kircher's map of the post-diluvial Earth in *Noah's Ark* still featured "the island once called Atlantis, now the Atlantic Ocean," as a peninsula that protruded from North America toward the western coasts of Portugal and Spain.[31] In spite of its attractiveness, Atlantis fell progressively out of favor as it suggested that the Americans were actually Europeans or at least belonged to the same branch of Noah's progeny. On the other hand, a land bridge between North America and Asia was physically more plausible, especially if it stretched across the Bering Strait. Even more importantly, it implied that the Americans stemmed from an Asian stock and were therefore further removed from Adamic perfection than white Europeans.[32]

The rejection of a global Flood and of biblical monogenism was certainly an important reason why Isaac La Peyrère's *Preadamites: or . . . the First Men Created before Adam* (1655) became one of the most confuted books ever.[33] A French Calvinist of possibly Jewish origin, La Peyrère contended that Adam was the ancestor of the Hebrews only, while the "gentiles" and the Americans descended from humans created much earlier. He still maintained that pre-Adamites were endowed with an immortal soul, and that Christ's sacrifice was meant to redeem everyone. While Christ's coming retained a universal value, La Peyrère turned the biblical narration into the story of the Chosen

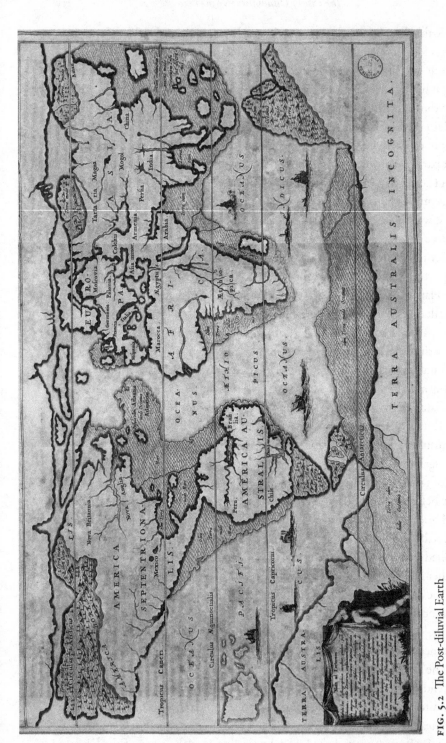

FIG. 5.2 The Post-diluvial Earth

Athanasius Kircher's conjectural map of the post-diluvial Earth, showing both ancient land bridges later lost to the ocean (like Atlantis between Europe and America) and new ones that emerged in the following centuries between North American and Asia, South America and Antarctica, and Indonesia and Northern Australia. Unnumbered plate between pp. 192 and 193 in Athanasius Kircher, *Arca Noë* (Amsterdam: apud J. Janssonium & E. Weyerstraten, 1675). Biblioteca Marciana, Venice.

People only and the Noetic Flood into a merely local event.[34] Within twelve months from its publication, *Preadamites* saw at least nineteen rebuttals. In 1656, the bishop of Brussels put the author under arrest. Sent to Rome, La Peyrère obtained Pope Alexander VII's pardon after a resounding recantation and a well-publicized conversion to Catholicism.

The space and tolerance for pre-Adamitism had clearly shrunk compared to previous centuries. Yet the doctrines of a local flood and of multiple creations cannot have been that shocking (especially the former). Some of La Peyrère's most ferocious detractors, like Anglican theologian Edward Stillingfleet, argued as well for the local nature of the Noetic Flood.[35] The Dutch scholar Isaak Voss (Vossius) shared the same view. When the French Benedictine Jean Mabillon examined his work on behalf of the Congregation of the Index in 1686, he saw no reasons why this opinion should be censored, since it did not conflict with the moral and theological significance of the biblical story.[36] La Peyrère's thesis of a local flood appears in the list of the statements he was asked to recant, but it would not have been so troubling if he had couched it in more hypothetical terms and in a less theologically charged work—that is, had he followed the rules for discussing such subjects.[37] Instead, he embedded his version of pre-Adamitism within a millenarian theology and a thoroughly unconventional reading of some passages of the Scriptures that alienated Catholic and Protestants alike. As various scholars have pointed out, most of La Peyrère's critics were concerned with his attacks on the traditional doctrine of the original sin (which pre-Adamitism compromised) and on the authenticity of the Pentateuch, whose attribution to Moses he disputed.[38] Absent such a controversial theological context, his denial of a universal Flood would still have raised strong opposition but hardly a resounding scandal.

The same was true of the ancient Creation required by his vision of human and Earth history. Normally this would not have been too problematic, but his world appeared so old as to lie dangerously close to the edge of eternity. La Peyrère contended that its past duration, like the date of its future demise, could not be known. He added that, when past philosophers wrote of the "eternity" of the world, they meant not that it was uncreated or co-eternal with God, but that its age was simply too great to reckon. Its duration was finite and known to God, but from the point of view of human time, the Earth could be considered and called "eternal."[39] La Peyrère's argument largely coincided with a standard assumption of medieval and Renaissance meteorology, but that battle had already been fought, and lost, one century earlier. The one thing that would no longer be tolerated was the reintroduction of a terminology that included the "eternity" of the world. Unsurprisingly, his recantation included his doctrine of the two different meanings of "eternity"—that of the created universe and that of God.

The barrage of confutations also reveals another of La Peyrère's feats: he exposed a deadly weakness of theories of human origins based on a universal Deluge that left but a handful survivors. Land bridges were necessary to explain the propagation of humankind across the post-diluvial globe, yet no one had ever brought solid evidence of their existence. Proposed to buttress the universality of the Flood, they were themselves a pure hypothesis. As Kircher acknowledged, his map of the world's coastlines as they appeared soon after the Flood was mere guesswork:

> It is impossible to determine them with any accuracy; so that it was necessary to conjecture them exclusively from similar mutations of the Earth that take place in our times, and from a number of historical accounts. We do not assert that this was the actual constitution of the Globe after the Deluge; we only define it as what it could have been, on the basis of our various conjectures.[40]

The physical reality of the land bridges assumed, rather than validated, that of a universal Flood. In the same way, Kircher's elaborate calculations on the size, tonnage, and internal structure of the Ark showed at best that a couple of all the Earth's land animals could fit and be fed into a man-made vessel without yet another miracle. They did nothing to prove that the global inundation actually happened. From the point of view of a natural philosopher, around 1650 the universal nature of the Flood was just another conjecture. Widely believed on account of Christian traditions, the letter of Genesis, and the urge to establish the unity of the human race, it was not supported by any compelling material evidence, nor was it needed to explain the geological appearance and the natural productions of the Earth's surface.

It is telling that La Peyrère never acknowledged he had been wrong. The text of his abjuration was written in such a way that he did not have to declare his theses contrary to reason and to historical documents. In a later apology, he explained that his pre-Adamatism was like the Copernican hypothesis to Catholic astronomers: they renounced it out of obedience to the Church, not because there was actual evidence against it. Whether right or wrong, it remained the theory that best explained the available data.[41] No physical proof existed of the young age of the Earth, nor of the universal Flood that could make it geologically plausible. Well versed in Nordic antiquities, he derided the hypothesis that the Americans descended from Viking settlers, pointing out that when they arrived in Greenland they found it inhabited already. In his endorsement of the Egyptian, Babylonian, Chinese, and Mexican antiquities, he compared the chronologists who clung to a recent Creation to those geographers who drew seas over every unexplored area:

> As is the Geographers' habit, who cover with seas the lands they ignore; Chronologists in the same way are used to obliterating the past centuries that they ignore. The former submerge the regions they are unaware of; the latter transfix unknown times with their cruel stylus, and deny what they cannot recognize.[42]

Leftovers of the Sea

The races to naturalize and rationalize biblical inerrancy, Christian universalism, imperial dominion, and racial exploitation finally converged onto the fulcrum on which everything hinged. Land bridges, mathematical demonstrations of the capacity of Noah's Ark, or hypotheses on how Eden did exist but then disappeared, all sought to avoid the endless miracles that were unacceptable to every reputable theologian. Yet they only served as beams and pillars of a building whose foundation was a recent universal Flood. That foundation was propped up by a number of traditions and historical documents, both sacred and profane, but it never rested on solid empirical ground. The transformation of a mainly theological and moral Flood into a physical event meant that it was no longer possible to overlook its natural causes and, above all, its material implications. As observation and experiment became increasingly central to the production of natural "facts," the search for irrefutable material proofs of a global inundation became all the more urgent. The examination of geological features and "figured stones" might have been able to anchor documents of uncertain reliability and interpretation and prove conclusively the reality of a universal inundation in historical times. However, failure to do so could have brought down the whole edifice it underpinned.

It is a common misconception that in pre-modern times the most typical explanation for the presence of marine fossils far from the sea would have been the Noetic Flood, or alternatively, "jokes of nature" (namely stones casually resembling actual living beings), or rocks molded into animal or vegetable shapes by astrological forces. These theories had the virtue that they did not require the Earth to be older than a few thousand years, a more ancient one being allegedly unthinkable. Early Christian authors, from Tertullian to

Isidore of Seville, mentioned of course marine fossils as plausible clues to the universal Deluge of the Bible, often drawing upon pagan sources like Plato and Ovid that used them to illustrate the past revolutions of the globe. In subsequent centuries, popular culture continued to associate marine fossils to the story of Noah, but by the thirteenth century (at the latest) very few natural philosophers seem to have subscribed to it. Even those who attributed fossil fishes to vast inundations did not necessarily identify the "deluge" with the global event described by the Bible, or considered it a unique occurrence. In the late thirteenth century, Restoro of Arezzo related them to any of the calamitous but still local floods periodically unleashed by astral conjunctions, and counter-balanced elsewhere by a great drought. The Flood of the Bible was simply the latest "big one." As Restoro argued, by moving dirt "from one location to another" the diluvial waters could sometimes form a mountain that incorporated the petrified remains of sea creatures. In the same decades, the Aristotelian natural philosophers that dominated university teaching excluded entirely the biblical Flood from the physical history of the Earth. Natural explanations for the origin of fossils included instead local floods, earthquakes in coastal areas, the upward push of the sphere of the Earth like in Buridan's model, and, above all, the slow crawling of the ocean around the globe that continuously uncovered lands long submerged by the waters.

Likewise, before the late Middle Ages there is little evidence that fossils were ever considered anything but real animals and plants turned into stone. The first influential work to discuss them as rocks molded into animal or veg-etable shapes by astrological forces was possibly Peter of Abano's *Conciliator* (ca. 1300).[43] Historians of geology have always found it difficult to understand how this theory could ever make any sense; instead, contemporary doctrines on spontaneous generation are illuminating. Hardly anyone questioned that lower animals could rise from decaying organic matter or mud. While the soil provided the matrix, astral influences acted like male sperm that gave the new living beings their "form" (in the Aristotelian sense of the term), namely their specific being. A few decades before Peter of Abano, Albertus Magnus's *Book of Minerals* explained how rocks were shaped by a "formative power" that "descends from the heavens through the influence of the stars, deter-mining the particular kind of mineral that will be formed at any particular time and place."[44] The philosopher did not extend this process to the forma-tion of stones with figures of animals or leaves; however, only a small step was required to interpret them as incomplete spontaneous generations or images impressed by the stars on soft and malleable material before it hardened.

One principle did not necessarily exclude the other. The power of the stars could still contribute to the process that turned once living organisms into stones. In his vernacular scientific poem *Acerba*, Cecco of Ascoli (ca. 1269–1327) described how a branch squeezed between two layers of muck left its impression on the contact surfaces as they hardened.[45] A 1516 commentary on this extremely popular work explained that the branch turned into stone owing to "the antiquity of time, and the operations of the Sun and other celestial bodies."[46] Cecco metaphorically described nature as a painter, but he did not imply that those shapes had never been alive. Leon Battista Alberti related in his treatise *On Architecture* the account of quarrymen who found in central Italy a snake that lived "inside a huge boulder, without there being any passage into it for air"; he had witnessed himself "the discovery of tree leaves within a piece of marble of the purest white." Detailed images of living beings carved into stones were ubiquitous and often reported.[47] Nature was definitely thought capable of producing works of art, but like in Cecco of Ascoli's *Acerba*, or in Bernard Palissy's pottery two centuries later, the starting point of the process could well be actual animals or plants.[48]

In the early 1500s natural philosophers pushed back with determination against the re-emergence of the Deluge, which they regarded not only as physically inconsistent but also as a threat to their traditional autonomy. The Deluge was peddled by an unlikely alliance that included biblical literalists as well as authors who supported the spontaneous generation of human beings and saw universal floods as the great annihilators and regenerators of earthly life. Tiberio Russilliano and Alessandro Alessandri mentioned the common observation of fossil fish on mountains as "reasonable" evidence that such floods actually happened.[49] Leonardo da Vinci argued vehemently against what he considered a baseless scheme that lacked any empirical or theoretical backing. Universal floods themselves could not happen short of a miracle, in which case they had nothing to do with natural philosophy. In the episode that opens this book, Girolamo Fracastoro defended the traditional Aristotelian doctrine, explaining that the Flood could not have produced fossil fish and seashells. In fact, he did not even mention specifically the biblical event, but rather "the dregs brought by waters" when they "overtook the mountains"—a much broader and more ambiguous definition. Even Aristotelian philosophers who did not share his view on the origin of fossils, like Cardinal Gasparo Contarini and anatomist Gabriele Falloppio (1523—1562), preferred to attribute marine fossils to incomplete spontaneous generations that took place inside the rocks.[50] Falloppio flatly stated that the "Deluge" was only invoked by "some unlearned people" ignorant of the principles of natural philosophy.[51] He ranked this belief in the same league as the "popular story" that made of petrified shark teeth "lightning fallen from the sky."[52]

While not unchallenged, the displacement of the ocean continued to be taught throughout the sixteenth century at Italian universities like Padua and Pisa. In 1596, Andrea Cesalpino held that "oysters or other shells that are sometimes found when cutting [stones or marbles], were left there by the sea when it withdrew. . . . Indeed, the sea once used to sojourn in every place that is now dry land, as Aristotle teaches."[53] The main alternative remained not the Flood, but the doctrine of spontaneous generation *in situ*, or the idea of the "playfulness" of a mimetic and animated nature.[54] Spontaneous generations or "jokes of nature" did not require an ancient Earth, but there is little indication that this had anything to do with their popularity. The diffusion of the Platonian notion that no sharp distinction existed between the mineral world and organic animals or plants was likely a much more significant factor. For philosopher Girolamo Cardano (1501–1576), what distinguished a live animal from a fish-like stone was only a different amount of the same generative force that pervaded the whole of nature.[55] Historian of geology F. Ellenberger noted that the diluvial theory found very few supporters in the sixteenth and early seventeenth centuries. The staunchest of them, Martin Luther, was hardly a quotable authority for Catholic scholars.[56]

The Flood received indeed only half-hearted support—at best—among Jesuit authors. In his commentary on Aristotle's *Loca Mathematica* (1615), Giuseppe Biancani attributed the marine fossils he had observed "both in the Apennines and in the Alps" to the "great Deluge." Not the biblical one, but Aristotle's and Ovid's cyclical inundations, which returned at intervals of many centuries causing the mixing of terrestrial and marine bodies. The startling conclusion that "we Christians must nonetheless refer [these objects] to Noah's flood" was a one-liner not supported by any reasoning or natural evidence.[57] As in similar cases, it might have been requested by his superiors against his own wishes. If so, it was not part of any consistent policy. Later in the century, neither Niccolò Cabeo nor Athanasius Kircher mentioned marine fossils as proofs of the biblical Deluge. On the contrary, Cabeo pointed out that "we have no records of this Deluge, outside of the Sacred history."[58] Stephen Jay Gould noticed that Kircher's primary concern was to distinguish between objects that could be readily accounted for by natural causes, and others that seemed to call for supernatural intervention. Kircher usually placed fossils of plants and animals into the first category, while stones impressed with religious images or symbols figured in the second.[59]

Among the most puzzling curiosities that nature could offer to a col-
lector, fossils always held a special place in Renaissance museums and their
illustrated catalogues that were influential and widely read.[60] The 1622 book
on the collection put together in Verona by pharmacist Francesco Calzolari,
certainly one of the most celebrated in Renaissance Europe, left the last
word on their origin to Fracastoro and Cesalpino.[61] Calzolari himself was
a self-avowed pupil of Fracastoro, and the Aristotelian tradition undoubtedly remained strong in the area.[62] His museum was later acquired by count
Ludovico Moscardo, a local gentleman who added those items to his large
and variegated cabinet.[63] In the catalogues he published in 1656 and 1672,
Moscardo acknowledged that opinions differed as to whether "shells, fish,
animals, plants, trees and many other objects made of stone . . . were ever
alive."[64] However, he considered Fracastoro's opinion "the one I truly deem
worthy of such a great philosopher: for it can be seen patently, that where
are now mountains, there used to dwell the sea."[65] In fact, Moscardo's idea of
the history of the Earth owed more to Biancani's linear conception than to
Fracastoro's cyclical one: he borrowed from the Jesuit scholar the vision of
a globe that used to be perfectly circular in the beginning and entirely covered by the ocean. The Almighty then raised the mountains, and dug the
cavities where the sea retreated, but erosion was slowly returning the Earth
to its primeval state.[66] As far as fossils were concerned, the net result did not
change: the dry land we inhabit was for a long time the bottom of the ocean.
One could scarcely wonder if its stones returned "fish, shells, wood, and other
petrified things."[67] Even though Moscardo was a good reader of Biancani's (or
maybe, exactly because of that), the Jesuit's admonishment on the Christian
duty to believe in the diluvial origin of fossils went unnoticed.[68]

The list of seventeenth-century naturalists who ignored or downplayed
the Flood is extensive. Ulisse Aldrovandi (1522–1605) remained unconvinced
that such petrifications had ever been living animals, arguing that the lack of
internal structures spoke, in general, against their organic origin. The editor
of his posthumous *Metallic Museum* (1648) definitely favored incomplete
spontaneous generations within rocks.[69] Among the members of the famed
Roman Academy of the Lynx-eyed, Francesco Stelluti (1577–1653) concluded
that the fossil wood he studied for decades resulted from the imperfect transformation of rock into wood, rather than the opposite.[70] Fabio Colonna
(1566–1640) determined on the contrary that the so-called "tongue stones"
could be nothing but petrified shark teeth turned into rock when "the withdrawal of the waters left them on dry land."[71] The biblical inundation was not
unique in this regard: Colonna reminded his readers that "the earth changed,

(a)

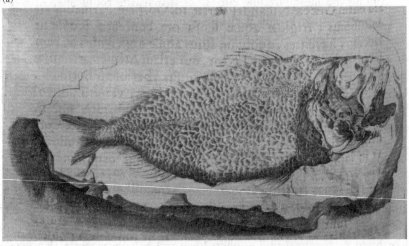

(b)

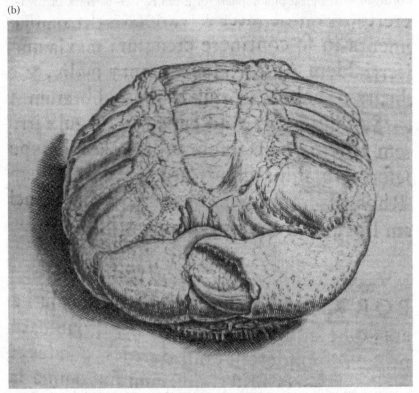

FIG. 5.3 Marine Fossils in Francesco Calzolari's Museum

Fossil fish (a) and crab (b) from Chiocco and.Ceruti's *Calzolari's Museum* (1622). The authors note that the crab has petrified in the same position that this animal assumes "when the sea throws it on dry land, and getting numb and scared it contracts its legs and remains still." Andrea Chiocco and Benedetto Ceruti, *Musaeum Calceolarianum Veronense* (Verona: Apud Angelum Tamum, 1622), 429.

and the sea as well, and they took each other's place not only in the time of the universal Flood, but even in other centuries and in other places."[72] When he set out to disprove the inorganic origin of fossils in his *Vain Speculation, Undeceived by Sense* (1670–1671), Sicilian painter Agostino Scilla (1629–1700) proudly touted his empirical expertise, founded on the study of Sicilian and Maltese fossils as well as his acquaintance with some of the best Tuscan, Roman, and Sicilian naturalists.[73] Together with the many authors he cited approvingly, Scilla recognized that the lands that yielded fossils of marine creatures had once been covered by the waves:

> I do not know whether it happened during the universal Flood, or in other particular inundations. Neither know I . . . if in a certain time the world, tired of lying on one side, turned on the other and exposed to the rays of the Sun the part that was underwater, packed with the leftovers of the sea.[74]

Some things, however, Scilla did know. For example, he was sure that many such inundations or revolutions of the Earth took place; that "the world is ancient"; and that the fishes whose petrified teeth littered the rocks of Malta could not have died all at once in a singular Flood. Antonio Boccone (1633–1704), a leading Sicilian naturalist and a close friend of Scilla's, was more open to the inorganic option, but he still maintained that "the Earth is far more ancient than us, and we do not know in what time mountains were formed."[75]

Archeologists of the Earth

Scilla's work appeared immediately after the two concise geological writings published by Danish naturalist Niels Steensen (Steno, 1638–1686) in 1667 and 1669.[76] Raised a Lutheran, Steno at the time was investigating the geology of Tuscany, where he arrived in 1666 preceded by a solid reputation as an anatomist.[77] His dense 1669 *Prodromus* should have been the outline of a larger work he never completed. Even in this preliminary form, it proved a crucial text in the history of geology and mineralogy that laid out, among much else, the main principles of stratigraphy as they are still taught today.[78] It is perhaps not equally appreciated that Steno's work was also a foundational text of diluvialism. Born out of the marriage of Protestant literalism with the vibrant environment of Italian natural history, the *Prodromus* placed the biblical Flood at the center of a cohesive natural history of the Earth as a unique event that largely explained its geological appearance. Moreover, Steno established that the reality of the universal inundation could be proved independently through the detailed study of Earth's surface, thus confirming the accuracy of the sacred and profane sources that supported the traditional story and chronology of the Flood.

Steno noticed that fossils were always embedded in rocks formed by deposition of sediments under water; yet they could be found everywhere, even at elevated places, thus suggesting a complete submersion of the globe. According to Scripture, nature, and profane histories alike, such an episode only happened twice in the history of the Earth: soon after its creation, and in the time of the Flood. Since Genesis says that plants and animals appeared after the emergence of dry land, the Flood remained the only possibility.[79] In Steno's time the vertical differentiation of groups of "soils" was already being used to demarcate major phases in human and Earth history, thanks to the "new art"—as the Dutch polymath Simon Stevin called it—of digging very deep wells.[80] Taking his cues from Stevin or from Varenius's *Geographia Generalis* (1650), Moscardo cited "stratigraphic" evidence to illustrate the progressive filling of valleys and plains due to the erosion of mountains, whose debris buried a more ancient crust untouched by human civilization:

> When architects dig the ground to build the foundations of their buildings, they first find the soil that they call "moved," which is mixed with wood, iron works, sometimes medals and ancient graves; then they find a soil that is firm, and hard because it was never moved, and pure, with no artificial objects.[81]

Unlike Moscardo's generic example, Steno strove to reconstruct an articulate history of the Earth's surface that could explain his first-hand observations of the inequalities of the Tuscan landscape. Strata, in particular, could only deposit horizontally, yet they were often seen bent and tilted at various angles. Steno inferred that Tuscany, and the crust of the Earth in general, must have undergone six different phases.[82] The first coincided with a time when the Earth was entirely submerged, the second with the first appearance of dry land: "Scripture and Nature agree in this respect . . . but of how and when it began, and how long it lasted as such, Nature says nothing, while Scripture speaks."[83] Steno declined, though, to provide any figures for the duration of these epochs. Scripture might have spoken, but whether humans could understand it was perhaps less clear.[84] The surface was then submerged again by the universal Flood, which Steno explained naturally with underground waters gushing onto the surface, pushed by internal fires that also lifted the seabed. Following the traditional chronology, Steno placed the event about four thousand years earlier. Finally, the "sixth aspect" coincided with the present Earth. While strata were all horizontal at first, some partially collapsed into the caverns dug by subterranean waters or fire; their outer face formed peaks and valleys, whose sides were composed of now inclined strata, while sediments filled the bottoms. The final result was the chaotic and tormented surface we inhabit.[85] No existing document or natural evidence, Steno claimed, contradicted the Genesis account and the conventional chronology of the Noetic Flood.[86] The "danger of novelty" he alluded to in the *Prodromus* consisted arguably in his unusual commitment to support the accuracy of the Bible through natural history, rather than the undetermined number of years he allowed between the Creation and the Flood.[87]

Steno's methodology and his conception of the Earth were clearly indebted to French philosopher René Descartes' system. Descartes placed the Earth into a modern universe where every star was another Sun and the Earth itself a "star" whose surface had cooled down while the core retained its primordial heat.[88] At a time when naturalists normally dismissed straight eternalism, it offered a naturalistic account of the evolution of the Earth that was independent from the Mosaic narration but still envisioned a beginning and an end. Steno criticized, though, his deductive and mechanistic approach.[89] Descartes reduced even living things to complex automata, and his universe was famously likened to a clock that did not need any further intervention after the clockmaker made it and set it in motion. Steno and many contemporaries were gravely concerned about the possible materialistic drifts of this philosophy, yet they also understood the apologetic potential of Descartes' method. A history of the Earth founded on natural reasons alone appeared not only possible, but even desirable. The biblical account of the Flood was accepted in early seventeenth-century scientific literature as a matter of faith, accompanied—if at all—only by the most generic evidence. Since the reliability of the earliest human records remained questionable and the Genesis account open to different interpretations, only the book of the Earth could provide incontrovertible proofs of its historical accuracy. If the biblical Flood explained the formation of strata and the fossil records, then other deluges and the cyclical revolutions of the meteorological tradition became unnecessary, together with the ancient Earth they entailed. Conversely, the common ascendancy of every human race from a single family could no longer be doubted.

Sparingly mentioned in the first half of the century, the association between strata, fossils, and the biblical Deluge began to loom large over

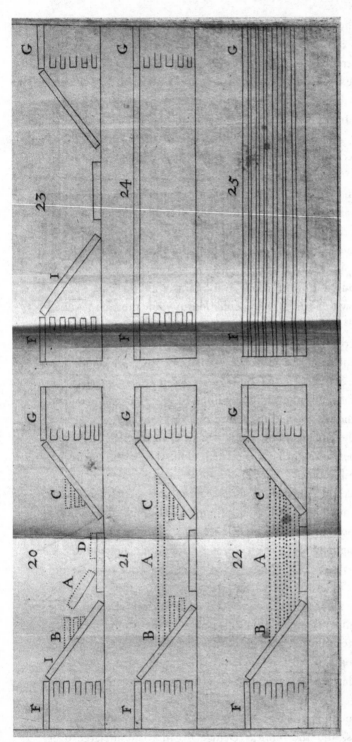

FIG. 5.4 Steno's Earth

The formation of the Earth's surface according to Steno. From the right bottom corner:

25. The primitive crust, formed by the deposition of sediments under the primeval universal ocean.
24. After the emersion of dry land, subterranean waters and/or fires dug cavities under the surface.
23. Some of their vaults collapsed, leaving valleys and elevations.
22. The universal flood left new layers of sediments over the primary crust.
21. Subterranean forces resumed their work.
20. The present state of the surface.

From Niels Steensen, *Nicolai Stenonis De Solido Intra Solidum Naturaliter Contento Dissertationis Prodromus* (Florentiæ: Ex typographia sub signo stellae, 1669). Medical Historical Library, Yale University.

natural history in its closing decades. In 1676 physician Jacopo Grandi, who dominated the anatomical theatre of Venice, co-authored together with Venetian nobleman Giovanni Quirini a *Letter on the Truth of the Universal Flood* (1676), whose title needs little clarification.[90] In the prefatory letter, Quirini provided an interesting overview of the ongoing controversies on the Mosaic story. The nobleman lamented that many questioned its literal sense, replacing it with symbolic readings in which Noah's Ark stood for the indestructible "seeds" of the living things that resume generation after every global catastrophe.[91] A less extreme interpretation, Quirini continued, was that the Flood had not been a global event because not enough water existed in the world to submerge every mountain.[92] While he believed both of these opinions to be wrong, he also doubted that the diluvial waters could bring sea creatures from the bottom of the sea to mountaintops in the little time they dwelled upon the surface.[93] How extremely detailed fossils could have survived for thousands of years in the bowels of the Earth was also problematic (a rather common objection). The only plausible alternative remained spontaneous growth within the rocks, given that the indestructible "seeds" of generation were present everywhere.[94]

Quirini did not contradict the Mosaic account of the Flood, but did not bring any evidence for it either. In his reply, Grandi took a further step by showing not only that the event was physically possible, but also that it took place as Genesis "accurately described."[95] "Do not get fooled," Grandi advised, by the "dreams of Rabbis and the babblings of the Kabbalists" with their extravagant exegeses: the literal truth of the biblical story was proved by the memories handed down by ancient historians and poets and, much more decisively, by the Earth itself.[96] In the footsteps of Steno, Grandi examined the geological strata exposed in the pits of his hometown of Modena; unlike his mentor, he tried to anchor each of them to an absolute chronology.[97] Under the so-called old town, he found soil mixed with ruins of buildings, then only soil, then more ruins, and so on. Armed with the ancient chronicles of the city, he argued that the first layer corresponded to the destruction of Modena during the barbarian invasions about one thousand years earlier. Assuming that strata accumulated uniformly over time, he could then determine when each of them formed. The lowermost, up to 24 feet thick, was made of a stone "entirely pure, with absolutely nothing mixed to it."

> It follows, that it was produced all at the once in a short span of time, and that an immense inundation took place from which a sediment of such a magnitude deposited. . . . There must have been then a great

Deluge, in every respect like that described and divulged by Moses, which occurred eight hundred years before his time.[98]

Grandi's *Letter* was widely read, but not always well received.[99] Bernardino Ramazzini (1633–1714), who taught medicine at the University of Modena and later Padua, severely criticized Grandi's observations and sources. What Grandi described as a single calcareous layer left by the biblical inundation was actually made of three different strata that deposited at different times and certainly well after the Deluge, he contended.[100] The vestiges of the latter event, Ramazzini hinted, may perhaps be found at greater depth.[101] Astronomer and antiquarian Francesco Bianchini (1662–1729) was on the contrary largely inspired by Grandi's *Letter*, when he used stratigraphic arguments to support the diluvial origin of fossils and the traditional chronology of the Flood.[102] His *Universal History*, published in 1697 in Rome, was an often criticized yet bold, admired, and widely influential work.[103] Bianchini presented himself as a firm supporter of a recent Creation, but while the account of the ages before the Flood needed to be anchored to the Bible ("so as not to abandon it entirely to fables"), he sought to create a history that was as independent as possible from Scripture precisely because he aimed to demonstrate its truthfulness.[104] Bianchini bolstered the supposedly converging traditions of different nations with two of the "more universal histories . . . that the same mutations of the Earth impressed in its body."[105] The first was the presence of "crustaceans" within the rocks of the highest mountains, which he found difficult to explain other than with "a general flooding of the terrestrial globe, caused by the Deluge."[106] The second came from an account of digs conducted a few years earlier at the footsteps of Mount Vesuvius near Naples.[107] Steno and Grandi assumed that strata were produced by slow sedimentation, but the Vesuvius area showed an alternation of sedimentary and volcanic "soils," down to a bedrock of solid limestone. Of the five volcanic strata that were counted, Bianchini assigned the uppermost to the great eruption of 1631, and the second to the explosion that buried Pompeii in 79 CE. In the reasonable assumption that nature is "uniform, in its works," the three volcanic layers that formed before the beginning of the current era spanned twenty-four centuries. The limestone bedrock coincided once again with the time of Noah's Flood—exactly where a chronologist would expect to find it.[108]

Historian of geology Martin Rudwick has argued persuasively that the Earth sciences became "historical" by borrowing the methods and approaches of chronology. A discipline that, at a superficial glance, stood in the way of a correct interpretation of the geological evidence, actually taught naturalists to think in terms of a world with a beginning and a directional development. Unlike Aristotelian eternalism, it assumed that its age could be reckoned; that it was "supremely important" to date historical records, be they human or natural; and that, when a precise dating proved impossible, at least they should be arranged in the correct relative order.[109] Rudwick placed these developments into the late 1700s (when chronology was a declining discipline, though still practiced), but their seeds seem to have been planted already at the end of the previous century, stemming from the program of proving the universal nature of the Flood and the young age of the Earth through natural evidence. Still, the sober and empirically grounded diluvialism of Steno, Grandi, or Bianchini was quickly overshadowed by its British versions, which placed the Flood at the center of grandiose but largely hypothetical reconstructions of world and human history.

As in Italy, in seventeenth-century England a history of the Earth based on Genesis was only one of various competing options. Together with Ovid's *Metamorphoses* and Plato's *Timaeus*, Aristotle's *Meteorology* continued to be one of the most cited sources for the textbooks that in the course of the century peddled a cyclical view of the Earth.[110] While straight eternalism was banished, many authors adapted Aristotle's cyclical Earth to a world with a beginning and an end. Noting that a single Flood could not explain the fossil records, in the 1660s Robert Hooke regarded Noah's Deluge as one of the many exchanges between dry land and sea that had taken place since the Creation.[111] As he put it, "a coin tells the archeologist that such and such place was ruled by such and such prince, but the petrified shells give much more convincing proof to the archeologist of Nature that such and such place was under water."[112] In the informal environment of the London coffeehouses, he debated with other fellows of the Royal Society the Creation, the local nature of the Flood, and pre-Adamitism.[113] In the 1690s, astronomer Edmund Halley held that the existence of the Earth should be measured over millions of years, and that Genesis only narrated the creation of a new world out of the old one.[114] As recent research has clarified, Halley cited his timescales as an argument against eternalism that remained the truly unacceptable option.[115] Divergences on how Scripture should be interpreted meant that there was, in fact, no such thing as a single biblically orthodox history of the Earth.

In *A Sacred History of the Earth* (1680–1681) Thomas Burnet sought to frame a biblical understanding of human history within the edifice of Descartes's physics.[116] Burnet expounded the imposing vision of an Earth whose geological revolutions deeply intertwined with the history of salvation, both of them having the biblical Flood as their central event. Like many medieval and Renaissance authors, Burnet supposed that the primitive Earth had no mountains nor valleys, and that Eden coincided with the whole of it. Nothing more was required to submerge it than the cracking of the crust, which released the subterranean waters onto the surface. The catastrophe left behind a harsh world of ruins and climatic extremes; in the future, other natural disasters would prepare the Earth for the second coming of Christ and its final conflagration. The evolution of the globe unfolded according only to natural law, without any need for supernatural intervention after the Creation. In spite of the book's title, Burnet (himself a clergyman) reduced the account of the six days to pure metaphor, whose purpose had been to inculcate in the minds of a primitive people the idea of a creator God. In his correspondence with Newton, he chastised "those Divines that insist upon the hypothesis of 6 dayes as a physical reality," while the Mosaic account had been "Ideal, or if you will, morall."[117] Newton acknowledged that Moses's account was "historical" rather than "philosophical," but he insisted that, while Scripture needed to be interpreted (and the first days, at least, could be made "as long as you please"), it could not be ignored.[118]

Burnet's approach to both Scripture and nature received much criticism. In his history, humanity lost Eden not with Adam's sin but much later, after Noah's Flood. To some, this questioned the same need for a Redeemer.[119] The Roman censors found Burnet's theory of the Earth "extravagant" rather than heretical, but for most of its Protestant readers the *Sacred History* was just not sacred enough.[120] His entirely deductive scheme, which did not even mention the problems raised by evidence like strata and fossils, also met with widespread disapproval. Yet many of his critics set out in turn to produce similar Flood-centered outlines of the Earth's history. Among them, John Woodward was certainly the most read and the most influential in early eighteenth-century Europe. His *Essay toward a Natural History of the Earth* (1695) replaced Descartes' physics with Newton's, positing that the primeval Earth had been very similar to the present one, only more fertile.[121] In order to turn the world into one that was more adapted to the fallen condition of humankind, God, the master of gravity, suspended or strongly reduced its attractive force so that waters could rise from the underground. Losing its cohesion, the surface dissolved into an undifferentiated muck. After the restoration of gravity, rocks hardened again and settled in layers according to their specific weight, trapping the petrified remains of once living organisms. Fossils of unknown or exotic animals (like elephant bones or molars), which puzzled British naturalists, actually reinforced the thesis of a global cataclysm: only such an event could have brought to those latitudes the bodies of species normally dwelling in remote parts of the world. Unlike Burnet, he made no concessions to the possibility of an ancient Earth.

Woodward's "paper world" maintained the primacy of the Flood over Adam's sin as the crucial event in the history of humankind, and it defied the basic observation that strata are not disposed according to the specific weight of the rocks they are made of. In a country where the inorganic origin of fossils was still upheld by leading naturalists such as Martin Lister, Woodward maintained at least that they had been living animals. Overall, his scheme proved appealing. Its adoption by influential naturalists in Switzerland had a considerable impact, especially because in the early 1700s they often acted as cultural mediators between Catholic and Protestant Europe. In 1704 Johann Jackob Scheuchzer published in Zurich an influential Latin translation of Woodward's *Essay* (originally published in English), thus making it available to a much larger audience.[122] A champion of the diluvialist cause, he became famous as the author of such works as the *Holy Physics*, the *Museum of the Flood*, and the *Herbal of the Flood*.[123] Calvinist Louis Bourguet (1678–1742), another firm supporter of diluvialism, in the early 1700s resided mainly in northeastern Italy for his family's silk business and built strong ties with local naturalists and scholars, religious differences notwithstanding.[124] Bourguet collaborated with a key cultural mediator, Jean Le Clerc, editor of a *Selected Library* (*Bibliothèque choisie*) that spread Italian literary and scientific culture across Europe, while facilitating the assimilation in Italy of British natural philosophy. After 1728, his work was continued by Bourguet with a new journal, the *Italian Library* (*Bibliothèque Italique*).[125]

Le Clerc enjoyed excellent relations with Francesco Bianchini, who in the years following the publication of his *Universal History* gained the confidence of the popes he served and used it to promote the diffusion of Newton's natural philosophy in Italy.[126] Bianchini and his Roman circles recognized that his system (based on a mysterious attractive force, which could only come from God's continuous intervention into nature) provided a modern yet theologically attractive alternative to Descartes's clockwork universe. The cultural alliance between papal Rome and schismatic England was sanctioned in January 1713 by warm meetings between Newton and Bianchini himself, who had been sent on a diplomatic mission in northern Europe. The fierce anti-papist and the Roman courtier agreed on almost everything, including (at least in principle) the historical and chronological issues in which they were both immersed.[127] Copies of the second edition of Newton's *Principia*, then being printed, were promptly sent to Rome. Bianchini made sure that the book would be well received and doubts on its theological orthodoxy promptly silenced.[128]

"Only God Knows When"

A number of Italian naturalists observed with alarm the growing influence of the international and interconfessional movement that sustained the tide of diluvialism. Its sharpest critic, physician Antonio Vallisneri (1661–1730) who taught at Padua University in the first decades of the new century, was widely regarded as the new leader of the Galilean experimental school in medicine and natural history.[129] Vallisneri did not question the existence of a Creation and of the biblical Flood, nor the authority of the Church over the interpretation of Scripture. Like his medieval and Renaissance predecessors, however, he maintained that a clear separation must exist between theology and natural philosophy. Therefore, the Flood must remain a miraculous event that escapes rational investigation, that did not leave any traces of its passage on the surface of the Earth, and that could not be treated as part of the Earth's physical history. Diluvialism represented an existential threat to what he considered the very foundations of local and Catholic scientific traditions, namely, a natural history unaffected by religious disputations and theological dictates, and the possibility of an ancient world. In his most influential geological treatise, *Of the Marine Bodies Found on Mountains* (1721), Vallisneri acknowledged unambiguously those long and uninterrupted intellectual roots:

> The highly praised Fracastoro [was close to the truth], when together with Cesalpino, Aristotle, Strabo and others of subtle and insightful mind, thought that the fish and shells of his mountains did not take origin from the universal Flood; but were cast there in the remotest and most obscure times (and only God knows when) by natural floods of the sea, [which later] withdrew to cover and fill other, more distant provinces.[130]

Only a great number of local inundations that succeeded each other over immense lengths of time could have laid the geological layers that enveloped marine fossils. As he argued in favor of an ancient Earth and against the biblical Flood as an active agent in its history, Vallisneri did not see himself as the herald of astonishingly new ideas but as the defender of a threatened tradition. Like Agostino Scilla, he confessed that he did not know how and when those "natural floods of the sea" rose and withdrew, but he was also not willing to erect a precarious scaffolding that lacked solid empirical bases. In line with much of contemporary Italian natural philosophy, Vallisneri's science was informed by a programmatic empiricism and by a deep mistrust of universal explanatory systems, be they Cartesian or Newtonian. Even though he could not assign a definite cause, he found it unquestionable that "the whole large plain around [river] Po used to be a marsh"; if so, "how can we be certain that the Adriatic Sea, and the whole sea that surrounds the booth of Italy, wasn't once as high as said mountains?"[131] He recognized that, while observation did not support any of the theories of the Earth elaborated up to then, it invalidated each one of them. In 1708 Vallisneri delivered in a literary journal a scathing review of Woodward's *Essay*:

> Finally, in the sixth part [of the *Essay*] he deals with the state of the Earth, and of his productions before the Flood; everything is founded on the author's imagination, and on his effort to conjecture over things of which we have nothing but most obscure relations, and founded more on our Saint and infallible Faith, than on natural history.[132]

Vallisneri did not seek to domesticate the Flood to make it plausible through natural means or to "explain Scripture with nature" as Bourguet suggested.[133] His intent was to keep "the useless miracle," as he called it, out of natural history altogether. Those of his correspondents, like mathematician Jacopo Riccati, who reduced the Flood to a local event while maintaining for it a role in the geological history of the Earth, received gentle but firm rebukes.[134] In his struggle to build a dam against diluvialism, Vallisneri strove to associate it with Protestant culture and with the "heretics" who, "enamoured with Scripture," stood by "the literal sound of the words, and they make [Scripture] say things the Holy Ghost never dreamt of."[135] The world, he wrote to Bourguet, "is older than many believe. . . . Apart from the reverence due to the Holy Scripture . . . who can assure us of this universal Flood?" Vallisneri did not deny that "many Italians too" increasingly accepted the Flood as the origin of marine fossils; but he portrayed the situation as the result of recent imports from non-Catholic countries and cultures.[136]

Vallisneri was indeed in an excellent position to assess the trends that were popular among Italian naturalists and amateurs. The vast majority of his scientific exchanges involved a regional web of physicians, pharmacists, surgeons, and city and country priests who made up the backbone of the naturalistic community in which he operated. His museum and his writings incorporated an astonishing amount of fossils, ideas, and suggestions he obtained from or elaborated alongside local naturalists.[137] Venetian apothecary Giovanni Girolamo Zannichelli held that the salty waters once covering the plain of river Po had been those of the biblical Flood, and Vallisneri rightly counted him among the Italian diluvialists.[138] Zanichelli suggested that as the world re-emerged from the catastrophe, those waters dwelled here and there like puddles after a storm, trapping the marine creatures that later turned into fossils.[139] In his correspondence, it was customary for him to call them "diluvian things."[140] In contrast, Vallisneri praised Bolognese naturalist Luigi Ferdinando Marsili for using the adjective "antediluvian": "They are not, then, trophies or sediments of the Flood, as everyone writes. They are 'antediluvian', and I infer that Your Excellency's system of the world resembles mine, positing that the sea once naturally was on mountains."[141]

In the same years, an obscure correspondent of Vallisneri, Ottavio Alecchi, entertained fascinating material and intellectual exchanges with Louis Bourguet on the origin of fossils, the Flood, the Creation, and the relation between faith and reason. The Protestant naturalist introduced Alecchi to British diluvialism and to Leibniz's philosophy and Earth history, but he was unable to shake the correspondent's skeptical attitude toward the Flood and its role in the history of the Earth. Alecchi insisted that "faith and human reason cannot get along; nor will a mathematician be able to demonstrate that, against every sound principles, one is three, and three is one."[142] The Flood and its asserted universality could not be explained by natural laws. Had it been appropriate to discuss the biblical account from a physical point of view (and he maintained that it was not), Alecchi still could not conceive how the whole world needed to be drowned in order to submerge the small area inhabited at the time.[143] Alecchi read the six days of Genesis as a moral fiction, accepting instead the Augustinian view that the universe was created all at once.[144] It was certainly not a coincidence that, during those years, the doctrine of the instantaneous Creation was being revived by Paduan theologians with Vallisneri's endorsement.[145]

Rather than foster a particular theory of the Earth, *Of the Marine Bodies* aimed to expose the inconsistencies of those formulated up to that point. Arranged as a loose conversation on the topic among different authors, the book was also representative to some extent of the diverse views of the community that contributed to it. Nonetheless, the final product reflected Vallisneri's own take on the subject, thanks to the firm control he maintained over the publication. *Of the Marine Bodies* made room for transparent allusions to the antiquity of the world, or for Riccati's conjectures on the local nature of the Flood, but it did not include any contributions from avowed diluvialists such as Scheuchzer, Bourguet, or Zannichelli (who figured otherwise among Vallisneri's closest and most respected scientific partners). Such strategies were not uncommon in the European "Republic of Letters," but in Vallisneri's case they were further complicated by the need to deal with religious issues that could no longer be circumvented.

To assert the old age of the Earth in a "philosophical" work, or to deny that the biblical Flood caused the formation of fossils, had never been considered heretical, dangerous, or even "extravagant" in Catholic culture. However, the increasing dependence of the truth of the Genesis on natural evidence stimulated powerful and dangerous connections between diluvialism and religious orthodoxy. The open-minded tone of the exchanges Vallisneri entertained with Scheuchzer, Bourguet, Zannichelli, or the learned ecclesiastics he corresponded with, could no longer be taken for granted. In 1720 he raged against a book he received from a German author, who claimed that every anti-diluvialist must be "a wicked and ungodly man, an atheist, a deist, and he calls for God's revenge on Earth and in Heaven. Oh what a German fool!"[146] The tendency to identify anti-diluvialism with libertinism and unbelief was spreading even in Italy. In 1719, Bolognese naturalist Giuseppe Monti published the description of a fossil that he excavated in a nearby mountain. Monti identified the finding with the head of a "cetacean" (apparently, a walrus) that according to naturalists well versed in Nordic fauna, like Conrad Gessner and Ole Worms, lived in remote Artic islands. The presence of such an animal in the Italian Apennines, Monti observed, was irrefutable evidence of the global upheaval caused by the Flood "against the strongest objections of the atheists."[147]

Vallisneri had to tread carefully, to the point of including a chapter composed by a theologian who certified that the Flood had been entirely supernatural. His collaborator attested that the "floodgates of Heaven" and the "springs of the great deep" could be metaphorical expressions, and that it was admissible to argue from a physician's point of view, as Vallisneri did, against the "giants" mentioned by Genesis and the incredibly long lives of the antediluvian patriarchs.[148] Vallisneri and the anonymous contributor did nothing but reinstate the traditional separation between faith and natural philosophy, which counter-Reformation cultural policies had restricted and redefined but never repudiated. It is difficult to determine whether such communication strategies were motivated by the need to strike the delicate balance his public position required, or by a genuine desire to protect Catholic culture against the kind of counterproductive overreach that produced the disastrous condemnation of Copernicanism. In fact, one may wonder whether he was a sincere Catholic at all. He certainly believed in a creator God, and he made sure that his public behavior would be as impregnable as his geological treatise. His private correspondence, however, often denigrated the "obstinate race of monks and priests, who do not want innovations,"[149] who "sold" as miracles what could be explained by the normal course of nature, and who battled mathematics and experimental philosophy in order to keep people "in the dark, always shrouded in mysterious mists."[150] Of course, anti-clericalism was not the same as irreligion, especially in a country like Italy. Toward the end of his life, though, he disclosed to Venetian philosopher Antonio Conti wholly heterodox opinions on the nature of the human soul (and more) that he had probably held for a long time: "Here is a confession that my beneficent Jesuit confessor never heard."[151]

Vallisneri knew how to navigate social and cultural conventions.[152] Lack of such skills and of adequate protections, status, or caution could still make the difference between approval or censure, tenure or discharge, and sometimes even life or death. On his way back to Padua from a trip to Florence, the naturalist heard that the Inquisition of Modena hanged an illiterate peasant who insisted that "all those things we are taught as infallible truths, could not be true according to nature."[153] Vallisneri commented with bitter sarcasm that he must have been "one of those lunatics, who openly and unwisely deny what they cannot understand," and that he "foolishly answered back to priests and monks." The peasant then listened to his sentence "with the utmost tranquility, and fearless faced death."[154] In that anonymous villager, the famous professor might have secretly admired the moral courage he never found.

6

The Invention of the History of Deep Time, 1700–1770

During the sixteenth-century several savants, among whom one can place Fracastoro ... did not doubt that it was the sea that brought them on the continent; and since they could see no more plausible cause for this phenomenon than the universal Flood, they attributed to it all of the marine bodies that are found on our globe, completely inundated by its waters.

ENCYCLOPÉDIE, "Fossils" (1757)

A FEW MONTHS after the publication of his treatise *Of Marine Bodies* (1721), Antonio Vallisneri could congratulate himself on the positive reception of his book, and on the powerless "mutterings of the friars."[1] Whether because of the soundness of his observations, his shrewd communication strategies, or because few were willing to challenge the author's powerful network of patrons and friends, "Mr. Monti, Mr. Zannichelli, and many more [diluvialists] have all remained silent so far." Writing to Louis Bourguet, Vallisneri was confident that even the Swiss naturalist would have found it hard to confute his scientific and theological arguments against the Flood.[2] At their core lay a consideration that he held as unquestionable: a young Earth had never been a meaningful tenet of Christian doctrine and Catholic traditions. As long as one acknowledged a creation in time (and steered clear of theological implications), it had always been licit to entertain the possibility of an ancient world and to exclude the Noetic Deluge from its history. History and tradition marched by his side.

One century later, that history and those traditions had all but disappeared from Europe's collective memory. In 1836, pious geologist and Anglican minister William Buckland assumed that his discipline had been considered

On the Edge of Eternity. Ivano Dal Prete, Oxford University Press. © Oxford University Press 2022.
DOI: 10.1093/oso/9780190678890.003.0007

"hostile" to revealed religion; it was then necessary to modify the "most commonly received" interpretation of Genesis in order to accommodate the immensely long geological eras that could no longer be denied.[3] A few years earlier, Charles Lyell held that an Earth only a few thousand years old had been "the consistent belief of the Christian world."[4] The material and the general framework for the historical introduction to his *Principles of Geology* were sourced from a work published in the Veneto, the area where Vallisneri taught and had been most influential. A new understanding of pre-modern Earth history, which cast an unfathomable past as inconceivable in traditional Christianity, had emerged in the course of the eighteenth century and appeared by then firmly established. It was to remain unchallenged for the following two hundred years.

This chapter explores the elaboration and consolidation of this narrative, tracing its origin back to the cultural and political tensions of Enlightenment Europe. While a number of naturalists tried to dissociate non-literalist positions on the age of the Earth and on the Flood from atheism and unbelief, immensely popular authors weaponized an ancient Earth as a tool of anti-Christian propaganda. *Philosophes* such as Voltaire, Boulanger, and d'Holbach turned the recent association of diluvialism and biblical inerrancy into a hallmark of historical Christianity and of the purported clash between "religious prejudice" and rational enquiry. In publications that were avidly read all over Europe, they outlined for the first time what remains to this day the received history of deep time and used it to undermine the divine origin of the Revelation. As they rearranged historical sources and natural evidence in ways that supported their agendas and their beliefs, they presented contemporary theories based on an ancient Earth not in their uninterrupted continuity with medieval and Renaissance natural philosophy, but as recent breakthroughs of the human reason finally freed from religious obscurantism. The attacks on the established religion, however, may not have been as consequential as their perceived political and social repercussions.

After 1750, a variegated coalition of "*anti-philosophes*," as they and their enemies referred to them, began to decry in apocalyptic tones the dangers of unbridled reason.[5] In their view, their opponents were sapping the moral and ideological foundations of a social order whose ultimate sanction was divine authority. Fears of social subversion and the associated rhetoric became rampant in the decades leading to the French Revolution, powerfully contributing to the politicization and polarization of Earth history. In their multipronged battle against "materialists," "atheists," and the dangerous ideas of democracy and equality they supposedly peddled, many conservatives embraced diluvialism and a young Earth as powerful arguments for the truth and historical accuracy of Scripture. Inevitably, politicization entailed simplification and radicalization. Friends and enemies had to be counted and categorized, nuances erased, and the public asked to take sides. A diluvialist orthodoxy was established, every deviation being regarded as a dangerous concession. In the end, they sought legitimacy in a version of the same narrative propagated by their opponents: a young Creation had always been an essential element of the Christian tradition, that the *philosophes* were questioningfor purely ideological reasons. Caught between their hammer and the conservatives' anvil, the complex and nuanced history reconstructed in the previous chapters was crushed.

Transmutations of Earth and Life

At the turn of the eighteenth century, a recent universal Flood had long been an ideological cornerstone of imperial expansion, global Christianity, and biblical inerrancy; yet, there was never any dearth of criticism and alternative approaches. Like Vallisneri, their proponents were perfectly aware of their continuities with uninterrupted traditions. Recent diluvialism did not appear out of the blue, either. Physician and naturalist Bernardino Ramazzini, for example, was remarkably unimpressed with Thomas Burnet's theory of the Earth, which he labeled a "fiction" and not even a good or an original one. He placed it firmly within a Renaissance lineage that assumed a fundamental discontinuity between a pre-diluvial and a post-diluvial Earth, and whose main representatives had been Scaliger, van Helmont, and Patrizi. The ample excerpts from Patrizi's 1562 dialogue that he published in 1691 circulated widely across Europe, particularly in Great Britain.[6]

Likewise, the idea that dry land slowly emerged from a primeval global ocean had been common in the Middle Ages among both Christian and Islamicateauthors. Discussed in vernacular literature and by authorities like Albertus Magnus, the inexorable desiccation of the globe had offered for centuries a plausible naturalistic reading of God's command that separated water from earth in Genesis. Leonardo da Vinci reflected seriously on this mechanism as a possible cause for the emergence of the surface in the past and the future extinction of terrestrial life when all of the water would be lost. One century later, Biancani envisioned the possibility that earth might instead be submerged as it was in the beginning, owing to the erosion of every height.[7] The subsidence of an early ocean gained considerable traction around 1700, as a doctrine traditionally used to explain the separation of waters from dry land in non-literal readings of Genesis. Yet its Christian appeal eroded steadily, when it appeared that it could also serve extremely unorthodox theories of geological and biological change.

The diminution of the ocean as a Christian alternative to diluvialism was the backbone of Wilhelm Leibniz's *Protogaea*, which was not published until 1749 but whose basic elements circulated in various writings that appeared between 1693 and 1710.[8] *Protogaea* was heavily indebted to Descartes in its pursuit of a miracle-free Earth history, whose course was determined by the unfolding of entirely natural processes. Unlike the French philosopher, though, Leibniz presented his model as the outline of real events that were compatible with the narration of Genesis and a finalistic Creation. In the beginning, the Earth was a molten mass "of a regular form, and has hardened from liquid, light or fire being the motive cause . . . [which cause] sacred history also takes as the beginning of Cosmogony."[9] The Earth's cooling resulted in enormous blisters under the crust filled with water or air, whose collapse produced the first valleys and mountain ranges. The release of subterranean liquids and vapors then covered the surface with a global ocean, as related "in the sacred books of our religion," the stories of ancient nations, and above all "the vestiges of the sea discovered inland."[10] As Leibniz reiterated in his 1710 *Theodicy*, Moses briefly hinted at those same phenomena:

> The separation of light from darkness indicates the melting caused by the fire; and the separation of the moist from the dry marks the effects of inundations. But who does not see that these disorders have served to bring things to the point where they now are, that we owe to them our riches and our comforts, and that through their agency this globe became fit for cultivation by us.[11]

It was hardly believable, of course, that natural forces could have accomplished the process during the second day of the Creation; unless miracles were invoked, those "days" must have been immensely long. As the waters slowly seeped back into the underworld, the first humans descended from the highest regions into the new lowlands. From time to time, the weight of the ocean could still crack the crust, unleashing catastrophic and even universal inundations. If the biblical Flood happened when the surface of the sea was much higher than in later eras, the submersion of the small stretches of inhabited land did not require any extravagant amounts of water. Not unlike Steno, whom he quoted extensively, Leibniz could then rejoice "in contributing, through natural arguments . . . to a belief in the sacred history and the universal Flood."[12] Yet the biblical Flood, placed at center of the diluvialists' narrations, was unnecessary to Leibniz's system. It needed instead an ancient Earth, which allowed for the great number of partial inundations and withdrawals indispensable for explaining the succession of rock strata and the origin of marine fossils. In spite of its deference to religion, Leibniz's attempt to sketch a theologically acceptable history of the Earth was not always welcome. While his system seemed the "least improbable" to Vallisneri, the well-informed Venetian philosopher Antonio Conti admonished him that German theologians had "rightly" exposed it as a form of concealed skepticism. A close confidante of Vallisneri's, and denounced in 1735 to the Inquisition of Venice for "atheism," Conti could hardly be troubled by large timescales or by the sidelining of the Flood.[13] He might have been worried, though, by the far more radical twists that Leibniz's concept was receiving in libertine milieus. A number of "philosophers" were using the diminution of the ocean not to foster a more Christian counterpoint to Aristotle's eternalism, but to undermine some of the most fundamental tenets of the religion. As Leibniz did not fail to mention, some unnamed authors speculated that terrestrial animals may not have been created anew when dry land first appeared. Instead, they could be the remote descendants of aquatic species that in the course of the ages adapted to an increasingly drier environment:

> There are those who take the freedom to conjecture so far that they have imagined how once, when the ocean covered everything, animals that now live on land were aquatic; then, as the water departed, these animals became amphibians, until their descendants eventually left that original home. But that conflicts with the sacred writers, with whom it is impious to disagree.[14]

In spite of Leibniz's restraint, *Protogaea* downplayed rather than confuted organic "transmutation." The author concluded that "throughout those great upheavals [of the Earth], animal species have remained mostly unchanged."[15] In 1710, he even acknowledged that "certain terrestrial animals which long ago lived in our regions are today extinct." As for the "unknown animals" of that ancient world, whose remains were continually dug out, one could think that "when [those lands] were once and for all abandoned by the sea, some finally advanced as terrestrial beings, such that, transformed by the prolonged effect of time, they were no longer able to withstand water."[16]

The possibility of both "transmutation" and that some species might go extinct over time were not stunning novelties. Aristotle's eternal Earth

required that the number and characters of the species of living things be immutable, but most other philosophical schools of the antiquity assumed that,
as the Earth evolves, species must show some degree of adaptation or disappear. In *On the Nature of Things*, Lucretius explained that the prodigiously
fertile loam of a young Earth produced every sort of creatures: the majority
of them, though, were incapable of finding food, reproducing, or adapting to
a changing environment. In the same way, species that thrive here might not
survive in the different conditions of other worlds. As the fertility of our globe
declined, humans too had to learn new ways to obtain food and shelter, invent language, develop agriculture, and form complex societies. Their bodies
underwent some changes too, while less adaptable species faded away.[17]
Renaissance physicians regularly invoked bodily changes in order to explain
discrepancies between ancient and recent descriptions of organs and diseases,
or the appearance of new syndromes (like syphilis). Did not Scripture itself
talk of the time when "giants" inhabited the Earth and the span of human lives
was measured in hundreds of years?[18] Noah's Flood took care of the giants
and, in the following ages, the bodies of Noah's descendants became smaller
and weaker accompanying the physical decadence of the Earth and their own
moral one. Colonial expansion and the need to justify the systematic exploitation of other races brought to the fore the problem of the origin of human
diversity. As spontaneous generation or multiple creations appeared less and
less viable, the influence of different climates became a common answer to
how races differentiated or decayed from the first couple or from Noah's
family (who were, of course, white). The Europeans who settled on other
continents worried that their bodies might also degenerate in those unfamiliar environments and become as vulnerable to old diseases as those of the
natives.[19] If a few hundred or a few thousand years seemed sufficient to bring
about such mutations, what could the ages of a vastly older world do? The inclusion of humans among the creatures that once abandoned the sea was just
a small step, albeit a giant leap away.

The most famous of those who did not hesitate to jump was French
diplomat Benoît De Maillet, who began to write his *Telliamed* in the late
1690s.[20] Manuscript versions did not circulate before ca. 1720, so that the
book appeared in many ways outdated when it was published posthumously
in 1748. Nonetheless, it seems to have been one of the most popular scientific
texts in France's libraries, and it loomed large over the debate on the history
of the Earth until the end of the eighteenth century.[21] In his fictitious dialogue between a French missionary to Egypt and an Indian philosopher, de
Maillet merged a Lucretian eternal universe with a Cartesian cosmology, in

which every world was cyclically either a hot sun or a cooling planet. In the course of those revolutions, the indestructible seeds of life that pervaded the universe developed and transformed as the environment allowed, without any need for supernatural intervention. On the Earth, the right conditions occurred when enough of the original global ocean was lost as to permit the existence of shallow waters and of the first dry land. The various species encased in those seeds first came to life in the sea, but as more and more of the sea withdrew and turned into land, some adapted to the new surroundings.[22] The numerous accounts and reported sightings of tritons, mermaids, and other "aquatic men" that inhabited Renaissance natural history attested that some of our cousins never left the seas where our species was born. De Maillet's theory did not represent a precursor of modern evolutionism, because in his cosmos species did not evolve and differentiate into something new: aquatic and terrestrial humans belonged to the same species, which always had the ability to adapt to life in the ocean or on land. Both had lungs and gills that atrophied when not in use, and they adjusted their skins or limbs to cope with the prevalent conditions. De Maillet maintained that different human races did not degenerate from an original archetype, but developed independently when "marine men" in different parts of the world abandoned their first habitat.[23] The problem was not whether humans descended from apes, but whether apes were a variety of humans that left the sea at a more recent time (de Maillet's answer was in the affirmative).[24] So much for the special creation of man, or the common descent of the human race from Adam and Noah.

Such conceptions of the origins and history of humankind depended on an immensely old Earth, whose transformations slowly affected the environment on a global scale. While species could adapt, they also needed the time to do so. Earth history, rather than biology, was the cradle of those theories. Taking a page from the diluvialists' book, de Maillet set out to buttress his claims by establishing a chronology of the epoch when dry land began to emerge from the global ocean, and humans might have been able to conduct a terrestrial life. The usual combination of historical and biblical sources, archeological findings, and stratigraphic observations could barely probe the surface of that bottomless pit. De Maillet relied instead on surveys of shorelines and shallow waters, which he conducted during his consular missions to Egypt and Leghorn (a Tuscan free port that was home to a vibrant and diverse naturalistic community), to determine that the Earth was losing about three inches of water per century.[25] Extrapolating that loss rate backwards, and assuming that it accelerated as the oceans became shallower, the diminution of the sea must have been in progress for a period that in some manuscripts goes as long as two billion years.[26]

De Maillet's biographers have constantly emphasized his originality, yet the redaction of his dialogue was a collaborative effort that involved prominent figures in the French cultural landscape. *Telliamed*'s editor, Jesuit abbot Jean-Baptiste Le Mascrier, modified considerably the original manuscripts in the arduous task of tempering the author's materialism and atheism. The reorganization of the original three parts in six "days" mimicked the six "evenings" of Bernard de Fontenelle's (1657–1757) *Conversations on the Plurality of the Inhabited Worlds*, possibly the most successful work of scientific popularization ever published. For fifty-eight years the secretary of the Paris Academy of Sciences, Fontenelle was likely the most influential of de Maillet's interlocutors. His prefaces to the yearly *Memoirs* of the Academy were crucial to the French diffusion of Leibniz's views on the history of the Earth. After de Maillet's return to France, Fontenelle suggested that he should expand the original geological treatise into a broader cosmological work, including the bold discussion on the aquatic origin of animals and humans that Leibniz declined to tackle.[27]

Fontenelle never laid out his own cohesive system of geohistory, but he shared with France's top naturalists, René Antoine de Réaumur (1686–1758) and Antoine de Jussieu (1683–1757), the willingness to reject both the diluvial and the inorganic origin of fossils. Instead, they envisioned an ancient Earth with a surface markedly different from that of their time, to which no human document could bear witness.[28] In 1703, the examination of some Italian petrified fish convinced him that "there is no reason to doubt that they were actual fishes, enveloped by a kind of sand that later solidified."[29] Three years later, Leibniz sent to the academy an essay on their origin based on his theory of the progressive lowering of the sea. Even as he endorsed Leibniz's and de Maillet's works, Fontenelle found it difficult for them to explain the existence in Europe of fossils of tropical species, nor did he think that the emergence of dry land was necessarily a slow and progressive phenomenon.[30] By 1715–1720, he came to conceive a past of catastrophic upheavals unrelated to the biblical Flood, when sea bottoms abruptly emerged from the waters while whole continents plunged into the abyss. Thomas Burnet turned to the tormented surface of the Moon for evidence that the inhabitants of other worlds might have suffered their own Deluges. The apparent instability of the surface of Mars, observed in those years by Paris astronomer Jacques Maraldi, suggested instead to Fontenelle that the Earth might have experienced in the past an epoch of similar convulsions:

Why such a great difference between bodies of a same nature? That would be a considerable problem of Physics, if it was not solved by the great and ancient Revolutions of our Globe shown by the seashells and fishes buried within the earth and on the mountains in every part of the world.[31]

"Either before or after the Flood," Fontenelle argued, "the surface of the Earth must have been arranged very differently from today." While the changes that have been known to us "since the first recorded histories, or from fables with some historical content" may not have been particularly significant on a global scale, "they let us imagine what longer ages could bring."[32] Fontenelle's reference to the Flood only paid lip service to scriptural authority, and none at all to diluvialism. Unlike Vallisneri, Fontenelle and de Maillet had no interest in saving the Christian Revelation from reason and physical evidence, or the other way around. On the contrary, they were willing to concede that an event similar to the biblical Deluge could have happened as part of the geological history of the Earth, leaving vague memories in some of the most ancient human records. To admit its universality, uniqueness, or supernatural origin was a different matter altogether. Unsurprisingly, de Maillet reduced the biblical Flood to a merely local event. When Scheuchzer discovered an alleged human skeleton that fossilized in the aftermath of the Deluge (his "witness of the Flood" was actually a giant salamander), de Maillet commented that he did not see any contradictions between that relic and more ancient "petrifications" also caused by the withdrawal of the sea.[33]

Theories of an eternal and cyclical Earth derived from the Aristotelian tradition were also common currency during the eighteenth century. When geological change was supposed to depend on celestial motions, many meteorological authors tied the rejuvenation of the Earth's surface to the period of the precession of the equinoxes, estimated at 36,000 years. Still, the obvious slowness of the geological processes called for much longer periods. Jean Buridan's approach to Earth history was vastly different from de Maillet's, yet the fourteenth-century Master of Arts would have been familiar with his timescales. French engineer Henry Gautier (1660–1737) shared with Buridan, and many more before and after him, the problem of showing that unremitting erosion did not rule out a timeless Earth. His theory descended directly from Renaissance meteorological traditions. As Louis Bourguet noted, it incorporated elements such as the "eternal circulation of earth between land and sea" and a hollow globe in which portions of the crust periodically fractured and crumbled into the underlying abyss. Gautier's remarkably accurate measurements of erosion and sedimentation rates in southern France suggested periods of millions of years, even though he finally settled for a figure of 35,000 years. Like Fontenelle or de Maillet, he did not need to deny the biblical Flood only its uniqueness and providential origin.[34]

As astrology faded away in the last decades of the seventeenth century, models of geological change inherited from the old astrological tradition were recast into modern molds. The period of the precession of the equinoxes was far too short, but the supposed variation of the obliquity of the ecliptic (the angle between the plane of the Earth's orbit and that of the equator, which is responsible for the seasons) seemed to offer a more promising alternative. Comparing ancient observations and more recent ones, in 1714 French astronomer Jacques Eugéne Louville reckoned that 400,000 years earlier the ecliptic would have been perfectly perpendicular to the celestial equator, as in the alleged Egyptian tradition handed down by Herodotus and Plato, the Babylonian records cited by Berossus, and Chinese and Indian calendars.[35] A full revolution of the ecliptic around the celestial equator would then take about two million years. In the heliocentric cosmos that was becoming the norm, what changed was of course the inclination of the rotation axis of the Earth, rather than that of the sky. Louville did not discuss the geological implications of his scheme or even if the Earth could actually be that old. A number of eighteenth-century naturalists, however, found his model alluring and conjectured that as the solid bulk of the Earth slowly revolved, liquid bodies remained in a fixed position with respect to the plane of Sun's orbit, gradually covering and uncovering the landmasses that passed underneath. Temperate regions yielded the fossils of both tropical and arctic creatures, because those continents traversed widely different latitudes in a remote past.[36] Fontenelle raised legitimate doubts about the accuracy of the ancient observations cited by Louville, but he was happy to enlist his theory in the anti-diluvialist front: "The physical derangements that Mr. Louville's hypothesis would one day bring to the Earth, may sound implausible to common people; but philosophers will digest them more easily."[37]

The Philosophes *and the History of the Earth*

Even the most daring geological and biological theories that circulated in the early eighteenth century, like de Maillet's, were not necessarily more disruptive than the Aristotelian eternalism Christian Europe had tolerated for centuries. That coexistence, however, was based on the assumption—or pretension—that the Earth was not actually eternal and that humankind (at least the one Europeans were part of) was created directly by God. As long as natural philosophers stayed in their lane and did not meddle with theology, theologians would do the same. The cultural and political environment, however, had changed drastically. The pursuit of a "pious" natural philosophy made Revelation far more vulnerable to rational arguments and physical evidence, a danger that was not lost to many clergymen, and lay Christians as well. Literary historian Maria Susana Séguin noted that the French low clergy tended to oppose the naturalization of the Flood, for the same reasons that had been considered valid centuries earlier: to explain and rationalize a miracle was to attack the dignity of Scripture, challenge divine wisdom, and, ultimately, submit faith to reason.[38] Apparently, Flood theories did not gather any more support among the reforming elites who made up the core of the so-called Catholic Enlightenment that peddled the compatibility of Christian doctrine and modern rationalism.[39] Even though Vallisneri railed against the "priests and friars" who opposed his disenchanted nature (he definitely resented their competition at the bedside of wealthy patients), diluvialism never received any institutional backing from the Catholic Church.[40] Clergymen involved with natural history seem to have been as likely as any other social group to support or deny the diluvial origin of fossils. One of the most influential exegetical works of the century, Dom Calmet's *Literary Commentary* (1707), presented the Flood as a phenomenon that could be perfectly explained trough a concurrency of secondary causes so that it would only make an enlightened faith stronger.[41] German naturalist Johann Gottlob Krüger (1715–1759) did not share Calmet's confidence. Around the middle of the century, he concisely voiced a believer's unease with diluvialism:

> I do not blame the intention; but experience teaches that this indiscreet zeal often leads to very dangerous consequences, because those who believed the fact [of the biblical Flood] without asking for evidence will continue to believe. And those who were disposed to doubt it, will remain in the same disposition and even confirmed into it, seeing how our arguments are not strong enough to convince them.[42]

Even worse, physical evidence against the reality of a recent universal Flood (or lack of convincing support for it) could turn diluvialism into a millstone around the neck of Scripture. Some of "those who were disposed to doubt"— and in many cases did not believe at all—were indeed forging Earth history into a powerful weapon to dislodge Christianity from the foundations of morality and society. While the geological models they adopted may not have been radically new, in the course of the century they put them in the service of a political project that was definitely unprecedented. In the early 1700s, a growing number of *philosophes* no longer limited themselves to defending "philosophical freedom" or the autonomy of natural philosophy from theology. Instead, they aimed to replace confessional identities, religious institutions, and the divine right of the sovereign, with freedom of thought and expression, equality of every man before the law regardless of status and religious affiliation, and a social pact between the government and the governed within a secularized state.[43] None of these objectives appeared possible so long as the Christian Revelation remained the unquestionable touchstone of truth and the ideological underpinning of political power.

More mainstream currents within the European Enlightenment refused to embrace social and racial equality and preferred to pursue human progress within the framework of the existing social order. Yet they also had little use for traditional religion except as an instrument of governance of the masses. Even as they rejected atheism and materialism, they turned not to Christianity but to deism and to a "natural religion" inspired by the necessary existence of the rational Supreme Being that governed the Newtonian universe.[44]

The history of the Earth and its antiquity were crucial to their efforts. Throughout the century, enormously influential authors such as Montesquieu, Voltaire, Buffon, Boulanger, and d'Holbach engaged deeply with the entanglements of Earth and human history, because they had long been considered inseparable. Their differences notwithstanding, *philosophes* of every persuasion intended to relegate Noah's Flood not to a different sphere of truth but to superstition, myth, or actual events that the Bible described as one of the human documents of the early stages of the current civilization. In the previous century, much emphasis had been placed on human agency on the evolution of the global environment through their sin toward the Christian God. The *philosophes* returned instead to a conception in which humans were but a cog in the inexorable mechanism of nature, who witnessed as powerless spectators the cosmic drama of the periodical renovations of the Earth. Their history was largely dictated by the course of nature, rather than the opposite. Many of the naturalists who dismissed diluvialism and a young Creation were unwilling or not in a position to question Christianity itself. Vallisneri's strategy hinged on the claim that diluvialism was in fact historically foreign to Catholic traditions. By charging their moderate opponents with irreligion and materialism, many diluvialists offered to more radical ones the opportunity to turn their arguments against them: Christianity and a young Earth could not, indeed, be separated. The falsity of the latter would entail the absurdity of the former. In the early 1700s, this was a low-hanging fruit that the *philosophes* were quick to grab.

The operation required a further element, namely, a history of deep time that was compatible with their representation of Christianity as a force against reason and humankind's progress. Such a history would forgo the continuities between recent ideas of an ancient Earth and their medieval and Renaissance antecedents. The complex and sophisticated pluralism of the previous centuries could simply not exist in a dark age supposedly dominated by religious intolerance, superstition, and the blind acceptance of the letter of Scripture. While they did not invent the equation between Christianity and a young Creation, those *philosophes* erased centuries of history by presenting it as the only imaginable and permissible belief in Christian Europe until recent times.

It would be simplistic, of course, to imagine them as a cabal engaged in willful deceit and determined to sell to the rest of Europe a narrative they knew was flawed. Before the eighteenth century, historical writing was conceived in terms of picking examples from the past to teach moral and civic lessons, rather than as a critical analysis and comparison of all the relevant sources.[45] Likewise, the *philosophes* rearranged and reinterpreted historical material in a way that corresponded to their own perception of history and to what they did maintain was true. Furthermore, a vast range of pre-modern scientific and philosophical literature was rapidly becoming obsolete, and the memory of it was fading. Educated persons were still familiar with the Aristotelian philosophy that made up the backbone of the instruction imparted in the Jesuit colleges, and many eighteenth-century theories of the Earth were clearly rooted in Aristotelian meteorology. Yet the physical, philosophical, cosmological, and methodological bases of that corpus were of little use in a time of increasing mathematization and mechanization of the natural world, and of widespread affirmation of the experimental method. Giovanni Maria Lancisi, personal physician to various popes, admitted that "Aristotelian philosophy is the disposition of our minds"; but he added that "experimental philosophy is the true disposition of the *human* mind."[46] Vallisneri and other proud empiricists could occasionally praise some Aristotelian theses, but as a whole, they distanced themselves from that tradition. In their rhetorical confrontations with Galenic physicians and the last defenders of spontaneous generation, they called those doctrines not only surpassed but altogether "rancid" and "rotten."

Many medieval and Renaissance works were by then forgotten and often difficult or impossible to find. Jean Buridan's commentary to the *Meteorology* never made it into a printed book. Vallisneri was unable to obtain a copy of Francesco Patrizi's dialogue, even though he lived in the same area where Patrizi worked and published. Like everyone else, he learned of his description of the Flood through the excerpts published by Ramazzini.[47] While he knew of Fracastoro's thesis on the origin of fossils, he never read either him or his contemporary sources directly.[48] Nonetheless, the existence of a continuous tradition was still obvious to him as he placed the Italian philosopher (and himself) within an uninterrupted stream. About thirty years later, the authors of the entry "Fossil" in the *Encyclopédie* completely misrepresented Fracastoro's position on the origin of marine petrifications and its context. To all appearances, they did so more because of their inability to grasp Renaissance Earth history and its meaning than as a deliberate distortion. By then, many of their conservative foes were neither better informed, nor less eager to rewrite the past in much the same way.

The invention of the history of deep time did not follow a unified pathway. The *philosophes* were themselves a loose galaxy marked by profound differences and enmities, rather than the compact conspiratorial block portrayed by their opponents.[49] While the new narrative was not articulated explicitly before the 1750s, the previous decades saw the systematic demolition of the biblical stories of the Creation and of the Flood, regardless of whether they were presented as inexplicable miracles or as the preordained results of plausible nature causes. The most brilliant and famous quill among the *philosophes*, François-Marie Arouet, best known as Voltaire (1694–1778), waded for decades through different theories on the formation of fossils, whose common denominator was the aversion to the idea that the Earth had ever been covered by a global ocean. As a deist convinced through reason of the existence of a rational Supreme Being, he lambasted fiercely not only the biblical universal Flood but even the progressive lowering of the ocean. Too indulgent toward the Genesis account in Leibniz's version, he saw it as too radically materialistic in the variant propounded by de Maillet.[50] Voltaire's first significant appraisal of the history of the Earth, included in his *Elements of Newton's Philosophy* (1738), adopted instead Louville's theory of the slow revolution of the ecliptic and proposed that this motion must be the engine of geological change.[51] The millions of years required by that model to explain the formation and characters of the marine fossils, Voltaire noted, "please human imagination"—or at least his own.[52] Even more importantly, they did justice to the true intuitions and the distant memories of the ancients that, in a narrative which was already taking shape, had been eclipsed by the long night of religious fanaticism:

Egypt and a part of Asia . . . once preserved an immemorial tradition, vague, uncertain, but that could not be without some basis.
They said that prodigious changes happened in our Globe, and in the
heavens as they related to our Globe. The simple observation of the
earth gave this opinion considerable weight. We can see that the waters have in succession covered and abandoned the seabed in which
they are contained. The plants and fishes from the Indies found among
European petrifications, the seashells piled up in the mountains, are
sufficient witness of this ancient truth, and the majority of these
seashells . . . show that they were deposited little by little by regular
tides, and during a long series of years.[53]

Likely written by a collaborator before Voltaire could thoroughly review it,
the chapter was still present in the 1741 edition. The new version added a
confutation of Burnet's and Halley's diluvial theories and the claim that the
Flood "cannot be explained with physical causes."[54] The *Elements* was a crucial work in the transition of the French scientific culture from a Cartesian
to a Newtonian framework, and it was arguably a major contribution to
the popularity of Louville's theory among eighteenth-century naturalists.
Voltaire's opinions on the history of the Earth wavered remarkably, however,
in the course of the following years. In his most organic geological work, composed in 1746 and addressed to the Academy of Sciences of Bologna, he not
only denied that the ocean ever covered the highest mountains, but he also
doubted that marine fossils originated from the sea.[55] In *Candide* (1759), the
character Martin, who usually incarnates philosophical wisdom, castigated
once again the thesis of the lowering of the ocean:

> By the way, said Candide, do you believe that the earth was originally
> all ocean, as they assure us in that big book belonging to the ship's
> captain? I don't believe any of that stuff, said Martin, nor any of the
> dreams which people have been peddling for some time now.[56]

In the entry "Inundation" of his *Philosophical Dictionary* (1764) Voltaire piv-
oted again toward Louville's theory, noting that the over two million years
required for a complete revolution of the terrestrial axis corresponded to the
historical rate of withdrawal of the sea from the shores of Aigues-Mortes,
Fréjus, and Ravenna. His argument and his examples had been used since the
Middle Ages, as was his insistence on the "physical impossibility" of a complete submersion of the earth.[57] What had never been stated openly until his

century was the exclusion of supernatural intervention. Indeed, he concluded with characteristic sarcasm, "everything about the history of the Flood is miraculous":

> But, since the deluge is the most miraculous story ever told, it would be mad to explain it; such mysteries are articles of faith; faith consists in believing what reason cannot believe: that, too, is a miracle.[58]

With the biblical narration publicly reduced to both a physical and metaphysical impossibility, a matter of amusement and laughter, what remained of it apart from myth? As some contemporaries pointed out, if Moses was not truthful in the account of a world-shattering event, how could he be trusted on anything else? The medieval separation of theology and philosophy, which shielded Scripture and the science of nature from each other, was still invoked around the middle of the century by naturalists and censors alike, but it appeared an increasingly formal tribute to faltering traditions and institutions. In 1749, the first volume of Georges-Louis Leclerc de Buffon's monumental *Natural History* included a general theory of the Earth that followed a plainly Aristotelian outline: the planet was the stage of continuous gradual change, brought by the forces of erosion and sedimentation. The same phenomena had been around throughout its existence, and there was no reason to think that the planet would be any different in the future—or that it ever was in the past. Buffon held a public position as director of the Jardin du Roi (the direct predecessor to the Paris Museum of Natural History), and in the employment of the king he needed to respect formalities and stave off possible allegations of atheism or materialism. While he did not mention either the Creation or the Flood, in other writings he reclaimed the old distinction between the book of nature and the book of Revelation in order to both protect his views and score points against diluvialists and British physico-theologians:

> Every time someone will be so bold as to try to explain theological truths with physical reasons; allowed to interpret in purely human views the divine text of the sacred books; and dare to rationalize the determinations of the Almighty, they will necessarily fall into the darkness and chaos in which the author of this system [William Whiston] has fallen.[59]

For good measure, in a different essay he also laid out a concept of how this cyclical world and the other ones in the solar system might have formed in an undetermined past.[60] Buffon's work was praised by the Jesuits but attacked by their theological and political rivals the Jansenists, and could not avoid the predictable charge of eternalism. The Sorbonne, that is, the Faculty of Theology of Paris, determined that the appearances had been saved when Buffon signed a declaration stating his belief in "all that is told [in the Scripture] about Creation" and that his theory was a pure "philosophical hypothesis." The statement was likely composed by the Faculty itself.[61] The much-publicized affair was never more than an annoyance for the already famous and well-connected naturalist, as was a second spat with the Sorbonne that took place thirty years later when he published his *Epochs of Nature* (1778). By then, Buffon had turned to a linear history in which the cooling of the primitive Earth to the present condition might have taken about 75,000 years (though he mentioned several millions in some manuscripts).[62] Citing theological opinions, he asserted that the six days of the Creation could be interpreted as "epochs" of the long pre-human history of the Earth and of the successive appearance of life forms on it. Finding no role nor evidence for the Deluge, he claimed again that the Flood was a miracle with no relation to natural history, so it did not need to be addressed.[63] Tellingly, the theologians of the Sorbonne did not denounce the theory nor its timescales, but they retorted that by discussing Genesis at all, Buffon overstepped the traditional disciplinary boundaries. It was important to reaffirm that those different epochs "had no relationship to the different days of the Creation, neither for the order of time nor for the circumstances of the facts." The public ignored them, but Buffon duly signed the retraction prepared by the Faculty of Theology, as many other philosophers had done for centuries before him.[64] He was probably the last one to do so.

Before Us, the Deluge

By the time he wrote his *Epochs of Nature*, even a man of the establishment like Buffon saw "a large movement coming," an inexorable swell that even without a clear leadership continued to rise, taking everything with it. With little appetite for revolutions or decisive modifications to the social hierarchy, he still hoped that the progress of science and reason would engender a more just, peaceful, and prosperous world. Human industry and ingenuity might even be able to slow down the inescapable death of the planet by progressive cooling, as he could see in Europe where the "clearing of woods" and the sheer number of men resulted in a warmer climate than Canada, even though Québec lay at the same latitude as Paris.[65] This new era would be not an unprecedented epoch in the history of humankind, but a return to a golden age of which only scattered debris remained. Since at least 1758, he had embraced the opinion that sciences "had been cultivated in the remotest antiquity, and perhaps perfected beyond what they are at the present day." Profound truths were then believed not only by "philosophers" but even by common people; yet they "could only preserve their purity in the centuries of light," and were lost almost entirely in the ensuing "revolution of darkness."[66] As he completed his *Epochs of Nature*, Buffon had under his eyes the recent *History of Ancient Astronomy* by Jean-Sylvain Bailly, who extolled the advanced astronomical knowledge of a mysterious pre-diluvial civilization.[67] Both were borrowed from far more radical voices, like those of Nicolas-Antoine Boulanger and Paul-Henry Thiry, Baron d'Holbach.[68]

Some of the most outspoken eighteenth-century proponents of materialism and atheism did little to disprove a universal (or quasi-universal) submersion of the globe in a not-too-distant past. On the contrary, they placed an exterminating Deluge at the very origin of the knowable human history, for the same reason once adduced by Plato, Aristotle, and the entire meteorological tradition: how could one explain otherwise the apparently recent appearance of arts, sciences, and human documents, if the Earth and humankind itself were immensely more ancient?[69] Since Avicenna and Albertus Magnus, a minority of meteorological authors had conceived of an Earth scourged by a succession of universal floods. Nonetheless, the local inundations of mainstream meteorology remained a sufficient explanation as long as the historians' horizon was limited to interconnected civilizations between Europe and the Middle East. In an age of colonial empires and worldwide cultural and economic exchanges, though, the great reset of natural and human history must be a global one.

Unlike the diluvialists, Boulanger rejected the Flood as the origin of fossils and, more importantly, he divested it of any providential meaning. As a civil engineer in the public works department of the French state, he amassed a considerable amount of first-hand observations on the geology of various regions, as well as on the oppressive reality of the forced labor that his administration exacted from the peasantry (the infamous *corvée*).[70] Even without fossils, "all you have to do," he wrote in his *Physical Anecdotes of the History of Nature*, "is lower your eyes towards the earth, and observe it with thoughtful attention, to recognize that it was built under water and by water."[71] The global inundation witnessed by the body of the Earth and perpetuated by the myths of so many nations was of course only the latest occurrence of a natural catastrophe that wiped out human civilization. The first origin of the human species was as unknowable as that of the universe, but humans managed to survive through the "revolutions" that followed. The periodical submersions of the globe were never complete, and many corners of the planet always harbored pockets of survivors.[72] The vast differences in "costumes, language, or polity" between the Europeans and "the Indians, the Chinese, the Mexicans, the Peruvian, and all the other people that inhabit the circle of the terrestrial hemisphere," proved that not every nation "originated from the countries watered by the Tigris and the Euphrates."[73] Boulanger's Deluge safeguarded neither scriptural monogenism nor its divine origin. To the contrary, religion itself rose from the traumatic experience of the near-extinction of humankind, as scattered and shocked survivors tried to make sense in supernatural terms of the inconceivable catastrophe that befell them:

> Humans finally, the most unhappy and terrified of every living being in those sad revolutions, were in all likelihood the last to leave the caves and rocks where the survivors took refuge. What a bitter and painful scenery, and well capable of outweighing the joy of having survived, must have astounded them at the first sight of the deplorable condition of their abode now cracked, ravaged, covered with immense ruins, and showing everywhere the markings of a divine curse."[74]

Rather than pure invention or metaphor, Genesis was an elaboration of the disaster from the point of view of its author, who followed and absorbed the traditions of other nations as well as of his own.[75] The "formless and empty" Earth stood for its state of general devastation and depopulation after the Deluge.[76] The Garden of Eden symbolized the wise government enjoyed before the disaster, just like other nations remembered a golden age "under great men that were called gods and heroes."[77] The Genesis distinction between the "children of men and the children of God" attested that different races escaped complete annihilation.[78] Boulanger even accepted the biblical chronology of the Flood, arguing that in spite of the many uncertainties and disagreements "the closest to the true one must be the less ancient which corresponds to the year 2348 before the Christian age."[79] Likewise, the Book of the Apocalypse used images from the destruction of the previous world to describe the end of the present, when another Deluge will once again submerge the globe: "the Sun will again be eclipsed . . . the Moon will no longer give its light, and the stars will vanish." From the center of the Earth, "vapors will rise like from a great furnace that will obscure the air and the Sun" with lightning, clatter and thunder; finally, an earthquake the like of which was never conceived will sink the islands, mountains will disappear, "and cities will fall into the sea like millstones, never to be found again."[80]

The *Physical Anecdotes* was in fact the indispensable geological premise to later works outlining the religious history of humankind, which descended directly from the tragedy of the Flood and determined its political structures. In the *Researches on the Origins of Oriental Despotism* and in the *Antiquity Unveiled through its Mores*, Boulanger developed the thesis that "those calamities were in those times their harsh Missionaries and powerful Legislators, who turned every eye towards the Sky, towards Religion, and towards Morality."[81] Terrorized and destitute, the first generations of survivors built their social institutions out of crude necessity and reinforced them with the fear of another heavenly punishment, in their life or in another one. "I have more than once noticed," Boulanger observed, "that most of the descriptions we have imagined of the other world come from impressions that the ancient revolutions of nature have left of themselves."[82] Mysticism was the survivors' psychological refuge. Even the positive aspects of the new religious sentiment were quickly outweighed by its degeneration. The Earth had not fully recovered from the devastation, and fierce competition for scant resources engendered inequality and conflict.[83] Monarchs soon posed as images of the dreaded God or as God themselves, rising above the wise laws inspired by the common calamity and subjugating the other men. Every government down to his times had been but a form of theocracy, enabled by the religious sentiment issued from an incomprehensible natural disaster. Religion replaced polity, reason was abandoned, and the fear of the unknown put in the service of despotism.

Boulanger's evocation of the scattered and shocked survivors is reminiscent of period literature, but also of much earlier narratives.[84] Two centuries earlier, Patrizi painted in similarly dark hues the condition of the "forever stunned" survivors, full of a "horrendous terror that shortened their lives, and those of their offspring. . . . Their children then inherited the same short life, and the same unremitting fear which is the true root of every evil." While vague memories remained of the sciences that had once been common knowledge, the powerful preferred to encourage the politically safer and less dangerous beliefs of the crowd, even knowing that they were false. They then "persecuted in every way, and mortally hated those who wanted to say the truth," so that the sciences had to be taught "in riddles, fables, numbers, esoteric mysteries, in silence, and one thousand more hidden ways."[85] Yet Patrizi's humankind had brought upon themselves the catastrophe with their arrogance and misbehavior. By making them morally and psychologically responsible for the flawed political organisms that oppressed them, the Church perpetuated their subjugation.[86]Boulanger's geological determinism rescinded even that last tie with the Christian tradition, absolving the human species from the crushing remorse for a fall they could not have prevented.

Completed around 1753, the manuscript of the *Physical Anecdotes* was never printed before modern times but it circulated widely among French naturalists.[87] The *Researches on the Origin of Oriental Despotism* certainly enjoyed a much wider readership, attested by the eight editions printed between 1762 and 1777 and the four more in 1790–1792.[88] Crucial to the diffusion of Boulanger's theses were his collaboration with Diderot and D'Alembert's immensely influential *Encyclopédie*, the manifesto of the French Enlightenment, and his close association with d'Holbach. The substantial entry "Political Economy" (*Oeconomie Politique*, 1765), based on the manuscript of the *Researches*, was largely dedicated to the establishment of the "primitive theocracy" of which the "Mosaic" represented a later evolution.[89] "Deluge" (1754), while not entirely by Boulanger, was presumably based on his original manuscript partially abridged by Diderot.[90] The article mounted a defense of the universality of Noah's Flood that may surprise the modern reader, if not for the assertion that there must have been local survivors in every place where the tradition of a great Deluge exists. After escaping, the inhabitants "passed on to their descendants what had happened to their countries in such and such a river, on such and such a mountain, and in such and such a sea."[91] Diderot wrote approvingly of how Boulanger's geological investigations, born "on the routes of France," turned into the foundation of the religious origins of humankind's woes.[92]

Most of the *Encyclopédie* articles dealing with geology, mineralogy, or Earth history were actually composed by d'Holbach, who, after Boulanger's premature death, published some of the latter's most significant writings. While his interventions were probably more limited than once thought, he used the deceased friend's name to conceal the true author of books like *Christianity Unveiled*, whose title immediately called to mind Boulanger's *Antiquity Unveiled*.[93] D'Holbach preferred the gradual revolution of the terrestrial poles to Boulanger's strict catastrophism, but otherwise he was profoundly influenced by his thesis on the intimate connection between the history of the Earth and the natural origin of religion and despotism.[94] The demotion of biblical monogenism, which underpinned the devastating critique of the European colonial empires delivered by d'Holbach, Diderot, and other radical thinkers, was more important than differences on what particular theory should replace it.[95] During the 1750s, decades of elaboration converged into a vision of the history of deep time that not only suited d'Holbach's political and ideological agenda, but proved appealing even to more moderate authors and readers. In this scenario the ancients, still uncorrupted or retaining memory of the scientific notions of the previous world, unanimously regarded the Earth as immensely old if not eternal. Their correct insights, however, were later suppressed by the "Mosaic theocracy" that imposed the idea of a recent Creation to the point that any alternatives became unconceivable. Only in the course of the previous century did some philosophers begin to shake the chains imposed on their minds. Guided by observation and reason, they rapidly rediscovered the evident truth of the great antiquity of the Earth long buried under the weight of theocratic oppression. D'Holbach expanded on this conception in his *System of Nature*, often cast at the time of the publication (1770) as the most systematic assault on religion ever perpetrated. Its essential elements, however, appeared already in the entry "Fossil" of the *Encyclopédie* published in 1757.[96] The relevant passages of that article spell out the core of a narrative that was to become plain common sense for the following two and a half centuries:

> It was observed since the remotest antiquity, that the Earth enclosed a great number of marine bodies; this led to think, that it must once have been the bottom of the sea.... This was also the opinion of Avicenna, and of the Arab *savants*; but even though it had been universally common among the ancients, it was later forgotten; and observations of natural history were entirely neglected among us, in the centuries of ignorance that followed. When observations resumed the *savants*, who adopted a quite bizarre way of reasoning owing to the peripatetic philosophy and to the subtleties of the scholastic, imagined that the shells, and other fossils foreign to the Earth, had been formed by a plastic force (*vis plastica*) or by a seed universally diffused (*seminium & vis seminalis*). Hence, one can see that they did not regard marine fossils if not as jokes of nature, without paying any attention to the perfect analogy that could be found between these same bodies extracted from the interior of the Earth, and other marine bodies, or belonging to the animal or vegetal kingdom; this analogy alone would have been enough to undeceive them.... Those imagined plastic forces and those explanations, as absurd and incomprehensible as they were, found and still find today their supporters.... Meanwhile, during the sixteenth-century several *savants*, among whom one can place Fracastoro, considering that fossil substances were extraneous to the earth, found

that they had such a perfect resemblance with other natural bodies, that they did not doubt that it was the sea that brought them on the continent; and since they could see no more plausible cause for this phenomenon than the universal Flood, they attributed to it all of the marine bodies that are found on our globe, completely inundated by its waters.[97]

D'Holbach correctly acknowledged that the ancients attributed marine fossils to a prolonged sojourn of the sea over those regions and that Avicenna and the Arabs believed the same. What followed was a caricature of medieval and Renaissance natural philosophy. In his narration, the true opinion on the origin of fossils was "forgotten" in the "centuries of ignorance." An "absurd" and "incomprehensible" philosophy, the bastard child of the unnatural marriage of Aristotelianism and Christian superstition, ignored even the most obvious evidence. In those dark ages, fossil fish or seashells were universally regarded as "jokes of nature" rather than the remains of once living creatures. Not until the sixteenth century did some begin to recognize that the resemblance between fossils and actual animals was too striking to be easily dismissed; yet biblical literalism dominated their minds to such a point that Noah's Flood was the only possibility they could come up with. D'Holbach managed to turn Fracastoro into both a diluvialist and a forerunner of a more enlightened age, instead of a representative of the mainstream scientific culture of his time. Inconsistencies and contradictions needed to be ignored or glossed over, if the core message was to pass: Christianity's grip on the European mind had been so complete, its suppression of reason so successful, that the antiquity of the Earth could not even be imagined until the recent appearance of a secularized science.

This simplistic but powerful representation of the past was meant for the consumption of the nascent "public opinion" rather than for a restricted number of specialists. The *Encyclopédie* addressed precisely that kind of readership, but the history of the Earth had always been a genre with a wide diffusion. This was also true of d'Holbach's and Boulanger's literary production, aimed at a large audience in the attempt to establish a new historical and geological narrative.[98] In the decade following Boulanger's death (1759–1770), the popularization of their approach to Earth history became central to d'Holbach's project. The frequent allusions to the resistance that their geological ideas would meet, apart from a few "physicists and philosophers," was little more than a rhetorical device meant to reinforce the narrative of a nation dulled by centuries of religious prejudices. The editorial success and the huge

popularity of their publications belied them every day.[99] Their counterpart
was not the technical works of diluvialist geologists like Bourget, but Abbé
Pluche's *Spectacle of Nature* (paradoxically, among the most cited authors in
Boulanger's *Anecdotes*), who aimed to teach the youth about the perfect compatibility of recent physics and natural history with Christian dogma. Pluche,
as opposed to the authors of great world histories, valued the simple wisdom
of "the most credible physicists, namely construction workers, miners, and
reasonable travelers," who would immediately attribute marine fossils to the
universal Flood. Boulanger answered with the words of the villager who,
shaking his head at the view of a marine fossil, told him in his dialect that "his
country must once have been a sea hole."[100]

The decade that saw the publication of the first tomes of the *Encyclopédie*
witnessed the rise of a forceful and vocal movement that denounced with
increasing anxiety the diffusion of materialism and irreligion. Initially, the
anti-philosophes were mainly enlightened Catholics or Protestants who opposed the real or alleged anti-Christian drifts of contemporary scientific and
philosophical literature.[101] Ready to battle those (often anonymous) authors
on intellectual grounds, they maintained that Christianity was fundamentally
rational, and invoked Newtonian gravity to demonstrate the existence of immaterial forces and entities in the universe and hence a supernatural Creator.
When the *philosophes* accused them of "religious fanaticism," Christian
apologists could point—not without sensible arguments—at the unreasona-
bleness of materialism.[102] As historian Jonathan Israel argues, their intent was
not to marginalize reason but to "capture 'reason' for their camp."[103]

Soon enough, though, emphasis within the movement began to shift to the
worrisome political implications of the "new philosophy." It was self-evident
to them that the altar could not be touched without shaking the throne: to
attack the first was to question the second.[104] If so, the differences between
deists like Voltaire and straight atheists seemed of little import. At their core,
the *philosophes* were all "republican," devoted to an ideology that replaced a
divinely ordained social system with the arbitrary faculty of establishing and
dissolving a government at will. Already in the late 1750s, the *anti-philosophes*
began to portray them as uniformly hostile to "monarchy, aristocracy, existing
morality, and all social subordination and patriotism as well as religion."[105]
Their frequent rejection of the conventional sexual morality offered cheap
ammunition to those who wanted to cast the *philosophes* as enemies of even
the most fundamental building block of society, the family. By 1770, a new
rhetoric lumped together *philosophes* of every conviction into a compact

conspiratorial group, eager to subvert an imperfect but well-ordered society into chaos.[106]

In this narrative, the individual free-thinkers, libertines, epicureans, and radical Aristotelians of the previous centuries who advocated the autonomy of human reason had coalesced into a far more numerous and threatening "sect" enabled by ineffective censorship and a weak state infiltrated at every level. Books like d'Holbach's *System of Nature* or *Christianity Unveiled* appeared all the more dangerous, in that they couched their philosophical premises in simple terms and maxims accessible to ordinary people.[107] They elicited similar, though often cheaper, responses designed to address the widest possible readership through emotional appeals and conspiracy theories.[108] By politicizing, polarizing, and oversimplifying the terms of the political and cultural debate, the *anti-philosophes* lived and spread the perception of an impending social apocalypse, fed by the distorted (yet compelling) image of the Enlightenment they constructed themselves. Through propaganda, repetition of simple concepts, direct appeal to ordinary people's feelings, and simple faith, they marginalized the moderate and more sophisticated voices of the religious Enlightenment. When such a determined foe threatened the very basis of civilization and human coexistence, no compromise was evidently possible; no middle ground could be admitted. As *philosophes* and *anti-philosophes* battled for the soul of Europe, moderation, conciliation, and dialogue became marks of weakness and treason.

Radical conservatives did not fight to preserve a present that they saw as utterly compromised. What they really evoked was a mythical past meant to prefigure and legitimize a future reality. In an era that had ended no more than a century earlier, no false intellectualism muddled the Christian teachings laid out in the Revelation and the social order that descended from them. Philosophers did not dream of using God's own creation to undermine the truthfulness of His word and question its origin, or even His existence. No one doubted what was, and must be again, plainly true: the fossil fish buried in the Earth demonstrated the truth of Genesis and the reality of the Deluge that a few thousand years earlier punished humankind's rebellion against authority.

7

Political Fossils, 1740–1800

*What greater and more telling proofs [than the existence
of fossil fish on mountains] could be brought to confirm the
ancient tradition, that all the waters of the globe once over-
came their levees, and flooded the earth in its entirety?*

S. VOLTA, *Fossil Fish of Verona*, 1796

IN EIGHTEENTH-CENTURY FRANCE, the *philosophes* tied the history
and the political future of humankind to an ancient or timeless Earth, and
to the denial or de-Christianization of the Flood. It is hardly surprising that
radicalized reactions should target the geological basis of the "new philosophy." While France was a constant source of either inspiration or scandal
across Europe, the characteristics and relative strengths of the opposing fields
varied considerably according to local cultural traditions and political and
social contingencies. This chapter focuses on Venice and its mainland in
northeastern Italy, as an especially valuable case study of how late eighteenth-
century conservative elites politicized Earth history in the service of their
agendas.

Louder, more effective, and more influential than historians have long
suspected, the voices of the French *anti-philosophes* were still muffled by
the brilliant quills and by the immense celebrity of their antagonists. Italian
conservatives, in contrast, were able to present themselves as the spokespersons
for public opinion and to exert considerable influence on the discourse on science in general and on the history of the Earth in particular, at a local level.
Firmly in control thanks to a combination of economic power, cultural prestige, and cautious reformism, the aristocratic circles that were the backbone of
the conservative camp never allowed the emotional and apocalyptic appeals
of more extreme elements to dominate their rhetoric. Their strategy rested on
a politically conservative science, meant to counter the agenda of a nascent
technocracy that in their eyes was the vehicle for radical ideas of freedom,

On the Edge of Eternity. Ivano Dal Prete, Oxford University Press. © Oxford University Press 2022.
DOI: 10.1093/oso/9780190678890.003.0008

equality, and democracy—in other words, social mass destruction. Slowly but surely, in the second half of the century they elaborated a diluvialist orthodoxy allegedly supported by "true philosophy" and "sane science," which appeared very different than the Earth history many enlightened Catholics conceived only a few decades earlier. By physically and culturally appropriating the outstanding "marine petrifications" provided by their territory, they turned those fossils into politically charged statements that reaffirmed the existing social order. Finally, like their French counterparts, they burned the bridges with the past and built new ones leading to a tradition that never existed, one in which a young Creation and the diluvial origin of fossils had always been the undoubted beliefs of Christian Europe, only recently questioned by "philosophical fanaticism." In the end, Vallisneri's legacy was undermined not by "priests and friars" afraid of losing their credulous customers but by the scientifically minded heirs of the aristocrats who applauded and patronized him.

Dams Against the Flood

For a few decades, the dam that Vallisneri and other Italian literati erected against the Flood looked like a resounding success. His moderate and conciliatory approach wooed many Catholics, while targeting Protestant naturalists, their Italian followers, and the part of Catholic culture that resented a secular natural philosophy. Even though he had not proposed a theory of the Earth that could replace the Deluge, new or renewed elaborations promised to fill that void. Italian naturalists did not take long to turn Louville's model into the engine of geological change. Mere months after the publication of *Of the Marine Bodies*, mathematician Giovanni Rizzetti wrote to Vallisneri that he had been thoroughly convinced by his criticism of the Deluge as a geological agent in the history of the Earth. He added two long letters, in which he discussed instead the hypothesis that the Earth was a spheroid whose major axis is that of the equator, and the minor that of the pole. As the position of its poles shifted in the course of the centuries, water, being a fluid, accommodated to that figure, while the solid earth could not. Different regions of the earth were then covered or uncovered by the waters at different times.[1] Vallisneri was impressed with the concept, which "if proved" seemed to solve all of the difficulties he had so painstakingly pointed out. However, since his mathematical and astronomical competencies were almost nil, he could only "applaud, without judging."[2] In the same months the Venetian philosopher Antonio Conti, whose opinion Vallisneri usually deferred to when it came to mathematics and astronomy, described to him a similar model in enthusiastic terms.[3] The dislocation of the poles remained indeed a popular theory until the end of the century.

The early diffusion of far more unorthodox doctrines, like de Maillet's organic transformism, remains difficult to assess with any accuracy because they were rarely mentioned directly. The existence of the marine creatures that the French diplomat considered the aquatic counterpart of humans was certainly believable to most of his contemporaries. Hardly prone to credulity, Vallisneri did not doubt the reality of tritons and mermaids "that resemble our most noble species not only in their external appearance, but even in many of its habits." He invited skeptics to admire the "beautiful hands of woman- and man-fish, with part of the forearms and a few ribs" that he exhibited in his museum.[4] Even without evolution, the borders between species remained as contested and porous as they had been for centuries. One of Vallisneri's correspondents, a physician who traveled throughout northern Europe while he served in the Imperial army, once had a discussion with a Danish nobleman who claimed that chimps were a race of humans rather than animals. Back home, the physician composed a booklet to dispel the dangerous error at least among his fellow citizens. Yet he insisted that a young mermaid caught recently by Dutch fishermen and a huge triton encountered by French sailors were absolutely real, even though they resembled human beings only superficially.[5] Disconcerted at the spread of those "insane philosophies" and especially of Cartesian materialism, regarded as the root of "modern atheism," he resolved to support the diluvial origin of fossils as the best physical evidence for the truth of Revelation.[6]

Anti-diluvialism, though, was backed by figures of much greater intellectual and social clout. After Vallisneri's death in January 1730 his son, Antonio Vallisneri Junior, was appointed to the newly created chair of natural history at Padua in exchange for the donation of the family museum to the university.[7] The son did not always follow the doctrines of his more famous father, but he never relinquished his approach to the history of the Earth.[8] Wary of public controversy, he published very little, yet he was for more than forty years a dedicated teacher, who influenced generations of students. One of his father's closest collaborators, the Veronese marquis Scipione Maffei, also turned into a staunch opponent of diluvialism. A leading intellectual, after 1720 Maffei became progressively more involved in natural philosophy. Being a devout Catholic did not prevent him from holding that Scripture, like every other document from the past, must be read and understood within the historical context that produced it. Maffei held that, just like economy, or the civic ethic he elaborated drawing upon the values of classical antiquity rather than Christian ones, natural philosophy should not be bound to unhistorical and myopic interpretations of the revealed text.[9] In fact, Maffei considered modern experimental science a crucial element of his project of cultural modernization of the aristocracies of the Venetian state. By the time the chair of Natural History was instituted in 1734, Maffei had lobbied for its creation for more than twenty years.[10] Meanwhile, he deployed his social and intellectual prestige to make of natural philosophy a worthy occupation for a nobleman, encouraging local initiatives and discussing chronology and experimental science in the "conversation" he held in his palace.[11] The journal of *Literary Observations* that he published in 1737–1740 was intended as a follow-up to the discontinued *Journal of the Italian Literates*; like its predecessor, it included a considerable proportion of scientific articles.[12] After 1732, Maffei could count on the collaboration of French naturalist Jean-François Séguier, whom he met in Nîmes during a European tour and who remained with him until Maffei's death in 1755. Séguier was the real mastermind behind Maffei's botanical garden, the small astronomical and meteorological observatory he installed in his palace, and his purchase of the fossiliferous quarry of Bolca.[13] Séguier also acquired most of the scientific texts in Maffei's library, including Fausto of Longiano's 1542 vernacular *Meteorology* (the French naturalist duly transcribed the pages on the Aristotelian displacement of the sea and the "eternity" of the world).[14] Even more importantly, he kept together the network of physicians, pharmacists, and priests that remained the backbone of the naturalistic research in the area, without any of Maffei's condescending paternalism toward the socially inferior.

Parish priests and their aides, in particular, represented crucial links in the circulation of naturalistic specimens, and sometimes they assumed an active role in the debate on the origin of fossils. Apparently, they were just as likely as physicians or apothecaries to either support or demote the role of the Flood in their formation. In spite of Vallisneri's invectives against priests and monks, ecclesiastic condition would be a poor indicator of one's standing in the controversy. According to Séguier, who visited a number of European museums including John Woodward's, around 1740, the collection of fossil fish belonging to a parish priest in the countryside of Verona was "infinitely above" any other of its kind.[15] The owner, Giovanni Giacomo Spada (1679–1749), put it together thanks to his physical proximity to the famous paleontological site of Bolca, and to his relations with the quarrymen and stonecutters that were numerous among his parishioners.[16] Backed by Maffei, in the late 1730s he entered the dispute on the Flood on the anti-diluvialist side, proudly touting his empirical expertise. Like Vallisneri senior, he did not try to elaborate yet another theory of the Earth; but he denied resolutely that his fossil fish originated from the Flood. His investigations confirmed that they could never be found on the highest mountains, but only at lower elevation.[17] Given this fact, the most plausible explanation was the progressive lowering of a global ocean. After the first emergence of dry land, the surface of the sea must have been much higher than its present level. As the waters slowly retreated, fishes, crustaceans, and even the deer whose bones he had personally excavated remained trapped in the saltwater lakes and pools that inevitably formed.[18] Spada made it clear that this must have happened long before the Flood, not because of it.[19] It is telling that his main criticism of Vallisneri senior was the Paduan naturalist's restraint, which he found unjustified: the priest could not see how it might be sinful to affirm without hesitation that the "Crustaceans, and other marine productions seen on mountains, were antediluvian."[20]

The anti-diluvialist theory elaborated by another low-ranking priest from the area, Anton Lazzaro Moro (1687–1764), had a far larger circulation and is still well known among historians of geology. Like other talented but penniless boys, Moro embraced the ecclesiastic career as the only means to obtain an education. After his ordination he spent most of his life in his hometown north of Venice.[21] The treatise *Of the Crustaceans, and of Other Marine Bodies found on Mountains* that he published in 1740 was bold and in some ways innovative.[22] Moro argued that there was no need to speculate that the surface of the sea had been much higher in the past; instead, the inner fire of the Earth could have lifted the sea bottom, in the same way as a volcanic island famously emerged from the Aegean Sea in 1707.[23] Even though the rising of the sea bottom was a common mechanism to explain the existence of mountains in medieval and Renaissance meteorology, Moro's treatise was among the first to emphasize the importance of volcanic forces (alongside water) in the formation of the Earth's surface. The process implied, of course, extremely long time scales that "some" judged contrary to Genesis. Moro spent half of his hefty book refuting Burnet and Woodward, but he dedicated only a few pages to those local critics who "think that they are citing the views of the Holy writers, while they just cite their own."[24] The final allegation that in any case, the emergence of "most dry land" could still have occurred during the "third day" was a formality that no reader could take seriously.[25] A preliminary draft composed in 1737 retained the biblical account as its starting point, but the feedback received from the naturalists he consulted convinced him to leave it entirely out of the picture.[26] In the final version, Moro dismissed the Flood as a "miracle of the Almighty: therefore, inexplicable; and natural and physical subjects have nothing to do with it."[27] Maffei embraced enthusiastically the anti-diluvialist stance of that "good priest." Even though he remained faithful to Vallisneri's empiricism, and continued to hold that "phenomena are facts, and their reasons just words," he regarded Moro's theory of the Earth as the most plausible among those formulated up to then.[28]

The early 1740s seem to have marked the high tide of anti-diluvialism in the region. In the following years, the moderation and the precautions displayed by Vallisneri, Maffei, Spada, and even Moro were increasingly unable to offset the concerns raised by less guarded writings imported from France. An Italian edition of Voltaire's *Elements* appeared in Venice in 1741, including the chapter with the discussion of the theory of the Earth based on Louville's work and its timescales.[29] Ecclesiastic censorship still demanded that readers be advised to consider the motion of the Earth as a hypothesis, rather than a proven fact, but no such disclaimer was required when it came to its age.[30] Even bolder texts followed suit, like *Telliamed* (1748). De Maillet's book, or at least his doctrine of transformism, were certainly known in the Veneto at an earlier time, but the availability of printed editions accelerated their dissemination and visibility. In 1753, Anton Lazzaro Moro confided to a correspondent that he had "long" mulled over its most radical elements:

> I dare to mention to you, that in the fables of the poets I have long envisaged a bottom [of truth] with which one could elaborate a system showing that humans had their origin in the water; and this could even be bolstered by recent events.[31]

Moro did not dare to quote his real source, mentioning instead the Greek philosopher Thales who made of water the principle of everything. Nonetheless, de Maillet's work circulated unhindered and was soon on Séguier's bookshelf.[32]

The "Abuses" of Reason

Vallisneri senior had insisted strenuously, and with excellent reasons, on the possibility of an ancient Earth as part of the Italian and Catholic scientific traditions. From the mid-1700s it proved increasingly difficult to defend this argument, as the opposite opinion became commonplace: anti-diluvialism represented a French import, and it went together with materialism and atheism. Fracastoro and Cesalpino remained glorious but fading memories; the names of Voltaire, de Maillet, d'Holbach, and Buffon were on everyone's lips, and their writings in everyone's hands. Language was hardly a barrier to the diffusion of French scientific and philosophical literature. According to the editor of a 1754 Italian compendium of Voltaire's *Elements of Newton's Philosophy*, there would have been little need to translate it, as "the French language is already common enough in our Italy, and the examples are unfortunately numerous."[33] The intent was instead to defuse its content for the most naïve readers, who could take Voltaire's alleged "demonstrations" at face value.[34] This was also the main purpose of the notes and supplements added to the Italian editions of the *Encyclopédie*, especially after the solemn condemnation issued by Pope Clement XIII in 1759. The papal edict did little, however, to thwart its diffusion.[35]

Diluvialists were emboldened by the appearance in 1739 of a Venetian translation of Woodward's *Essay*.[36] Woodward had been known for decades, and Latin or French versions were accessible to educated readers; yet the impact of the Italian edition was immediate. Moro, whose geological work was published in 1740, used it alongside a 1735 French edition.[37] Authors, who had never mentioned Woodward in 1739, in 1741 made his system the fulcrum of their conception of the physical universe and of the history of the Earth, turned abruptly from Cartesians to Newtonians, and called Woodward a "classic" author under whose guidance one can never err.[38] The readers of the Venetian edition learned that Woodward's work was motivated, first and foremost, by a desire to confound unbelievers: "A spirit of skepticism reigns in the world, which tends to subvert ideas, and to reject the most universal principles." Possessed by this spirit, some deduced from the fixity of natural laws the eternity of material things, "so that God is unnecessary."[39]

Unlike other local diluvialists, Paduan lawyer Gianfrancesco Pivati (1689–1764) was hardly an unknown. Beginning in 1730 he worked for decades as an overseer of the publishing industry for the government, developing close relations with Venetian magistrates, printers, booksellers, and University of Padua professors. His resounding (albeit contentious) experiments with medical electricity, conducted between Padua and Venice in 1746–1747, gained him wide renown.[40] Among many other works, he was also the author of an ambitious illustrated encyclopedia published in ten volumes between 1746 and 1751.[41] The long entry "Deluge" (1746) consisted almost entirely of a resolute defense of the universality of the Noetic inundation, of the chronology that placed the Flood "one thousand three hundred years after the creation," and of the diluvial origin of marine fossils. Even though human beings should not submit divine omnipotence to their limited understanding, Pivati strove to show that the submersion of the world was compatible with natural laws.[42] The most puzzling aspect of the article, however, is that Pivati cited copiously from British or French diluvialists while overlooking entirely the Venetian debate, to the point of not even mentioning Vallisneri senior or Moro. Simple ignorance is inconceivable in an author with his acquaintances and relations. More likely, Pivati tried not to hurt local sensibilities or raise the ante, even as he was obviously taking sides. His approach might have been dictated by his public position, but it was also in line with the tone of the controversy on the Flood, which up to that point had been respectful of the other side's opinions. In particular, Venetian diluvialists usually refrained from questioning the piety or good faith of their opponents. During the same years, another well-known essayist, Giuseppe Antonio Costantini (1692–1772), showed none of Pivati's caution and opted for a frontal attack.

Costantini was a familiar figure to Venetian conservatives, thanks to his widely read *Critical, Playful, Moral and Scientific Letters*.[43] According to his own account, one day he immersed himself in the reading of Vallisneri's *Of the Marine Bodies* in search of quotable evidence for the "undeniable relics of the Universal Inundation." To his bitter surprise, he discovered instead the doubts, uncertainties, and perplexities of the great naturalist. Vallisneri couched his work in such terms that some diluvialists even pretended he was one of their number, but Costantini did not let himself be fooled:[44] without questioning the Bible directly, Vallisneri spread the even more venomous seed of doubt. His method turned the Mosaic story into a "fairy tale one can tell to ignorant women, or use to frighten illiterate peasants."[45] Had Christianity not been perceived as under attack, the stakes might have been lower, but this was not the case. In the fourth volume of his *Letters* and in a ponderous *Defense of the Universal Flood* (1747), Costantini framed the controversy in binary terms: if Moses was not reliable in the narration of that grandiose event, which must have left imposing scars on the face of the Earth, how could he be trusted on anything else? Conversely, "if Moses is truthful in such a resounding case, then even the minor events that he narrates to us are true."[46] Fossils were, and must be, "proofs of the truths that we believe."[47] To deny their diluvial origin was to deny Christianity itself. While Costantini condemned Vallisneri and Moro with unprecedented virulence, they were only the closest targets in a list of "impious" philosophers that included Aristotle, Fracastoro, Cesalpino, Leibniz, and, surprisingly, even Woodward, who invoked God's use of natural causes when simple divine will should suffice.[48] Against them, and in particular against Moro, Costantini used a stinging sarcasm that was the opposite of the polite civility typical of the early eighteenth-century "Republic of Letters."[49] Just as the second aimed to discipline confessional and cultural differences and made their coexistence possible, the former was meant to divide and inflame. Only a man "without civility and without letters," a disgusted Maffei wrote, could use Costantini's language.[50] Yet Costantini was more a symptom than a cause of the rising stakes of the controversy and of the creation and consolidation of a diluvialist orthodoxy.

The government began to take notice. The Venetian elite had become extremely wary of anything that could agitate the public or even remotely perturb the stability of the state. In 1718, the Republic concluded—and lost—the last of three centuries of wars against the Ottoman Empire. Defeated as a Mediterranean and commercial power, and threatened at its northern borders by the expansion of the Austrian monarchy, Venice developed a deep awareness of its fragility. For the rest of the century, the once proud republic adopted a stance of rigorous neutrality, fearful that any major modification of the European status quo would lead to its partition among bigger players. Alliances and military readiness were replaced with the myth of the stability and wisdom of the Venetian government. The dynamics of internal politics were equally frozen. While administrative and economic reforms often found support, calls for changes in the political system were ruthlessly crushed by those who believed that the Venetian state was so fragile that if even a single element was disturbed, the whole edifice would crumble.[51] The Catholic religion and its institutions remained a pillar of this precarious balance. Conflicts with Rome always aimed to protect the prerogatives of the secular government and the political independence of the city. Religion itself and the Venetian Church, whose high hierarchies were staffed by members of the great aristocratic families, were a different matter, and there was no desire to see them weakened or questioned. When non-scriptural theories of the Earth began to be perceived as a danger to the credibility of established institutions and to the ideological foundations of the state, the controversy on the Flood turned into a political problem.

The politicization of Earth history in northeastern Italy did not wait for the French anti-philosophes to make their voices heard. In the late 1740s, Vallisneri junior withdrew from the debate on the origin of fossils in favor of less controversial subjects. His decision was driven mainly by the attitude of the Venetian senate that, according to the Paduan naturalist, rejected every attempt to interpret the existence of marine fossils "without the light of the Holy Faith." Surprisingly, senators seemed ready to condemn as "heretics" those who did not explain "the origin of these fishes with the Universal Flood."[52] Vallisneri junior may have exaggerated the influence of the most conservative faction within the assembly: little evidence exists that such drastic initiatives were seriously proposed, let alone implemented, but it is significant that they were mentioned. Venetian state censorship (not to be confused with the ecclesiastic censorship) remained extremely tolerant and concerned primarily with the protection of the thriving publishing industry. Between the early 1720s and the early 1740s, the censor was Franciscan friar Carlo Lodoli, who held the office with great moderation and enjoyed the full support of leading intellectuals. In spite of the changing cultural and political climate, at the end of his tenure he was replaced with equally open-minded censors.[53] Vallisneri junior's orthodoxy, however, was even more suspect than his father's. In a manuscript *On the Necessity of Free Thought*, he stated that as long as "our Christians" remained attached to their superstitions, they would never be able to "see their mistakes." In fact, his most important biographer suggests that he did not differentiate at all between Christianity and "superstition."[54] Wary of general explanatory systems, Vallisneri leaned nonetheless toward a cyclical Earth of undetermined age, drawing explicitly upon medieval and Renaissance models including the notion of a cyclical history of humankind. Like many others in his time, he regarded recent theories of the Earth (in particular, Buffon's early one) as the updated versions of long traditions.[55] Always shy of public controversy, Vallisneri never published the numerous tomes of his *Natural History*. A prudent line prevailed, too, in Venetian intellectual circles. The numerous "worthy men" who met every night at publisher Pasquali's "conversation" did not love Costantini; yet they did not dare either to uphold Moro's theses as "true and sure things," even though they seemed "likely and philosophically sound."[56]

In the following years, even provincial elites began to voice their concerns about the "excesses" of modern natural philosophy. Once again, the composition of aristocratic museums is telling. A well-furnished gallery was a must-have for a prestigious household, but until the 1760s natural specimens did not receive any of the attention devoted to antiquities, manuscripts, drawings, or paintings. Little social prestige was attached to fossils and other "natural curiosities," which were left, with few exceptions, to the care of physicians, apothecaries, and country priests.[57] The situation changed radically in the last decades of the century. According to historian Krysztof Pomian, the ratio of natural objects in Parisian museums was around 15% in 1720, but raised to 21% by 1750 and stabilized at 40% in the course of the following forty years.[58] The parallels between Paris and the Veneto can be explained in part with the increasing prestige of scientific and technical expertise throughout Europe. There can be little doubt, though, that they also reflected the growing involvement of the elites in politically fraught scientific controversies. As alarm heightened for the social and political implications of "modern philosophies" in the 1760s, the need for a bulwark against "materialists," "atheists," and "pyrrhonists" (used more or less as interchangeable labels) became urgent. Scholars of eighteenth-century Italy have often noticed that radical materialism does not seem to have been nearly as strong or threatening in the country as local conservatives portrayed it. Not even with the wildest imagination could one see the Italian *anti-philosophes* as an embattled group besieged by the triumphant atheism of the century, as their French counterparts often depicted or conceived themselves.[59] Yet, they adopted an analogous albeit less heated rhetoric and created a similar perception of danger fueled by the diffusion of libertine and irreligious literature—or even more often, by the same writings of the French *anti-philosophes*, which at a certain point became a major vehicle of diffusion of the radical ideas they fought.[60] Even though criticism of the political, legal, and philosophical works of authors such as Spinoza and Hobbes overshadowed scientific issues in those years, conservatives normally assumed that such systems must be underpinned by a materialist philosophy of nature.

Their responses were rarely a straight dismissal of "modern" science. More often, they opted for an appropriation of its methods and language so as to put it in the service of their agenda and defuse its political impact. Their influence should be evaluated not only in terms of the literature they produced, but for the social credibility and support they lent to the theories they favored. Around 1764 Giovanni Arduino, perhaps the most significant geologist in eighteenth-century Veneto, reckoned that the strata of the nearby mountains could not have been produced by either inundations or volcanic fire, but only by their alternating action over immensely long eras. A few decades earlier, academies and "conversations" of Vicenza—the city between Verona and Venice where Arduino resided—would have met such ideas with benevolent curiosity or, at worst, indifference. Writing in the mid-1760s, however, he felt that he should keep them for himself lest they engendered "excessive uproar":

> I greatly fear that this fire, and this Sea, that in so many different and distant occasions produced their effects, leaving in each of them their Inscriptions and Medals, will excite excessive uproar and derangement in certain malignant microcosms.... When it comes to terrestrial and marine volcanoes, this could be made tolerable. But what was that most terrible fire which, by vitrifying the core of the Earth, produced the primary mountains? Maillet, Whiston, Leibniz, Buffon, freely discuss it; but we are in Vicenza, where the wise man must never tell "the truth that seems a lie / for not believed, it turns to shame."[61]

The "malignant microcosms" Arduino dreaded were conservative circles that promoted biblical literalism in natural history. Arduino's biographer has found it perplexing that this situation should coexist with the surge of interest for geological and paleontological research that took place in the area in the same years.[62] This is only paradoxical, however, if one assumes that empirical evidence should automatically undermine diluvialism. On the contrary, the fossils and minerals that increasingly made their way into the aristocratic galleries were often presented as the ultimate proof of a young Creation, upset by a global Flood four thousand years earlier. Born in a modest family and often employed as a surveyor and mining consultant, Arduino did not fear the ecclesiastic inquisition, only the "soft power" that could orchestrate public ridicule and exclude him from qualified frequentations and remunerative jobs.

A younger and more combative geologist, Alberto Fortis (1741–1803), found himself enmeshed in a series of scientific disputes with the conservatives who had a strong foothold in the academies and coffeehouses of Verona. After visiting a large natural bridge in the Alps north of the city, Fortis deduced that it must have been the product of the slow action of water erosion over time. "And when I say over time," he reiterated, "I do not mean months, or years, but [whatever length of time] is compatible with its interior structure, upper loads, and the resistance of the materials that are more or less prone to alteration." From the pages of a widely read Venetian periodical, the *Journal of Italy*, he criticized the secretary of the Academy of Agriculture of Verona who believed that the bridge formed when "in the beginning, waters separated from the earth and the latter was still malleable."[63] The scriptural echoes in those lines, and the implicit preoccupation to protect the biblical chronology, irritated Fortis: "The opinion of those, who believe that the Bridge was made by nature in the first creation, does not even deserve refutation."[64] Fortis embraced instead the idea of the slow revolutions of the rotation axis of the Earth, which remained common in the area.[65] A former pupil of Vallisneri discussed the same hypothesis at length in 1761, openly citing de Maillet and stating ambiguously that he "wished the World were older, of what we know it is."[66] The Academy of Agriculture, however, was an organ of the local landed aristocracy.[67] Its members closed their ranks in defense of their man and of his scriptural geology, which had come to coincide with that of the social order. In coffeehouses and in conservative pamphlets, Fortis was portrayed as a mercenary "journalist" who, lacking personal wealth, could not be trusted as a dispassionate seeker of truth. The lack of civility that Maffei once imputed to Costantini was then turned against Fortis and his party.[68]

FIG. 7.1 A "Deluge Machine"

A 1767 "Deluge machine" designed to illustrate one of the possible physical causes of the universal Flood. According to the accompanying pamphlet, "The obvious shortcomings of all of these systems [to explain the Flood with physical means] have been used by some modern *Philosophes* as a pretext to question the Story of the Deluge." The machine showed that "the system of Moses is consistent with the Laws of Nature." A terrestrial globe filled with water was enclosed within a larger sphere of glass. As an operator spun the machine faster and faster, centrifugal force pushed the water out of the smaller sphere through some valves. An increase in the rotational speed of the Earth would have been "a quite simple cause for the ascension of the waters that formed the Deluge." *Expérience physique, servant à expliquer les causes du Déluge* (Paris: 1767)

The Diluvialist Orthodoxy

The patriciate of the capital shared the concerns of their aristocratic subjects. An "irreversible fracture" took place in those years between some of the most prestigious intellectuals of the state and the conservative majority of the senate that came to consider them politically unreliable.[69] When Vallisneri passed away in 1777, Fortis hoped to obtain his chair of natural history at the University of Padua, but the assembly vetoed his appointment citing his lack of respect for the authority of Scripture in natural philosophy.[70] In 1780 he was attacked even by the *Literary Ephemeris* of Rome for discussing the age of the world without the necessary precautions.[71] Fortis had to adopt softer tones, assure that his theses were compatible with Scripture, and reject any association with libertinism and eternalism.[72] In private, he vented his frustration by calling natural history "a whore, with little business left in Italy," and the writings of Italian naturalists "toilet paper."[73]

In the controversy on the natural bridge and in many others, Fortis sided with mathematician Anton Mario Lorgna, who directed the school for military engineers of the Venetian army established in Verona in the late 1750s.[74] Modeled after the French military schools, it admitted boys of modest origins, providing them with excellent scientific and mathematical training. Unfortunately, their hopes that merit and technical expertise might become a tool for social advancement were regularly frustrated. Under Lorgna's long tenure, the school became a hotbed of political dissent: inspections caught pupils reading Machiavelli and Voltaire, while their overseers incited "irreligious acts" and spread "ideas of freedom."[75] Masonic lodges were discovered in the school in 1785 and 1792; in both cases, officers and teachers were heavily implicated.[76] The scientific academy that Lorgna founded in town, which initially recruited a good number of patricians, did not survive its profound divisions. Lorgna and his collaborators, some of whom were ecclesiastics, lectured on Newtonian attractions, discussed meteorology, or defended the instantaneous creation of the world against the six days of Genesis. In the same rooms, counts Marioni and Buri thundered instead against the recent "abuses" of reason" and the "ill-advised souls of our days" who tried to penetrate the secrets of nature "without the true light of solid, and purged philosophy."[77]

Count Luigi Torri's philosophy ended up being "purged" as well. Well-traveled and well-read, Torri represented a new generation of moderate conservatives who tried to acquire an up-to-date scientific culture and put it in the service of the "public good"—or their idea of what it should be. In 1780 he delivered two geological lectures before the exclusive musical academy of the *Filarmonici*, in which he strove to reconcile the Flood with Buffon and with the succession of inundations and volcanic activities described by Arduino and Fortis.[78] Since he found it impossible to fit the history of the Earth within the canonical 6,000 years, he added at least the couple thousand more that even some Jesuit authors admitted.[79] The reception was glacial. Torri's manuscript was later examined by Orazio Rota, an ecclesiastic and an amateur naturalist who enjoyed a solid reputation as a die-hard conservative. Rota explained to the noblemen that to extend the age of the world by "almost two thousand years" was a "most absurd" and "extremely dangerous" proposition. If two thousand more years were acceptable, why not as many as modern materialists could dream of, thus leading to eternalism and atheism? Tellingly, Rota held that Torri's chronology did not coincide "with the opinion embraced by the Catholic Church" (which never had one, apart from the fact that the Earth is not eternal).[80] Once so amended, the reviewer concluded, the dissertation would be unassailable by modern unbelievers and well received by the lovers of "sane philosophy." The lectures were never published, but the fourth and final redaction included each and every one of the proposed revisions.

Meanwhile, fossils conquered a place of honor in patrician houses where they served to demonstrate the truth of the Flood. In 1779 Marquis Ottavio Canossa, whose family was among the most ancient and prestigious in the region, acquired for a "considerable amount" of money the collection of a pharmacist.[81] The large room devoted to those specimens hosted at least 400 petrified fishes, as well as less prestigious items. Noblemen who wanted to enlarge their collections often asked Maffei's heirs, who still owned the quarry of Bolca, for the permission to conduct digs. The actual work was carried out by local quarrymen, and, given the economic value of the specimens, illegal extractions or smuggling of fossils became common. In 1778, the discovery of enormous and mysterious petrified bones sparked a rush to the best specimens among collectors. The first to arrive was the wife of a pharmacist, who skillfully dug them out from the hard rock that encased them. It did not take long, however, for her hard-won fossils to move to a patrician household.[82] The largest cabinet of fossil fishes in the area, and arguably throughout Europe, was put together by Count Giambattista Gazola. A dynasty of physicians, the Gazolas forsook their considerable scientific interests—together with their fossils—after the family was ennobled in the mid-1700s. The reconstruction of their scientific cabinets began in earnest around 1784 and culminated a few years later in the acquisition of the last great local collection still not in aristocratic hands.[83] By then, the Gazolas owned 1,200 petrified fish of the highest quality, plus the usual mollusks and shells. A drawing of their gallery sketched around 1795 shows slab after slab arranged in glass cases, where they could be admired by amateurs, gentlemen, and gentlewomen.[84] The courses of chemistry and natural history that Serafino Volta, a firm diluvialist, began to teach around 1789 to forty "noblemen, and citizens," moved as well from the house of an apothecary to Count Marioni's palace.[85]

The Flood and the biblical chronology might have been presented as a stark choice between Christianity and atheism (or conversely, between "reason" and "religious fanaticism"), but rhetoric should not be mistaken for a reality that remained much more nuanced. Recent scholarship has swept

FIG. 7.2 A Diluvialist's Museum

Count Giambattista Gazola's museum of fossil fishes in Verona, ca. 1794. Verona, City Library, Sez. Stampe.

away the notion that eighteenth- and nineteenth-century controversies on the history of the Earth were fundamentally a confrontation between "science" and "religion," although many individuals certainly experienced personal conflicts.[86] Likewise, the social and political divide was real, but describes broad tendencies. Diluvialism was widespread in every class, not only among the aristocracy, even though the influence of the conservative patriciate remained crucial. Pharmacist Vincenzo Bozza, whose collection after his death made up the bulk of Gazola's gallery, was himself a diluvialist (as well as an accomplished chemist) who teamed up with the aforementioned Orazio Rota to defend the diluvial origin of Bolca's fossils.[87] On the other hand, the customers and acquaintances who frequented his shop may have feared public "shame," but in private they continued to entertain the most varied opinions. When he showed them his fossils, only some agreed that they were evidence of the "universal inundation." Others pointed instead to the slow revolutions of the sea, or to the climate changes brought by the alterations of the Earth's rotation axis "over countless centuries," or to the slow cooling of the globe.[88] Very few, if any, could have been actual "atheists." Most of them simply refused to subscribe to the rhetoric that associated an ancient Earth with irreligion.

The landed aristocracy was not a compact bloc, either. To find among them sympathizers of the radical Enlightenment, and later even of the French Revolution, would take little effort. Many of those identified here as "conservatives" would be described more accurately as enlightened Catholics, actively engaged in economic, legal, and technical reforms, as long as they did not endanger the established social hierarchy. A cautious top-down reforming process was actually deemed essential to its preservation. Neither had they identified their morality and religiosity with the mundane interests of the papacy: the restriction of the ecclesiastical prerogatives pursued by the Venetian government, which culminated in the late 1760s, received widespread support among them.[89] So did Jansenism, a theological and moral current explicitly condemned by the Catholic Church. Unlike their French counterparts, those moderates claimed the right to speak for the "public" and never allowed the heated rhetoric of more extreme fringes to push them to the margins.

Their cultural and political project, however, hinged on the continuing relevance of an early modern conception of science, performed in city salons and country villas with the participation of the wider social body, and in which scientific credit went together with social status.[90] Lorgna and Fortis represented instead a nascent technocracy that demanded exclusive control over the scientific discourse to the exclusion of "amateurs" and dilettantes. Their attitude was deemed arrogant and dangerous by the aristocracies of the Venetian state, which were determined to take part in and retain control over the elaboration and diffusion of natural knowledge at a local level. Indeed, the cultural and political significance of Earth history is best understood in the context of the larger conservative offensive against the "abuses of reason." Alongside geology and natural collecting, they appropriated Newtonian gravity, electricity, and embryological theories that invoked the agency of immaterial forces. While diluvialism validated the historical accuracy of the revealed text, other scientific pursuits could demonstrate, against modern materialism, that the universe was pervaded with imponderable entities that could act upon matter—be they gravity, electric fluids, or souls.[91]

The beginning of the French Revolution confirmed their worst fears, steeled their resolve, and radicalized their convictions. The events of the subsequent years proved shocking but not surprising to many of them. Like the French *anti-philosophes*, they had warned for decades that social subversion would be the natural outcome of skepticism toward the established religion and of the "fanatical" belief that nothing existed in the universe unless it could be seen, counted, and weighted.[92] Astronomer Antonio Cagnoli (1743–1816), an enlightened Catholic and a moderate but sure reformer, resided for ten years in Paris where he built close relations with scientists such as Lalande, Condorcet, and Bailly. As the heads of his colleagues, mentors and friends began to roll, it is little wonder that the mild-mannered astronomer erupted against "the brutish system" of democracy.[93] In 1791, the Venetian censorship finally noticed *Telliamed*'s "extreme impiety" and moved to prohibit it after de Maillet's dialogue had been imported, read, and cited for more than forty years.[94] In the same months, the informants of the government reported that in Lorgna's school some teachers, including the deputy director, spread "Jacobinism" among the students.[95] Count Gazola was signaled instead as politically trustworthy, which meant, in that context, strongly opposed to revolutionary ideas. The brother of the beheaded king of France, who reigned as Louis XVIII after Napoleon's fall, must have been of the same opinion. Between 1794 and 1796 he sojourned as a guest and personal friend at Gazola's suburban villa that included his botanical garden and hosted his electrical experiments.[96] Gazola himself became a leader of the conservative naturalists in the Veneto, even though he refrained from extreme positions or from placing his signature under the aggressive essays he probably wrote. The monument of late eighteenth-century Italian diluvialism, the superbly illustrated *Fossil Fish of Verona* (1796–1809), showcased his and other aristocratic families' collections of fossils as the ultimate evidence in support of the Flood.[97] The associates who backed the costly enterprise, in some cases down to the brink of financial ruin, entrusted the text to the diluvialist naturalist Serafino Volta, who did not leave any doubt as to its purpose:

> The analysis and description of so many remnants of fossil fishes . . . would be useless, or the object of mere curiosity, if it did not lead us to the realization of a most important truth. . . . Namely that fishes of different waters, and different seas gathered in the famous mountain of Vestena close to Bolca, and were later buried there. This fact, authenticated by the analysis and illustration of the fossil fish here examined, seems to prove beyond any doubt the event of a general inundation of our Globe.[98]

Fossil Fishes of Verona aimed to represent the moderate wing of Italian diluvialism, allegedly founded on the best empirical evidence and on the latest advancements in geology and chemistry. The dedication to the Royal Society of London and the bilingual text (Italian and Latin) were meant to boost not just its international visibility, but first and foremost its authoritativeness before a local public. Yet while Volta hardly mentioned the Bible at all, the

Il Pipistrello Aquatico. ——— *Chætodon Vespertilio*

FIG. 7.3 The "Sea Bat"

"The seas of Japan are the abode of the Sea Bat, where today it lives and reproduces its species. Should we say that the same climate and the same sea once covered the continent of Italy, where now the fossils of this fish are found? Or rather, that it was carried there with many more ... when in the course of the multiple floods of the terraqueous globe marine waters raised above the tallest mountains?" Giovanni Serafino Volta, *Ittiolitologia veronese del Museo Bozziano ora annesso a quello del Conte Giovambattista Gazola e di altri gabinetti di fossili veronesi: Con la versione latina,* vol. 1 (Verona: Stamperia Giuliari, 1796), XXVII, table VI. Beinecke Library, Yale University.

"certain knowledge" that mountains existed even before the Deluge could only come from the description of Eden in Genesis.[99] The author conceded that some of the local heights must have been produced by volcanic forces, but he relegated their activity to a time after the Flood. His simplification and misrepresentation of pre-modern Earth history was as radical as in the writings of the *philosophes*, but with a typically localistic twist: Aristotle and all the ancients had uniformly believed that fossil fishes were "jokes of nature" until Veronese naturalists and collectors began to unearth (quite literally) their true origin.[100]

On a different front of the same war, Gazola and his circles lent full support to the French physician Pierre Thouvenel, who fled to the Veneto at the onset of the Revolution with many of his aristocratic patients. Thouvenel toured the region with a water diviner who, he claimed, was especially sensitive to the imponderable electric fluid that permeated the universe. The experiments organized by Gazola in his botanical garden and in other patrician mansions, in front of a large and vocal public, seemed to confirm those assertions against the skepticism of Lorgna and of a large part of the Italian scientific establishment.[101] Thouvenel did not hesitate to place them among the "incredulous men, commonly called materialists," who regarded every "occult cause" as "unnatural, or against nature."[102] The anonymous editors (possibly Gazola himself) of *The Ten Years' War*, a collection of writings that exposed the history of the controversy from a conservative standpoint, denounced the "culpable monopoly" those scientists wanted to impose on natural history, concluding that they were one and the same with "free masons," "Jacobins," and "modern philosophers."[103]

Eighteenth-century Veneto is a compelling example of the formation of the narrative that construed diluvialism and a young Earth as defining elements of historical Christianity. Its elaboration took place not because of official pronunciations or cultural pressure from the Church (which would have been of little consequence in non-Catholic countries), but through cultural and political offensives that converged from opposing sides to promote the same perception. To what extent the Venetian case is an indicator of larger trends remains an open question. The area might have been a political and social backwater compared to France or Great Britain, yet the same could be said of most of Europe. Responses to perceived threats to the social order must have varied considerably, but the Veneto was not the only place where the landed aristocracy anxiously watched the challenges to the ideological bases of its supremacy. In the same years in which the Venetian conservatives concluded their diluvialist offensive, René de Chateaubriand (1768–1848) touted a fiercely literalist view of the Deluge and of the age of the Earth in his militant *Genius of Christianity* (1802).[104] Discussing late eighteenth-century England, Martin Rudwick came to similar conclusions: the "perceived threat" lay not in non-literal readings of Genesis in and of themselves, but in the social instability they could engender.[105] Even in the British islands, Earth history was often employed in simplified and radicalized forms in the service of opposing political agendas.[106] By the late 1700s the identification of a young Earth with the tradition of Christian Europe, never challenged until the recent emergence of a secular science, was complete even in contexts where the *philosophes* were unable to present themselves as the champions of the "public opinion."

Conclusion

Q. Mr. Governor [Mike Huckabee], I think the specific question is: do you believe that literally, it was done in six days and it occurred six thousand years ago?
A. No I didn't answer that, what I said is I don't know. My point is: I don't know, I wasn't there, but I believe in what God did in six days. Whether He did it in six days that represent periods of time, He did it: and that's what's important.

US PRESIDENTIAL PRIMARIES, televised debate
hosted by CNN, June 5, 2007

HISTORIOGRAPHICAL COMMON SENSE has long placed the discovery of "deep time" in the decades between the late eighteenth and the early nineteenth century that Martin Rudwick called "the Age of Revolution"—both in politics and in the Earth sciences.[1] This book argues instead that the idea of an immensely old Earth circulated openly in medieval and early modern times; that for most of those centuries, it was largely unproblematic; and that the notion of a deep fracture between a pre- and a post-nineteenth-century Earth history (one bound to a young world, the other to an ancient one) was a product of the cultural and political tensions of the Enlightenment. Far outliving the circumstances that created it, in the subsequent two hundred years this historiographical myth has nourished the perception of an inescapable conflict between the Christian tradition and modern science. It has powerfully contributed to the shaping of political and cultural landscapes, particularly in the United States, and provided ideological ammunition to both religious conservatives and their opponents; it has fed culture wars and contributed to the growing mistrust of the academic establishment. Persuaded that their religious traditions uniformly supported a literal reading of Genesis, conservative Christians and state legislatures alike have touted a recent Earth against

On the Edge of Eternity. Ivano Dal Prete, Oxford University Press. © Oxford University Press 2022.
DOI: 10.1093/oso/9780190678890.003.0009

the millions of years required by evolution. While the implications will not be lost to the readers, it seems appropriate to conclude this book with a brief re-assessment of the immense influence the Enlightenment narrative of the history of deep time exerted on the following centuries, down to our own time.

The conservative reaction that swept Europe in the decades following the French Revolution paradoxically increased the appeal of a more moderate diluvialism. The transformation of revolutionary France into Napoleon's socially conservative empire largely voided the controversy on the history of the Earth of its previous political stakes: in the end, a materialistic and technocratic science had not begotten the dreaded democratic and egalitarian society. The entanglement of Scripture and Earth history was far from dead, but the association of non-literal readings of Genesis with social subversion became much less compelling.

Moderate diluvialists could muster evidence that was often persuasive even on purely scientific grounds. The idea that the Noachian deluge and a recent humankind could be embedded within a much longer history had always been widespread, but at the turn of the 1800s it became commonplace. While the Age of Revolution did not create the idea of an almost immeasurable pre-human past, it did lead to its widespread acceptance as the physical and historical reality of the Earth, rather than as a matter of speculation or probability. Apart from the fossils of tropical or unknown species found in European quarries, geological features (such as erratic blocks or U-shaped valleys) attributed to the last ice age were often interpreted as evidence for a recent global inundation. Genevan naturalist Jean-André de Luc (1717–1817) claimed that Genesis contained "a true history of the world," but he thought that human presence marked only a small part of it. A wealth of geological evidence demonstrated the reality of a recent "sudden Revolution" that initiated the present world.[2] During his vast survey of the Russian Empire, commissioned by Empress Catherine, Prussian geologist Simon Peter Pallas (1741–1811) noticed that the region of the Ural Mountains featured a thick layer of sediments laden with the remains of trees and exotic animals. Again, an abrupt inundation of the globe in a not-too-remote past seemed the most plausible cause.[3]

In Napoleon's Paris, Georges Cuvier (1769–1832), one of the most famous and powerful European naturalists, recognized that the large mammals whose skeletons he painstakingly reconstructed disappeared suddenly from the face of the Earth, rather than migrate to still-unexplored climates. A watery catastrophe at the dawn of human history, arguably preceded by similar ones, seemed to best explain those great extinctions. His insistence on an inconceivably vast, albeit not eternal pre-human history never interfered with his lifelong Lutheran faith.[4] According to Italian naturalist Giambattista Brocchi (1772–1826), who was from the Veneto and wrote only a few years after the completion of *Fossil Fishes of Verona*, geologists had established but one certain fact: that present continents emerged from the sea a few thousand years earlier, after a long history of progressive withdrawal of the waters interspersed with devastating inundations. Humankind must have appeared recently too, preceded by Cuvier's now-extinct creatures.[5]

In Great Britain, geological fieldwork became a fashionable activity for educated men and women during the "heroic age" of geology in the early nineteenth century. The discipline was evolving at breakneck pace, and lectures on the topic were certain to summon large and interested audiences. The foundation of the "Geological Society" in 1807 provided a venue for heated disputes, which contrasted sharply with the gentlemanly formalism of the Royal Society.[6] The exploration of prehistoric caves in different parts of Europe, once used as dens by hyenas and bears, provided Anglican Reverend William Buckland (1784–1856) cogent evidence for Noah's flood.[7] Buckland went on to become the first professor of geology at Oxford, Dean of Westminster, and author of the sixth of the famous *Bridgewater Treatises*, which established for the Anglo-Saxon public the compatibility of reason and Revelation. His voluminous contribution, *Geology and Mineralogy Considered with Reference to Natural Theology*, was introduced by a chapter on the "Consistency of Geological Discoveries with Sacred History."[8] Buckland shared the already general belief that geology had been "for a while considered hostile to revealed religion."[9] It was essential, then, to avoid strict biblical literalism, so as not to fall into the same mistakes as Galileo's prosecutors. This entailed the modification of the "most commonly received and popular interpretation of the Mosaic text" that would prevent it from marching in agreement with modern science.[10]

"Christian geology" proved indeed a popular genre in those years, but it was not without vocal critics.[11] The contemporaries of Scottish naturalist James Hutton (1726–1797) had good reasons to suspect him of eternalism. Hutton devised an Earth without a clear beginning, whose continents were endlessly washed away to the sea only to be replaced by new ones, thanks to the expansive force of its internal heat.[12] The same natural causes always operated at a slow, uniform rate, without any need for catastrophic revolutions. Hutton has traditionally been hailed in English literature as the true discoverer of deep time, but fourteenth-century natural philosophers like Jean Buridan or Albert of Saxony could have found his Earth almost familiar.

In the early 1800s Hutton found himself relegated among the obsolete "world makers" of the previous century, until his "uniformitarian" principles and his cyclical Earth were rescued by another Scotsman, Charles Lyell (1797–1875). Lyell took it for granted that the young age of "this planet" and the dominant role of the Flood in its geological history had been "the consistent belief of the Christian world."[13] Their undoing by modern geology was part of the general path of enlightenment of the human mind.[14] Lyell's dismissive picture of pre-modern Earth history, and his tribute to Fracastoro as the great unheeded precursor, drew almost word by word from Giambattista Brocchi's *Fossil Conchology* (1814).[15] Long isolated in his rejection of periodical catastrophes, Lyell was later extolled as the greatest hero of a heroic age. In the process, the larger context of his work came to be misrepresented as the epic confrontation between a young and an old Earth, rather than a controversy on the causes of geological change (slow, uniform forces vs. sudden, catastrophic "revolutions").

By the 1820s, naturalists of every religious and scientific persuasion had come to accept that a number of past species went extinct. However, the origin of the living ones that did not appear in the fossil records remained controversial.[16] The medieval and Renaissance doctrine of spontaneous generation had long lost any credibility. As more of the world outside Europe was explored, the idea of large migrations of animals to and from remote corners of the Earth became implausible. On the contrary, the possibility that old species "transmuted" into new ones was gaining ground. Organic transmutation had been often discussed in the previous centuries, especially in the context of an ancient Earth that allowed enough time for the process to carry out its effects. The transformation of aquatic species into terrestrial ones offered a natural mechanism for the inhabitance of the continents if—as it was often held—dry land emerged gradually from a global ocean. At the turn of the nineteenth century, the idea proved especially attractive to those who favored uniform geological processes over global catastrophes.

In 1800 Jean-Baptiste Lamarck (1744–1829), a senior colleague of Cuvier's at the Paris Museum of Natural History, publicly embraced the theory of the continuous "transmutation" of species: rather than die out, they unceasingly change; since life forms were in constant flux, species were ultimately arbitrary, and fossils recorded just a singular point in a continuum.[17] More conservative than Lamarck, Brocchi still held that animal species were not only born and died just like individuals, but could also modify their physical traits in the course of their existence.[18] Lyell extended to the species of living things his conception of a cyclical Earth: the famous 1830 cartoon that showed "prof. Ichthyosaurus," discussing with other ancient reptiles a newly found human skull, played with his idea that one distant day, dinosaurs might indeed return.[19]

Skeptics of transmutation could bring many, often compelling objections. Rather than arbitrary constructions as claimed by Lamarck, species appeared quite real and distinct from one another.[20] Like earlier doctrines on the spontaneous generation of human beings, transmutation raised troubling questions about God's special creation of humankind. Neither could account for the deafening silence of human documents on the time that preceded the last few thousand years. On the one hand, new evidence undermined the association of the latest "revolution" of the Earth with the biblical Flood, as Buckland himself acknowledged. On the other, it suggested that even the most ancient animals had been sophisticated organisms carefully designed to suit their environment rather than crude first steps toward fully functional animals, as Lamarck presumed.[21] Yet, when the first human fossils and prehistoric artifacts began to be unearthed in the early 1830s, they came not from recent strata supposedly deposited by the Flood, but mixed with the much older remains of extinct mammals. Non-human primates also began to emerge from even deeper layers.[22]

In 1844, Robert Chambers's *Vestiges of the Natural History of Creation* delivered to a mass audience an evolutionary view of the universe, in which a providential design guided the progress toward ever-increasing complexity. After reassuring his public of the self-evident truth of a supreme Creator, the narrator enquired freely on the details of their development according to nature's God-given laws.[23] Chambers's work provoked an enormous sensation among readers of every social class, both in Great Britain and in America. While the author did not seem to be a traditional Christian, his work could not be associated to eighteenth-century materialism, either. As could be expected, a number of groups, such as Evangelicals or the Anglican establishment of Oxford and Cambridge, accused the *Vestiges* of atheism and of erasing "all distinction between physical and moral."[24] The overall reception of the *Vestiges*, though, shows that many Christians were willing to accept not only an ancient Earth but even organic evolution. After reading the *Vestiges*, future British Prime Minister William Gladstone (1809–1898) came to regard the creation of Adam as the culmination of God's plan, whose unraveling may have passed through generations of pre-Adamic humans.[25] The controversy seems to have been largely perceived as one within Christianity itself and its definition, rather than one between "religion" and "science." Provided that the Creator's hand remained the guiding principle, evolution could still be part of a Christian worldview.[26] A few years later, though, Charles Darwin (1809–1882) removed even that last prop by replacing the Almighty's benevolence with natural selection, namely, blind chance and the ruthless survival of the fittest.

Darwin devoted many more pages to geology than to biology in the notebooks of his voyage around the world (1831–1836).[27] By the time he set foot on the *Beagle*, he had long been convinced of the old age of the Earth and that the biblical Flood had no recognizable part in its history.[28] Neither attitude was particularly noteworthy (though the second more so), but in the course of his travel he came to adopt Lyell's uniformitarianism without renouncing the Earth's directional history.[29] Evolution could indeed replace, with an initial single act, the many separate creations needed to repopulate the Earth after its recurrent catastrophes. At the end of *On the Origin of Species* (1859), Darwin famously envisioned a Creator breathing life into the members of the very first species.[30] Any such reference disappeared, though, in his later works, replaced by the explicit rejection of divinely guided evolution.

In spite of gaps and problems in the fossils record, Darwin provided the plausible natural cause for speciation that had been lacking until then. Whether the underlying motive force was design or chance continued to be ferociously debated, but within twenty years from the appearance of *On the Origin of Species* it had become difficult to find reputable scientists willing to deny the reality of evolution. It is hardly a coincidence that one of the foremost early opponents of evolution, Louis Agassiz (1807–1873), was also one of the last "catastrophists." Yet his catastrophes no longer had to do with the biblical Flood, which he actually helped remove once and for all from natural history. In 1837 he proposed that the geological features often related to a recent global inundation could be attributed instead to an older "ice age" that once gripped most of the Earth.[31] The penetration of evolutionism in mainstream culture was difficult, but no more than that of heliocentricism, which also defied, perhaps more than religious precepts, what seemed plain common sense.

An immensely old Earth was certainly not an issue, and it was easily conceded even by those American Protestant leaders who vehemently rejected organic evolution. Opinions on how to interpret the Genesis narration varied greatly among them, but around 1875 virtually no one thought of compressing the history of the Earth within a few thousand years, or of denying the possibility of a long pre-Adamic history.[32] At the turn of the twentieth century, evolution was largely accepted in American high school textbooks.[33] As for the Catholic Church, the age of the Earth had never been a matter of faith.

Evolutionism too was handled with extreme caution at an official level, including in its Darwinian incarnation. Neither Lamarck nor Charles Darwin
ever made it onto the list of the forbidden authors. Pious IX's 1864 *Syllabus
of Errors*, largely perceived as a sweeping condemnation of liberalism and modernity, omitted any explicit reference to Darwinism, which remained thus
open to debate.[34]

The flip side of the coin was the decline of the tradition of natural theology, which had long represented a space for accommodation. Since religion
and science no longer sought mutual support, radicals of either camp could
ignore, dismiss, or fight the other. The late nineteenth-century struggle between "religion" and "science" involved both a revival of biblical literalism
(alongside the more conciliatory approaches) and a new unwillingness to
compromise by a large part of the scientific establishment.[35] Beginning in the
1870s, science became a weapon often wielded in attacks against religion.[36]
An enduring outcome of that rhetoric was the "conflict thesis."[37] The most
influential enunciations of the doctrine were John William Draper's *History
of the Conflict between Religion and Science* (1874) and Andrew Dickson
White's *A History of the Warfare of Science with Theology in Christendom*
(1896). Considering the crucial role played by nineteenth-century geology
in defining the boundaries of science and religion, it is hardly surprising that
both authors dedicated substantial parts of their works to the "controversy
regarding the age of the Earth."[38] Draper specifically targeted the Catholic
Church, in spite of the fact that it had been less prone historically to biblical
literalism than its Protestant counterparts. However, Draper considered it the
religion whose "demands are the most pretentious" and that "sought to enforce
those demands by the civil power."[39] It was also the religion of the poorest and
the most prolific of the immigrants, the Irish and the Italian, often regarded in
those decades as a cultural and demographic threat to the traditional Anglo-
Saxon, Protestant supremacy in America.[40] Not even Draper, however, could
deny that discussions on the age of the Creation "might within certain limits
be permitted" in Catholic Europe; and that Rome, "instructed by the outcome of its opposition to Copernicanism," did not exhibit the same kind of
active resistance "when the question of the age of the world presented itself
for consideration."[41] He assumed, of course, that the biblical chronology was
never up for consideration until recent times, which to him coincided not
even with the 1700s but with "the present century" (he presumably referred
to the 1830s).[42] Beforehand, "marine shells, found on mountain tops far into
the interior of continents, were regarded by theological writers as an indisput-
able illustration of the Deluge."[43]

Draper's book became immensely influential in spite of its extreme scientific weakness. Actually, its representation of the history of deep time might have appeared as one of its most solid parts, since its basic premises were largely shared. Twenty years later, Andrew White tried to establish Draper's thesis on firmer ground with a more nuanced, better researched, and academically more compelling work.[44] White claimed that he did not intend to attack revealed religion per se, but only the "decaying mass of outworn thought which attaches the modern world to medieval conceptions of Christianity." Science and Christianity could coexist; the conflict was between science and "Dogmatic Theology."[45] In the inspired opening of the book, he compared his project to the patient toiling of the Russian peasants who, in the month of April, broke up with picks and shovels the ice sheet still imprisoning river Neva in Saint Petersburg. Everyone knew that, sooner or later, the ice had to yield to springtime; but if the frozen barrier lasted too long, it could break all of a sudden, causing floods and misery. In the same way, an outmoded Christianity violently swept away by modernity could bring down with it not only "outworn creeds and noxious dogmas, but cherished principles and ideas."[46] In theory, White's stated program of saving Christianity by historicizing it might have looked attractive even to a number of modern believers; in practice, he contributed to the propagation of many of Draper's less substantiated arguments. White maintained too that until the late 1600s "to doubt [that man was created from four to six thousand years before the Christian era], and even much less than this, was to risk damnation."[47] A largely forgotten list of heresies composed in the fourth century that anathematized any deviation from Isidore of Seville's reckoning of the time of the Creation became, in White's book, a "guide to intolerance throughout the middle ages."[48]

Catholic scientists like Pierre Duhem and James J. Walsh began immediately to question the simplistic dualism of the "conflict thesis," at least in the form cast by Draper and White.[49] The fact that their followers continued to do so throughout the twentieth century is telling of its endurance and of its lasting appeal to sizeable sectors of the Western societies. Historians may have dropped the Draper–White thesis as a meaningful tool to understand the interaction of science and religion (while sometimes retaining more nuanced versions[50]); much of the public and influential authors like Richard Dawkins still think in similar terms, though, all the more so as they face the global resurgence of religious radicalism.[51] Christians had even more reasons to worry about modern science, as at the turn of the twentieth century Darwin's evolutionism morphed into social and political Darwinism. Elevated to the rank of universal laws governing the evolution of human society as well as of biological forms, the notions of struggle for life and survival of the fittest were used to justify social inequalities, colonialism, nationalism, and the extermination of "inferior" races or individuals. A new secular religion, whose central value was the preservation and increase of one race at the expense of the others, seemed poised to replace the traditional Christian ethics.

It is not difficult to understand why William Jennings Bryan (1860–1925), the three-time presidential candidate who figured among the protagonists of the "Scopes Monkey Trial" (1925), regarded Darwinism as a mortal threat not just to Christianity but to civilization itself, and came to mistrust the academic elites who theorized it. The horrors of World War I cemented Bryan's convictions, not unlike the French Revolution confirmed those of eighteenth-century diluvialists. Yet, the portrayal of Bryan as a biblical literalist popularized by *Inherit the Wind* is often inaccurate.[52] The probing of Bryan's reading of Genesis is the climax of the theatrical drama and of its Hollywood adaptations. During the dramatic showdown, lawyer Drummond/Darrow has a sudden intuition: since the Sun was created on the fourth day, there can be no way to tell the actual length of the previous ones. He then drags Brady/Bryan to the reluctant admission that the first "days" might have been longer than twenty-four hours. If so, Drummond/Darrow retorts in triumph, why not "thirty hours! Or a month! Or a year! Or a hundred years! . . . Or ten million years!"[53]

That exchange never happened. In the actual debate, Bryan opined that the Earth must be much older than a few dozen centuries and that the creation could have continued for "millions of years."[54] Prodded by the interrogators, he emphatically refused to concede that the Bible taught otherwise.[55] Even though he believed in a recent exterminating Flood, he accepted that fossil records formed much earlier, and that the "days" of the creation symbolized vast periods of time; to say that the Bible taught a creation in six literal days was akin to saying that it taught the flatness of the Earth.[56] Likewise, the sentence that a "day" could not be defined before the creation of the Sun was never pronounced at the trial. Far from the sudden stroke of genius of a witty barrister, it represents a classic argument formulated countless times before, from Thierry of Chartres's twelfth-century cosmological treatise to Thomas Burnet's correspondence with Isaac Newton.

Bryan's stance was typical among early-1900s conservative Christians. Even the fundamentalists who animated the anti-evolution crusade of the 1920s preferred to reconcile Genesis with the evidence for an old Earth, rather than oppose it.[57] Those who today believe in the possibility of a recent

Earth do not cling to a corpus of medieval beliefs that somehow resisted modernity. On the contrary, according to historian of American creationism Ronald Numbers, the movement grew in the second half of the twentieth century from small beginnings among the Seventh-day Adventists, a millenarian group that attaches particular importance to the seven days of Genesis.[58] There seems to have been no continuity between them and the critics of transmutation who, in the 1840s, touted a young Earth against Chambers's *Vestiges*.[59] The founder of the so-called "Flood geology," George McCready Price (1870–1963), had to defend his interpretation of Genesis not only against professional geologists but also against more mainstream religious leaders. Yet, his lectures and his most ambitious work, *The New Geology* (1923), found sympathetic audiences even though he later shifted focus onto a recent appearance of life on Earth.

As the movement expanded outside the initial Adventist circles and began to count members with actual scientific training (including in geology), its different strands became difficult to reconcile. Some creationists gave up trying to obtain recognition by established academic circles and formed their own societies instead, where they shaped an alternative culture in which "true" science confirmed their interpretation of the Bible. However, others began to accept at least limited evolution and pointed out that there was no need to conflate evolution and an old Earth.[60] A new direction was also necessary because of crucial tactical changes. In the 1920s, only three US states passed anti-evolution laws, but creationists obtained the de facto disappearance of evolution from high-school textbooks.[61] Starting from the 1960s, its reintroduction in new standard textbooks caused a firm reaction. Instead of trying to ban evolution, creationists lobbied for the teaching of creationism alongside it. In order not to infringe the US constitutional separation between state and religion, though, "creation science" had to stand on its own scientific merits, avoiding any reference to the Bible. In the 1980s, legislation to this effect was passed in Arkansas and Louisiana, but when challenged in court it proved impossible to produce a strong case for the scientific viability of creationism. In 1987, the Supreme Court ruled "creation science" a religious doctrine aimed to promote a particular faith.[62]

In hindsight, it is no longer surprising that moderate creationists failed to thwart the rejection of evolution (and often, even of an old Earth) among more radical religious denominations.[63] At the dawn of the new millennium, allegiance to group orthodoxies even in face of the most obvious evidence has come to define political and social, as well as religious, identities. Refutations from academics and "experts," held as corrupt or intrinsically unable to renounce their skepticism, have come to constitute corroborating evidence. On the one hand, the alleged continuity of Young Earth creationism with historical Christianity has remained a significant factor for its association with "medieval" obscurantism. On the other hand, the same postulate has contributed to validating a literal reading of Genesis among conservative religious groups. There can be little doubt that in the last 150 years the real issue has been evolution, rather than the age of the Earth. By and large, the teaching of biology, not geology, has been the arena of the various attempts to introduce creationism in American public schools.[64] Still, it has proved tempting to try to undermine evolution by denying or casting doubt on the millions of years it requires. The Arkansas law that in 1981 mandated "balanced treatment" in teaching creation and evolution listed among the six main tenets of "creation science" the "Explanation of the earth's geology by catastrophism, including the occurrence of a worldwide flood" (n. 5) and "A relatively recent inception of the earth and living kinds" (n. 6).[65] In 1999, the Kansas State Board of Education voted for the removal of references not only to evolution, but even to the age of the Earth in statewide standardized textbooks.[66]

During a televised debate in the 2007 presidential primaries, Mike Huckabee, then governor of Arkansas and an ordained evangelical minister, openly rejected evolutionism while refusing to take a clear stance on the issue of the age of the Earth. The presidential hopeful contended that the essential point of his faith was that somehow, at some time, "[God] did it"—not when. Yet it is hard to understand why he would find it difficult to accept the overwhelming geological evidence for a billion-years-old Earth, unless it is associated to evolution. Indeed, a question on his views concerning specifically the time of the creation ("Do you believe that literally, it was done in six days and it occurred six thousand years ago?") prompted an answer that shifted immediately to evolution.[67] Some of the most visible advocates of a biblical geology heavily emphasize the supposed uniformity and continuity of the Christian doctrine of a recent world, deprecating compromises with non-literalist interpretations of Genesis as a recent "mistake."[68]

Mainstream scholarship has long shared the same fundamental assumption on the Western history of deep time. Since that narrative fitted comfortably within the broader account of the scientific revolution and of the rise of modern scientific rationalism, it was rarely questioned in spite of clear indications to the contrary that had long been available. More than a historical thesis, it represented a crucial element of a worldview that was too resilient and widespread to be seriously disrupted by scattered and apparently incoherent evidence. It just made sense, and documents that seemed to state the opposite could be ignored, downplayed, or placed among the writings of "precursors" and lonely geniuses. That such a flawed account of our past managed to stand unchallenged for centuries is perhaps the most important lesson for the future one could draw from this book.

Notes

INTRODUCTION

1. Torello Saraina, *Torelli Saraynae Veronensis . . . De origine et amplitudine ciuitatis Veronae* (Verona: ex officina Antonii Putelleti, 1540), 5–8. Translation is mine.

2. Baron d'Holbach, "Fossile," in *Encyclopédie, Ou Dictionnaire Raisonné Des Sciences, Des Arts et Des Métiers*, ed. Denis Diderot and Jean le Rond d'Alembert (Paris: Briasson, 1757), 210.

3. Sir Charles Lyell, *Principles of Geology: Being an Attempt to Explain the Former Changes of the Earth's Surface, by Reference to Causes Now in Operation*, vol. 1 (London: J. Murray, 1832), 27.

4. François Ellenberger, *History of Geology*, vol. 1 (Rotterdam: A. A. Balkema, 1996), 95.

5. Paolo Rossi, *The Dark Abyss of Time: The History of the Earth & the History of Nations from Hooke to Vico* (Chicago: University of Chicago Press, 1984).

6. See for example, his outstanding scientific poem on syphilis: Girolamo Fracastoro, *Hieronymi Fracastorii Syphilis sive morbus gallicus* (Basel: Bebel, 1536). The disease is named after the main character of the poem, Syphilis.

7. Girolamo Fracastoro, *Hieronymi Fracastorii Homocentrica. Eivsdem De Cavsis Criticorvm Diervm per Ea Qvæ in Nobis Svnt* (Venice: 1538), 11. On Fracastoro's cosmology, see: Enrico Peruzzi, *La Nave Di Ermete: La Cosmologia Di Girolamo Fracastoro* (Florence: L. S. Olschki, 1995).

8. On Fracastoro's medical consultation that contributed to the displacement of the council from Trent to Bologna (a city in the State of the Church): Alessandro Pastore, "Il consulto di Girolamo Fracastoro sul tifo petecchiale (Trento, 1547)," in *Girolamo Fracastoro*, ed. Alessandro Pastore and Enrico Peruzzi (Florence: Olschki, 2006), 91–101.

9. Craig Martin, *Renaissance Meteorology: Pomponazzi to Descartes* (Baltimore: Johns Hopkins University Press, 2011), 25.

10. A notable exception is the first volume of: Ellenberger, *History of Geology*.

11. Stephen Toulmin and June Goodfield, *The Discovery of Time* (New York: Harper & Row, 1965), 74.

12. Toulmin and Goodfield, 75.

13. Toulmin and Goodfield, 74.

14. M. J. S. Rudwick, *The Meaning of Fossils: Episodes in the History of Paleontology*, 2d rev. ed. (New York: Science History Publications, 1976), 37.

15. Gabriel Gohau, *A History of Geology*, trans. Albert V. Carozzi and Marguerite Carozzi (New Brunswick: Rutgers University Press, 1991), 28.

16. Pascal Richet, *A Natural History of Time* (Chicago: University of Chicago Press, 2007), 43.

17. William J. Connell, "The Eternity of the World and Renaissance Historical Thought," *California Italian Studies* 2, no. 1 (January 1, 2011): 1.

18. Connell, 10.

19. Ann Blair, "Mosaic Physics and the Search for a Pious Natural Philosophy in the Late Renaissance," *Isis* 91, no. 1 (2000): 32–58.

20. John William Draper, *History of the Conflict Between Religion and Science*, 2nd ed. (New York: D. Appleton, 1875), 182.

21. Andrew Dickson White, *A History of the Warfare of Science with Theology in Christendom* (New York: D. Appleton, 1896), 250.

22. See in particular, Pierre Duhem's stalwart defense of medieval science in the early 1900s, and Robert H. Merton's claim (first formulated in 1938) that modern experimental science would be rooted in 17th-century Puritan ethics and values. Both theses gave origin to enduring historiographical trends. Among Duhem's most recent followers, see in particular Edward Grant, *God and Reason in the Middle Ages*

(Cambridge; Cambridge University Press, 2001). For an overview of the critical fortune of the "Merton thesis" in the twentieth century: I. Bernard Cohen, ed., *Puritanism and the Rise of Modern Science: The Merton Thesis* (London: Rutgers University Press, 1990).

23. See in particular: Pamela O. Long, *Artisan/practitioners and the Rise of the New Sciences, 1400–1600* (Corvallis, OR: Oregon State University Press, 2011); Pamela O. Long, *Openness, Secrecy, Authorship: Technical Arts and the Culture of Knowledge from Antiquity to the Renaissance* (Baltimore: Johns Hopkins University Press, 2001); Pamela O. Long, *Technology, Society, and Culture in Late Medieval and Renaissance Europe, 1300–1600* (Washington, DC: Society for the History of Technology and the American Historical Association, 2000); Pamela H. Smith, *The Body of the Artisan: Art and Experience in the Scientific Revolution* (Chicago: University of Chicago Press, 2004). Also: Deborah E. Harkness, *The Jewel House: Elizabethan London and the Scientific Revolution* (New Haven: Yale University Press, 2007).

24. Martin, *Renaissance Meteorology*; Joëlle Ducos, *La Météorologie En Français Au Moyen Age (XIIIe-XIVe Siècles)*, (Paris: Honoré Champion, 1998); Claude Thomasset and Joëlle Ducos, *Le temps qu'il fait au Moyen âge: phénomènes atmosphériques dans la littérature, la pensée scientifique et religieuse* (Paris: Presses Paris Sorbonne, 1998). In recent years, the universities of Venice and Warwick have launched two consecutive joint projects for the study of Aristotelian philosophy in Renaissance vernacular publications. See: http://www2.warwick.ac.uk/fac/arts/ren/projects/vernaculararistotelianism/, and http://www2.warwick.ac.uk/fac/arts/modernlanguages/research/italian/projects/aristotle/

25. Federico Barbierato, *The Inquisitor in the Hat Shop: Inquisition, Forbidden Books and Unbelief in Early Modern Venice* (Farnham; Burlington (VT): Routledge, 2012).

26. Anthony Grafton, *Defenders of the Text: The Traditions of Scholarship in an Age of Science, 1450–1800* (Cambridge, MA: Harvard University Press, 1991); Anthony Grafton, *Joseph Scaliger: A Study in the History of Classical Scholarship* (Oxford: Oxford University Press, 1983).

27. C. Philipp E. Nothaft, *Dating the Passion: The Life of Jesus and the Emergence of Scientific Chronology (200–1600)* (Leiden: Brill, 2012); C. Philipp E. Nothaft, "The Early History of Man and the Uses of Diodorus in Renaissance Scholarship: From Annius of Viterbo to Johannes Boemus," in *For the Sake of Learning. Essays in Honor of Anthony Grafton*, ed. Ann Blair and Goeing Anja-Silvia, vol. 2, 2 vols. (Leiden: Brill, 2016), 711–28; C. Philipp E. Nothaft, "Walter Odington's De Etate Mundi and the Pursuit of a Scientific Chronology in Medieval England," *Journal of the History of Ideas* 77, no. 2 (2016): 183–201; C. Philipp E. Nothaft, "Origen, Climate Change, and the Erosion of Mountains in Giles of Lessines's Discussion of the Eternity of the World (c. 1260)," *The Mediaeval Journal* 4, no. 1 (2014): 43–69.

28. Lydia Barnett (Baltimore: Johns Hopkins University Press, 2019).

29. Pratik Chakrabarti, *Inscriptions of Nature: Geology and the Naturalization of Antiquity* (Baltimore: Johns Hopkins University Press, 2020), 5.

30. Ivano Dal Prete, "The Ruins of the Earth: Learned Meteorology and Artisan Expertise in Fifteenth-Century Italian Landscapes," *Nuncius* 33, no. 3 (2018): 415–41.

CHAPTER 1

1. Alden A. Mosshammer, *The Chronicle of Eusebius and Greek Chronographic Tradition* (Lewisburg, PA: Bucknell University Press, 1979), 31.

2. Mosshammer, 41–42. On the tradition and reconstruction of Eusebius's text, see also: R. W. Burgess, *Studies in Eusebian and Post-Eusebian Chronography: 1. The "Chronici Canones" of Eusebius of Caesarea: Structure, Content and Chronology, AD 282–325; 2. The "Continuatio Antiochiensis Eusebii": A Chronicle of Antioch and the Roman Near East during the Reigns of Constantine and Constantius II, AD 325–350* (Stuttgart: Franz Steiner, 1999), 21–27.

3. Mosshammer, *The Chronicle of Eusebius and Greek Chronographic Tradition*. See also: Anthony Grafton and Megan Hale Williams, *Christianity and the Transformation of the Book: Origen, Eusebius, and the Library of Caesarea* (Cambridge, MA: Belknap Press of Harvard University Press, 2006).

4. On Joseph Scaliger, see the classic work by Anthony Grafton, *Joseph Scaliger: A Study in the History of Classical Scholarship* (Oxford: Clarendon Press; Oxford University Press, 1983).

5. C. Philipp E. Nothaft, *Dating the Passion: The Life of Jesus and the Emergence of Scientific Chronology (200–1600)* (Leiden: Brill, 2011), 9–11.

6. Nothaft, 12.

7. Eusebius Caesariensis, *Chronicon Bipartitum Nunc Primum Ex Armeniaco Textu in Latinum Conversum Opera P. Jo. Baptistae Aucher . . .*, ed. and trans. Johannes Baptista Aucher (Venice: Typis Coenobii PP. Armenorum in insula S. Lazari, 1818), 6–9.

8. Eusebius Caesariensis, 97–98; "Itaque placet Aegyptiis, priscis temporibus, quae paecesserunt diluvium, se jactare ob antiquitatem": Eusebius Caesariensis, 98. It is worth remembering that the Latin text I quote is a modern version published together with the original text in Armenian.

9. Eusebius Caesariensis, *Chronicon Bipartitum Nunc Primum Ex Armeniaco Textu in Latinum Conversum Opera P. Jo. Baptistae Aucher . . .*, 54.

10. "Etenim ab ipsa habitatione in Dei, ut nominatur, Paradiso, nemo tempora designare ullo modo poterit. Et quidem, ut mihi videtur, ipse quoque mirabilis Moyses Divino Spiritu de meliore quodam habitaculo, quam nostri orbis, de felicissimae, iniquam, mansionis divina viatae ratione annuit: namque Paradisum nuncupat primam generis huamani habitationem; erat enim in Paradiso Adam, dulci ac felicitatis plena fruens vita supra humanam conditione: quo omnino quidem de universo humano genere declarasse compertum est." Eusebius Caesariensis, *Chronicon Bipartitum*

Nunc Primum Ex Armeniaco Textu in Latinum Conversum Opera P. Jo. Baptistae
Aucher . . . , 54.

11. Eusebius Caesariensis, 54.

12. "Primam vero [chronologiam], quaecumque ea fuerit historia, quasi ineffabilem, a
successivis temporibus continuatis disjunctam relinquet": Eusebius Caesariensis, 55.

13. On Origen, and particularly Origen's library, see: Grafton and Williams,
Christianity and the Transformation of the Book, chapters 1 and 2.

14. Mark J. Edwards, "Origen," in *The Stanford Encyclopedia of Philosophy*, ed. Edward
N. Zalta, Spring 2014 (Metaphysics Research Lab, Stanford University, 2014),
https://plato.stanford.edu/archives/spr2014/entries/origen/.

15. As quoted in: Michael J. Crowe, *The Extraterrestrial Life Debate, Antiquity to
1915: A Source Book* (Notre Dame, IN: University of Notre Dame, 2008), 15.

16. The condemnation was issued by a local synod summoned in Constantinople
at the behest of Emperor Justinian. There is little evidence, though, that it was
ever approved by a pope in an ecumenical council. The "Origenism" that was
anathemized at the fifth ecumenical council (553) seems to have had little to do
with Origen's original doctrines.

17. Edwards, "Origen."

18. Augustine, *On the City of God*, XI, 8. From: William A. Christian, "Augustine on
the Creation of the World," *The Harvard Theological Review* 46, no. 1 (1953): 4.

19. Christian, "Augustine on the Creation of the World," 5; Scott A. Dunham, *The
Trinity and Creation in Augustine: An Ecological Analysis* (Albany, NY: State
University of New York Press, 2008), 71.

20. Dunham, *The Trinity and Creation in Augustine*, 64. Christian, "Augustine on the
Creation of the World," 16.

21. Dunham, *The Trinity and Creation in Augustine*, 64.

22. Christian, "Augustine on the Creation of the World," 17.

23. Dunham, *The Trinity and Creation in Augustine*, 60.

24. Dunham, 60; Christian, "Augustine on the Creation of the World," 14.

25. Dunham, *The Trinity and Creation in Augustine*, 60.

26. Theologians have even argued that the doctrine of the creation from nothing
(*ex nihilo*) emerged in early Christianity around the second century, but is not
demanded by the biblical text itself. See: Gerhard May, *Creatio Ex Nihilo: The
Doctrine of "Creation out of Nothing" in Early Christian Thought*, trans. A. S.
Worrall (Edinburgh: T&T Clark, 1994).

27. Richard C. Dales, *Medieval Discussions of the Eternity of the World* (Leiden: E. J.
Brill, 1990), 4.

28. In: Richard Sorabji, ed., *The Philosophy of the Commentators, 200–600 AD: Physics*,
vol. 2 (Cornell University Press, 2005), 138. The editor argues that a cushion would
have been an even more apt comparison, since the causality would be established by
the fact that "the cushion *would* spring back to shape, if the foot ever *were* removed"
(emphasis by the author).

29. On Calcidius and his commentary, see the recent edition by John Magee, *Calcidius, On Plato's Timaeus*, ed. and trans. John Magee (Cambridge, MA: Harvard University Press, 2016).

30. In: Dales, *Medieval Discussions of the Eternity of the World*, 10.

31. J. Magee has found no conclusive evidence that Calcidius was a Christian. Calcidius, *On Plato's Timaeus*, x–xiii. See also: Anna Somfaj, "Calcidius' 'Commentary' on Plato's 'Timaeus' and Its Place in the Commentary Tradition: The Concept of 'Analogia' in Text and Diagrams," *Bulletin of the Institute of Classical Studies. Supplement*, no. 83 (2004): 204.

32. Dales, *Medieval Discussions of the Eternity of the World*, 13.

33. Dales, 16.

34. Dales, 14. On eternity in Boethius, see also: Brian Leftow, "Boethius on Eternity," *History of Philosophy Quarterly* 7, no. 2 (1990): 123–42.

35. Dales, *Medieval Discussions of the Eternity of the World*, 16.

36. Dales, 17.

37. Kenneth B. Wolf, ed., *Chronicon, Isidore of Seville, c. 616*, Medieval Texts in Translation, 2008, para. 122, canilup.googlepages.com., accessed November 5, 2021.

38. "Alia est heresis, quae dicit incertum numerum esse annorum a mundi origine, et ignorare homines curricula temporum: cum ab Adam usque ad diluvium decem sint generations, et annorum MMCCXLII." Philastrius of Brescia, *De haeresibus liber*, ed. Johannes Albertus Fabricius (Hamburg: Theodor Christopher Felginer, 1721), 228.

39. Andrew Dickson White, *A History of the Warfare of Science with Theology in Christendom* (New York: D. Appleton, 1896), 251. On Augustine's poor opinion of Philastrius, see: Allan Fitzgerald and John C. Cavadini, *Augustine Through the Ages: An Encyclopedia* (Cambridge, UK: Wm. B. Eerdmans, 1999), 369.

40. Peter Hunter Blair, *The World of Bede* (Cambridge University Press, 1990), 267. On Venerable Bede's chronology and the ages of the world, see, above all: Peter Darby, *Bede and the End of Time* (Farnham, Surrey; Burlington, VT: Ashgate, 2012), 17–34. Also: Nothaft, *Dating the Passion*, 81–88; R. A. Markus, *Bede and the Tradition of Ecclesiastical Historiography*, Jarrow Lecture 1975 (Jarrow on Tyne [Eng.]: St. Paul's Rectory, 1975).

41. In: Dales, *Medieval Discussions of the Eternity of the World*, 18–19.

42. C. Philipp E. Nothaft, "Walter Odington's De Etate Mundi and the Pursuit of a Scientific Chronology in Medieval England," *Journal of the History of Ideas* 77, no. 2 (2016): 187.

43. Jon McGinnis, *Avicenna* (New York: Oxford University Press, 2009), 181.

44. McGinnis, 196.

45. See also: Dales, *Medieval Discussions of the Eternity of the World*, 43.

46. Jon McGinnis, "Arabic and Islamic Natural Philosophy and Natural Science," in *The Stanford Encyclopedia of Philosophy*, ed. Edward N. Zalta, Fall 2015 (Metaphysics

Research Lab, Stanford University, 2015), https://plato.stanford.edu/archives/fall2015/entries/arabic-islamic-natural/.

47. Frank Griffel, "Al-Ghazali," in *The Stanford Encyclopedia of Philosophy*, ed. Edward N. Zalta, Winter 2016 (Metaphysics Research Lab, Stanford University, 2016), https://plato.stanford.edu/archives/win2016/entries/al-ghazali/.

48. Hillier, "Ibn Rushd (Averroes)," *Internet Encyclopedia of Philosophy*, http://www.iep.utm.edu/ibnrushd/#H3.

49. "It is well known that Arabico-Latin translators were assisted by Mozarabs" (that is, Iberian Christians who lived under Muslim rule and spoke a romance language). Pieter L. Schoonheim, ed., *Aristotle's Meteorology in the Arabico-Latin Tradition: Critical Edition of the Texts, with Introduction and Indices*, trans. Pieter L. Schoonheim (Leiden: Brill, 2000), xxvii–xxviii.

50. For an overview of the translation movement of the twelfth century, see: Maire-Therese d'Alverny, "Translations and Translators," in *Renaissance and Renewal in the Twelfth Century*, ed. Robert Louis Benson, Giles Constable, and Carol Dana Lanham (Cambridge, MA: Harvard University Press, 1982), 421–62.

51. Brian Stock, *Myth and Science in the Twelfth Century; a Study of Bernard Silvestris* (Princeton University Press, 2015), 11. Among the ample literature on Bernard Silvestris's *Cosmographia*: Mark Kauntze, *Authority and Imitation: A Study of the Cosmographia of Bernard Silvestris* (Leiden; Boston: Brill, 2014). Mark Kauntze, "The Creation Grove in the 'Cosmographia' of Bernard Silvestris," *Medium Ævum* 78, no. 1 (2009): 16–34. Bernard Silvestris, *Cosmographie*, ed. and trans. Michel Lemoine (Paris: Cerf, 1998). Also the above mentioned and recently reprinted classic by Brian Stock: Stock, *Myth and Science in the Twelfth Century; a Study of Bernard Silvestris* (first published in 1972); Bernardus Silvestris, *Cosmographia*, ed. Peter Dronke (Brill, 1978); Peter Dronke, "Bernard Silvestris, Natura, and Personification," *Journal of the Warburg and Courtauld Institutes* 43 (1980): 16–31. Silvestris's work has been recently translated into English: Bernard Silvestris, *Poetic Works*, ed. and trans. Winthrop Wetherbee (Cambridge, Massachusetts: Harvard University Press, 2015).

52. From the unpublished translation by Katharine Park: Thierry of Chartres, "Treatise of the Work of the Six Days," trans. Katharine Park, https://www.academia.edu/31388090/Thierry_of_Chartres-Treatise_Six_Days-trans._Park.pdf. Park's translation is based on the edition of the Latin text, which I have also used, published by N. Haring, "The Creation and Creator of the World According to Thierry of Chartres and Clarenbaldus of Arras," *Archives d'histoire Doctrinale et Litteraire Du Moyen Age* 22 (1955): 137–216.

53. Dales, *Medieval Discussions of the Eternity of the World*, 35.

54. Helen Rodnite Lemay, "Science and Theology at Chartres: The Case of the Supracelestial Waters," *The British Journal for the History of Science* 10, no. 3 (1977): 228.

55. Thierry of Chartres, "Treatise of the Work of the Six Days," 2.

56. Thierry of Chartres, 4, translator's emphasis; Haring, "The Creation and Creator of the World According to Thierry of Chartres and Clarenbaldus of Arras," 187. The editor of the Latin text agrees that the first days of the creation may not be conventional days: "Thierry [. . .] does not explicitly claim that Moses uses the word to designate a 24-hour period." Haring, 156.

57. Lemay, "Science and Theology at Chartres," 228.

58. Dales, *Medieval Discussions of the Eternity of the World*, 27–31.

59. R. Dales does not mention Thierry de Chartres in his 1965 edition of this treatise; however, in his 1990 survey of medieval discussions on the eternity of the world he points out that *On the Elements* appears "in many ways an elaboration of Thierry de Chartres's *De sex dierum operibus* ('Treatise on the Work of the Six Days'), with which it has numerous points of contact." Dales, 31. See also: Richard C. Dales, "Anonymi De Elementis: From a Twelfth-Century Collection of Scientific Works in British Museum MS Cotton Galba E. IV," *Isis* 56, no. 2 (1965): 174–89.

60. Manuscripts of Lucretius's *On the Nature of Things* were listed in the libraries of Murbach and Bobbio in the eighth century, and in three more (all in France) in the twelfth century. In spite of that, the text remained basically unknown before the 1417 rediscovery by Poggio Bracciolini. Alison Brown, *The Return of Lucretius to Renaissance Florence* (Cambridge, Mass: Harvard University Press, 2010), 1–3.

61. In Edward Grant, *A Source Book in Medieval Science* (Cambridge, MA: Harvard University Press, 1974), 25.

62. Griffel, "Al-Ghazali." In both *On the Elements* and in Al-Ash'ari's theological school (to which even Al-Ghazali belonged), the atoms that compose material things have no attributes or qualities, like geometrical shapes. The treatise also discusses the possibility that the Earth could spins on its axis—although this is considered unlikely—a problem treated in Islamic sources such as Al-Biruni (973–1048).

63. Dales, "Anonymi De Elementis," 179.

64. Dales, 177.

65. Dales, 182. Part of the ambiguity comes from the Latin expression *ab eterno*, which can mean either "eternally" or "from what is eternal."

66. Dales, *Medieval Discussions of the Eternity of the World*, 32–33.

67. Dales, 33.

68. Kauntze, *Authority and Imitation*, 94. Kauntze's assessment is arguably the most reasonable, as well as the most recent, in a range that has seen the *Cosmographia* oscillate between "one of the last religious experiences of the late-pagan world" (Ernst Robert Curtius) to a work that simply employs "pagan cosmologies to elucidate the creation story of Genesis" (Étienne Gilson). Dronke, "Bernard Silvestris, Natura, and Personification," 16; Kauntze, "The Creation Grove in the 'Cosmographia' of Bernard Silvestris," 16–17.

69. From: Dales, *Medieval Discussions of the Eternity of the World*, 34.

70. Stock, *Myth and Science in the Twelfth Century; a Study of Bernard Silvestris*. On Silvestris's readers and imitators, see, above all: Kauntze, *Authority and Imitation*, 133–71.

71. On Silvestris and his relation with the school of Chartres, see: Kauntze, *Authority and Imitation*, 25–49.

72. Dales, *Medieval Discussions of the Eternity of the World*, 35.

73. Kauntze, *Authority and Imitation*, 91.

74. Silvestris, *Cosmographia*, 2; Kauntze, *Authority and Imitation*, 91.

75. For an overview of twelfth-century historiography, see: Peter Classen, "Res Gestae, Universal History, Apocalypse," in *Renaissance and Renewal in the Twelfth Century*, ed. Robert Louis Benson, Giles Constable, and Carol Dana Lanham (Cambridge, MA: Harvard University Press, 1982), 421–62.

76. Nothaft, *Dating the Passion*, 114.

77. Nothaft, 16.

78. Nothaft, 118.

79. On the elaboration of the modern category of "religion," see: Brent Nongbri, *Before Religion: A History of a Modern Concept* (New Haven: Yale University Press, 2013). For a critical review of Nongbri's book, and of religion as a modern concept: James Broucek, "Thinking About Religion Before 'Religion': A Review of Brent Nongbri's 'Before Religion: A History of a Modern Concept,'" *Soundings: An Interdisciplinary Journal* 98, no. 1 (2015): 98–125. A more concise discussion, with particular reference to Christianity and Western civilization, is in: Peter Harrison, "'Science' and 'Religion': Constructing the Boundaries," in *Science and Religion: New Historical Perspectives*, ed. Thomas Dixon, G. N. Cantor, and Stephen Pumfrey (New York: Cambridge University Press, 2010), 23–49.

80. Richard Dales concludes his study of medieval eternalism by arguing that "It was not the reacquisition of Aristotle's natural philosophy which occasioned disputations about the eternity of the world; rather it was Lombard's *Sententiae*, whose sources were patristic, not Aristotelian" (Dales, *Medieval Discussions of the Eternity of the World*, 259). I believe that the author does not intend to deny the impact of Aristotle's works, but rather to emphasize the continuing influence, throughout the thirteenth century, of traditional sources that previous historiography may have understated. See for example, p. 85: "During the 1240s and 1250s Aristotle's natural philosophy became a much more noticeable feature of the debates [on the eternity of the world], but the older arguments and the replies to them still maintained their place."

81. Dales, 39–42.

82. Luca Bianchi, *Censure et liberté intellectuelle à l'université de Paris: XIIIe-XIVe siècles* (Paris: Belles lettres, 1999), 92 and following. On censorship at the University of Paris in the Middle Ages, see also: Johannes M. M. H. Thijssen, *Censure and Heresy at the University of Paris, 1200–1400* (Philadelphia: University of Pennsylvania Press, 1998).

83. "Creator omnium invisibilium et visibilium, spiritualium et corporalium, qui sua omnipotenti virtute simul ab initio temporis utramque de nihilo condidit creaturam, spiritualem et corporalem, angelicam videlicet et mundanam ac deinde humanam quasi communem ex spiritu et corpore constitutam." English translation in: Norman P. Tanner, ed., *Decrees of the Ecumenical Councils*, vol. 1 (London; Washington: Sheed & Ward; Georgetown University Press, 1990), 230.

84. Dales, *Medieval Discussions of the Eternity of the World*, 50.

85. The list was first translated into English by Lynn Thorndyke and later reprinted in: Grant, *A Source Book in Medieval Science*, 44. For a detailed study of censorship in the twelfth and thirteenth centuries at the university of Paris, see: Bianchi, *Censure et Liberté Intellectuelle à l'université de Paris*.

86. Dales, *Medieval Discussions of the Eternity of the World*, 57, 63–64.

87. On Christian readings of Aristotle, see in particular: Luca Bianchi, ed., *Christian Readings of Aristotle from the Middle Ages to the Renaissance* (Turnhout: Brepols, 2011).

88. Dales, *Medieval Discussions of the Eternity of the World*, 71.

89. J. D. North, "Chronology and the Age of the World," in *Cosmology, History, and Theology*, ed. Wolfgang Yourgrau and Allen D. Breck (Springer US, 1977), 319.

90. "Those masters who perceived anything dangerous in Aristotle's natural philosophy were in the clear minority, and most intellectuals of whatever stamp cited these works as authorities, with little or no hint that they were in any way dangerous." Dales, *Medieval Discussions of the Eternity of the World*, 259.

91. In Dales, 76.

92. The elaboration and reception of this doctrine has been a major theme in the history of medieval philosophy, and it enjoys a vast literature. See in particular: Dales, *Medieval Discussions of the Eternity of the World*; Luca Bianchi, *L'errore di Aristotele: la polemica contro l'eternità del mondo nel XIII Secolo* (Firenze: La Nuova Italia, 1984); Luca Bianchi, *Il Vescovo e i Filosofi: la condanna parigina del 1277 e l'evoluzione dell'aristotelismo scolastico* (Bergamo: Lubrina, 1990); Boethius of Dacia, *Sull'eternità del mondo*, ed. and trans. Luca Bianchi (Milano: UNICOPLI, 2003); Luca Bianchi, *Pour une histoire de la "double vérité"* (Paris: Vrin, 2008).

93. Dales, *Medieval Discussions of the Eternity of the World*, 147. On Boethius, see also: Boethius of Dacia, *Sull'eternità del mondo*.

94. Boethius of Dacia, *Sull'eternità del mondo*, 38–42. Dales, *Medieval Discussions of the Eternity of the World*, 107.

95. For a reconstruction of the historiographical debate, see: Bianchi, *Pour une histoire de la "double vérité."*

96. Jozef Wissink, ed., *The Eternity of the World in the Thought of Thomas Aquinas and His Contemporaries* (Leiden; New York: E.J. Brill, 1990), 1–8.

97. The problem of whether or not the beginning of the world is rationally demonstrable was fiercely debated. Bonaventura of Bagnoregio, most Franciscans, and even some Dominicans sided against Aquinas. See in particular: Bianchi, *L'errore*

di Aristotele, 115–61. R. Dales points out, however, that it is impossible to speak of a Franciscan, a Dominican, or a secular position: Dales, *Medieval Discussions of the Eternity of the World*, 258–59.

98. Bianchi, *Il vescovo e i filosofi*, 21–25.
99. Andrew E. Larsen, *The School of Heretics: Academic Condemnation at the University of Oxford, 1277–1409* (Leiden; Boston: Brill, 2011), 26–41.
100. Bianchi, *Il vescovo e i filosofi*, 27.
101. Dales, *Medieval Discussions of the Eternity of the World*, 176. For a more conservative view of the 1277 condemnation, see: Thijssen, *Censure and Heresy at the University of Paris, 1200–1400*, 40–41.
102. Dales, *Medieval Discussions of the Eternity of the World*, 261.
103. Dales, 196. See also: Richard Clark Dales and Omar Argerami, *Medieval Latin Texts on the Eternity of the World* (Leiden: Brill, 1991), 41–42.
104. Bianchi, *Il vescovo e i filosofi*, 29–30. Thijssen, *Censure and Heresy at the University of Paris, 1200–1400*, 52. Bianchi and Thijssen point out that the 1277 condemnations were only revoked "insofar they touch, or are believed to touch upon the doctrine of Saint Thomas." Bianchi, *Il Vescovo e i Filosofi*, 28. This does not mean that they were acknowledged as true, but they could certainly be taught.
105. On the use of *novitas mundi* as opposed to its eternity in Thomas Aquinas, see Patricia Clare Ingham, *The Medieval New: Ambivalence in an Age of Innovation* (Philadelphia: University of Pennsylvania Press, 2015), 32–36.

CHAPTER 2

1. Only 3 of the 4 books of the *Meteorology* deal with properly meteorological phenomena. The fourth book discusses the structure of matter, and is usually considered a separate treatise that for some reason was bundled together with the meteorological books. See: Malcolm Wilson, *Structure and Method in Aristotle's Meteorologica: A More Disorderly Nature* (Cambridge: Cambridge University Press, 2013).
2. Aristotle, *Meteorologica*, trans. H. D. P. Lee (Cambridge, MA: Harvard University Press, 2014), 5.
3. Aristotle, *Meteorologica*, 108–9.
4. "As proof of former submersion by the waters of the sea, there are deposits of pebbles, shells and various objects of the type that are usually thrown up with the foam on seashores." Theophrastus, fragment XXX, 3; in François Ellenberger, *History of Geology*, vol. 1 (Rotterdam; Brookfield, VT: A.A. Balkema, 1996), 15.
5. Aristotle, *Meteorologica*, 352a. Emphasis mine.
6. On the origin of life, mankind, and civilization in Lucretius, see: Gordon Lindsay Campbell, *Lucretius on Creation and Evolution: A Commentary on De Rerum Natura, Book 5, Lines 772–1104* (Oxford: Oxford University Press, 2003);

7. Titus Lucretius Carus, *On the Nature of Things*, trans. W. H. D. Rouse and Martin Ferguson Smith (Cambridge, MA: Harvard University Press, 2014), 403–04.

8. Plato, *Timaeus*, trans. Benjamin Jowett (Blacksburg, VA: Virginia Tech, 2001), 5.

9. John Sallis, *Chorology: On Beginning in Plato's Timaeus* (Bloomington: Indiana University Press, 1999), 42.

10. Aristotle, *Meteorologica*, 351b.

11. On the sources of Diodorus's chapters on the creation of the universe, see: Kenneth Sacks, *Diodorus Siculus and the First Century* (Princeton, NJ: Princeton University Press, 1990), 55–66.

12. Diodorus Siculus, *Diodorus Siculus: Library of History, Volume I, Books 1-2.34*, trans. C. H. Oldfather (Cambridge, MA: Harvard University Press, 1933), 37.

13. Siculus, 73.

14. On the origin and early transmission of Noah's story: Norman Cohn, *Noah's Flood: The Genesis Story in Western Thought* (New Haven, CT: Yale University Press, 1996), 1–31.

15. On the influence of Lucretius in Ovid's poetry, see: K. Sara Myers, *Ovid's Causes: Cosmogony and Aetiology in the Metamorphoses* (Ann Arbor: University of Michigan Press, 1994); Stephen M. Wheeler, "Imago Mundi: Another View of the Creation in Ovid's Metamorphoses," *The American Journal of Philology* 116, no. 1 (1995): 95–121; Philip Hardie, "The Speech of Pythagoras in Ovid Metamorphoses 15: Empedoclean Epos," *The Classical Quarterly* 45, no. 1 (1995): 204–14.

16. Ellenberger, *History of Geology*, 1:12. I chose Ellenberger's version over more philologically authoritative editions, because of the special attention paid by the author to passages relevant to the history of geology.

17. Lucius Annaeus Seneca, *Natural Questions*, trans. Harry M. Hine (Chicago: University of Chicago Press, 2010), 44; Seneca, 50. On the dating of the *Natural Questions*: Seneca, 10.

18. Seneca, *Natural Questions*, 49.

19. Ellenberger, *History of Geology*, 1:59. The English text read "buccins," possibly an inadvertence in the translation from the French original.

20. Ellenberger, 1:60.

21. Ellenberger, 1:60.

22. "The fables of the Greeks say that humankind was recreated from stones by Deucalion, because of the ingrained stoniness of the human heart." Isidore of Seville, *The Etymologies of Isidore of Seville*, trans. Stephen A. Barney (Cambridge: Cambridge University Press, 2006), 283.

23. "Cuius indicium hactenus videmus in lapidibus quos in remotis montibus conchis et ostreis concretos, saepe etiam cavatos aquis visere solemus." Translation from: Isidore of Seville, *The Etymologies of Isidore of Seville*, 282.

24. In: Ellenberger, *History of Geology*, 1:61.

25. On the availability of Seneca's *Natural Questions* in the Middle Ages: Harry M. Hine, "Seneca and Anaxagoras in Pseudo-Bede's De Mundi Celestis Terrestrisque Constitutione," *Viator* 19 (January 1, 1988): 113.

26. See: Barbara Obrist and Irene Caiazzo, eds., *Guillaume de Conches: philosophie et science au XIIe siècle* (Florence: SISMEL edizioni del Galluzzo, 2011).

27. William of Conches, *Philosophicarum et astronomicarum institutionum, Guilielmi Hirsaugiensis olim abbatis, libri tres* (Basel: Heinrich Petrus, 1531), 16. See also: Pierre Duhem, *Le Système Du Monde; Histoire des Doctrines Cosmologiques de Platon à Copernic*, vol. IX (Paris: A. Hermann, 1958), 269–70. For a modern edition: William of Conches, *Philosophia mundi: Ausgabe des. 1. Buchs von Wilhelm von Conches' "Philosophia" mit Anhang, Übersetzung und Anmerkungen*, ed. Grealgor Maurach (Pretoria: University of South Africa, 1974). W. of Conches also dealt with the creation of the elements in the philosophical dialogue *Dragmaticon Philosophiae*, available in English translation: William of Conches, *A Dialogue on Natural Philosophy (Dragmaticon Philosophiae)*, ed. Italo Ronca and Matthew Curr (Notre Dame, IN: University of Notre Dame Press, 1997), 13–29.

28. William of Conches, *Philosophicarum et astronomicarum institutionum, Guilielmi Hirsaugiensis olim abbatis, libri tres*, 23.

29. William of Conches, 23.

30. "Et quomodo communi philosophorum sententia fuit, terrena modo diluvio, modo exustione finiri, unde utriusque contigat videamus." William of Conches, 63. This sentence is a quotation from the chapter on the destruction of the world by water and fire in Seneca's *Natural Questions*.

31. Pieter L. Schoonheim, ed., *Aristotle's Meteorology in the Arabico-Latin Tradition: Critical Edition of the Texts, with Introduction and Indices*, trans. Pieter L. Schoonheim (Leiden: Brill, 2000), viii, xv. Lettinck, 7.

32. Lettinck, *Aristotle's Meteorology and Its Reception in the Arab World*, 3, 7–9. Unfortunately, Lettinck's work deals only marginally with meteorological topics.

33. Pierre Duhem, *Le Système Du Monde; Histoire des Doctrines Cosmologiques de Platon à Copernic*, vol. IX (Paris: A. Hermann, 1958), 253–55. See also: Ellenberger, *History of Geology*, 1:64. The writings of the "Brethren of Purity" are currently being edited and published by Oxford University Press. See in particular: Carmela Baffioni, ed., *Epistles of the Brethren of Purity: On the Natural Sciences: An Arabic Critical Edition and English Translation of Epistles 15–21*, trans. Carmela Baffioni (Oxford: Oxford University Press in association with the Institute of Ismaili Studies, 2013). For a general introduction: Ian Richard Netton, *Muslim Neoplatonists: An Introduction to the Thought of the Brethren of Purity, Ikhwān al-Ṣafāʾ* (London; Boston: G. Allen & Unwin, 1982).

34. The precession of the equinoxes was first discovered by Hipparcus of Nicaea (ca. 190–120 BC). By comparing his observations with those of some predecessors, he suspected that that fixed stars placed around the plane of the Sun's orbit drifted slowly at an estimated rate of 1° per century. Hipparchus's findings were confirmed three centuries later by Ptolemy (ca. 100–170 CE), who attributed the phenomenon to a movement of the whole of the sphere of the fixed stars (the eight sphere in his

system). In modern terms, the rotation axis of the Earth slowly draws a double cone in the sky as it spins, not unlike a spinning top and for similar reasons.

35. Baffioni, *Epistles of the Brethren of Purity*, 231–32.

36. Baffioni, 233. Emphasis mine. Ellenberger's version, which is entirely independent from Baffioni's, also translates "layer upon layer." Ellenberger, *History of Geology*, 1:64.

37. Baffioni, *Epistles of the Brethren of Purity*, 258–59.

38. Duhem, *Le Système Du Monde; Histoire des Doctrines Cosmologiques de Platon à Copernic*, IX:256. Duhem followed the Latin translation in a collection of works erroneously attributed by the Renaissance editors to Aristotle: [Ḥunayn ibn Isḥāq al-ʿIbādī], "Liber de proprietatibus ælementorum," in *Hoc in uolumine continentur infrascripta opera Aristotelis uidelicet: in principio: Vita eiusdem etc.* (Venice: Gregorio de Gregori, 1496).

39. Duhem, *Le Système Du Monde; Histoire des Doctrines Cosmologiques de Platon à Copernic*, IX:256.

40. Duhem, IX:256–57.

41. Avicenna, *Avicennae De Congelatione et Conglutinatione Lapidum; Being Sections of the Kitâb al-Shifâ'*, ed. Eric John Holmyard and Desmond Christopher Mandeville (Paris: P. Guethner, 1927), 8.

42. Avicenna, 27.

43. Avicenna, 17–26.

44. Avicenna, 28.

45. Avicenna, 28–29.

46. Avicenna, 28.

47. Avicenna, 29. See also: Frank Dawson Adams, *The Birth and Development of the Geological Sciences* (Baltimore: Williams & Wilkins, 1938), 334.

48. Gherard of Cremona only translated the first 3 books of the *Meteorology*; the fourth book—of lesser interest for this study—was translated by Henry Aristippus (d. 1162). On Gherard's version and translation techniques, see: Schoonheim, *Aristotle's Meteorology in the Arabico-Latin Tradition*, xvii–xviii, xx–xxviii.

49. Schoonheim has counted 190 surviving manuscripts for Moerbeke's version, against 107 for the "old translation" (other studies give slightly different figures); Schoonheim, xviii. For a modern edition of Moerbeke's translation see: William of Moerbeke, *Meteorologica. Translatio Guillelmi de Morbeka*, ed. Gudrun Vuillemin-Diem, 2 vols. (Turnhout: Brepols, 2008). See in particular vol. 1 for a detailed history of the Aristotelian text and of Moerbeke's translation. On W. of Moerbeke: J. Brams and W. Vanhamel, eds., *Guillaume de Moerbeke: recueil d'études à l'occasion du 700e anniversaire de sa mort (1286)* (Leuven: Leuven University Press, 1989).

50. Dag Nikolaus Hasse, "Latin Averroes Translations of the First Half of the Thirteenth Century," in *Universality of Reason, Plurality of Philosophies in the Middle Ages*, ed. A. Musco (Palermo, 2012), 149–78.

51. Duhem, *Le Système Du Monde; Histoire des Doctrines Cosmologiques de Platon à Copernic*, IX:266–68.

52. This work was also known as *De Congelatione et Conglutinatione Lapidum* ("On the Congelation and Agglutination of Stones"), with reference to the two main modalities of rock formation according to Avicenna: solidification of a fluid by heat (agglutination) or by cold (congelation).

53. See the introduction in Albertus Magnus, *Book of Minerals*, ed. Dorothy Wichoff (Oxford: Clarendon Press, 1967).

54. Aristotle, *Meteorologica*, I–7.

55. For the debate on "probable opinions" in pre-modern Europe, see: Rudolf Schuessler, *The Debate on Probable Opinions in the Scholastic Tradition* (Leiden: Brill, 2019). As Schuessler notes, "an opinion conflicting with a truth of faith would not have been considered probable"; Schuessler, 9.

56. Darrel H. Rutkin, "Astrology and Magic," in *A Companion to Albertus Magnus* (BRILL, 2013), 459. For a systematic treatment of the subject, see now: Darrel H. Rutkin, *Sapientia Astrologica: Astrology, Magic and Natural Knowledge, ca. 1250–1800: I. Medieval Structures (1250–1500): Conceptual, Institutional, Socio-Political, Theologico-Religious and Cultural*, vol. 1, 3 vols. (Springer International Publishing, 2019). Chapter 3, "The Natural Philosophical Foundations for Astrological Revolution" etc., is especially relevant.

57. Aristotle, *Meteorologica*, I–14.

58. D. Rutkin emphasizes Al-Kindi's *On Stellar Rays* as a crucial source for medieval astrology: Rutkin, *Sapientia Astrologica*, 1:72–75.

59. Rutkin, 1:113–15.

60. Alfred of Sareshel, *Alfred of Sareshel's Commentary on the Metheora of Aristotle: Critical Edition, Introduction, and Notes*, ed. James K. Otte (Leiden: Brill, 1987), 22; 28–31. J. Otte notes that while A. of Sareshel "allows a creator of water, earth, vapor and everything included, he is not the Augustinian 'creator ex nihilo.'" Alfred of Sareshel, 81. See also: James K. Otte, "The Role of Alfred of Sareshel (Alfredus Anglicus) and His Commentary on the Metheora in the Reacquisition of Aristotle," *Viator* 7 (January 1, 1976): 197–210; James K. Otte, "The Life and Writings of Alfredus Anglicus," *Viator* 3 (January 1, 1972): 275–92.

61. On Albertus Magnus's physics, medicine, and natural philosophy in general, see the relevant chapters in: Irven Michael Resnick, ed., *A Companion to Albertus Magnus: Theology, Philosophy, and the Sciences* (Leiden: Brill, 2013).

62. Duhem, *Le Système Du Monde; Histoire des Doctrines Cosmologiques de Platon à Copernic*, IX:239. See: Alexander of Aphrodisias, *Alexandre d'Aphrodisias commentaires sur les Météores d'Aristote*, ed. A. J. Smet (Louvain: de Wulf-Mansion Centrum, 1968), vii. By 1275, Alexander of Aphrodisias's commentary was already integral part of the university teaching in Paris; Alexander of Aphrodisias, vii.

63. Albertus Magnus, *Liber Methauroru[m]* (Venice: Impressi p[er] Renaldum de Nouimagio theotonicum, 1488), Liber II, Tractatus III, Capitulum I. See

also: Duhem, *Le Système Du Monde; Histoire des Doctrines Cosmologiques de Platon à Copernic*, IX:121. Pierre Duhem called this theory "puerile," oddly attributing it to Albert even though he rejected it.

64. Albertus Magnus, *Liber Methauroru[m]*, Liber II, Tractatus III, Capitulum I.

65. Albertus Magnus, *On the Causes of the Properties of the Elements (Liber de Causis Proprietatum Elementorum)*, ed. and trans. Irven Michael Resnick (Milwaukee: Marquette University Press, 2010), 48–48 and n. 99. See also: Albertus Magnus, *Opus Nobile de Causis Proprietatum Elementorum* (Magdeburg: Jacob Winther, 1506), chaps. III, "De improbatione opinionum que dicit mare mutari de loco ad locum"; Duhem, *Le Système Du Monde; Histoire des Doctrines Cosmologiques de Platon à Copernic*, IX:272.

66. Albertus Magnus, *On the Causes of the Properties of the Elements (Liber de Causis Proprietatum Elementorum)*, 50–51. Albertus Magnus, *Opus Nobile de Causis Proprietatum Elementorum*, chap. III, "De improbatione opinionum que dicit mare mutari de loco ad locum." In the late Middle Ages the Belgian and Dutch coastline underwent impressive modifications (like the formation of the Zuiderzee) and catastrophic floods that were duly noticed. See: Jean-Pierre Leguay, *Les catastrophes au Moyen Âge* (Paris: Editions Jean-paul Gisserot, 2005), 53–54.

67. Albertus Magnus, *On the Causes of the Properties of the Elements (Liber de Causis Proprietatum Elementorum)*, 117. See also his description of fossils and their origin (which is entirely taken from Avicenna) in: Albertus Magnus, *Book of Minerals*, 52–53.

68. Albertus Magnus, *On the Causes of the Properties of the Elements (Liber de Causis Proprietatum Elementorum)*, 117.

69. Albertus Magnus, 87.

70. According to Henryk Anzulewicz, Albertus Magnus's work was in fact a milestone in the demarcation of the external and internal boundaries of theology, and in its differentiation from philosophy and the other sciences. Henryk Anzulewicz, "The Systematic Theology of Albertus Magnus," in *A Companion to Albertus Magnus: Theology, Philosophy, and the Sciences*, ed. Irven Michael Resnick (Leiden: Brill, 2013), 18.

71. Albertus Magnus, *On the Causes of the Properties of the Elements (Liber de Causis Proprietatum Elementorum)*, 71.

72. Albertus Magnus, *Opus Nobile de Causis Proprietatum Elementorum*, "Capitulum Nonum." For a more detailed study of the astrological causes of floods of water and fire in Albertus Magnus: Rutkin, *Sapientia Astrologica*, 1:100–104; 110–13.

73. Albertus Magnus, *On the Causes of the Properties of the Elements (Liber de Causis Proprietatum Elementorum)*, 85. Albertus Magnus, *Opus Nobile de Causis Proprietatum Elementorum*, "Capitulum xii."

74. Albert held that Deucalion's flood was the same as Noah's, not a later and much smaller one as established by most Christian chronologists. Albertus Magnus, *On the Causes of the Properties of the Elements (Liber de Causis Proprietatum*

Elementorum), 71. See also: Albertus Magnus, *Opus Nobile de Causis Proprietatum Elementorum*, "Capitulum Nonum".

75. Albertus Magnus, *On the Causes of the Properties of the Elements (Liber de Causis Proprietatum Elementorum)*, 91–93. Albertus Magnus, *Opus Nobile de Causis Proprietatum Elementorum*, "Capitulum xiii."

76. Albertus Magnus, *On the Causes of the Properties of the Elements (Liber de Causis Proprietatum Elementorum)*, 93.

77. Bruno Nardi argued instead that: (a) medieval Christian philosophers saw a contradiction between the "Aristotelian doctrine that places water and the other elements concentric to the earth, and the biblical statement that God himself gathered all of the water in a single place"; and (b) the problem was in fact nonexistent for the Greeks and the Arabs. Bruno Nardi, *La caduta di Lucifero: e l'autenticità della "Quaestio de aqua et terra"* (Turin: Società Editrice Internazionale, 1959), 434. I believe Duhem provides ample evidence that (b) is not accurate. As for (a), I would argue that the contradiction was not between Aristotle's theory of natural place and the Bible, but between Aristotle's theory of natural place and the obvious observation that dry land does exist—whatever or whoever the reason.

78. Aristotle supposed—just supposed—that it may take ten measures of air to make one measure of water, and vice versa; many commentators inferred that the proportion between water and earth, air and water, and air and fire should be of the same order. Duhem, *Le Système Du Monde; Histoire des Doctrines Cosmologiques de Platon à Copernic*, IX:91–96.

79. The subject was considerably more complex than the summary I give here. Chapters XVI, XVII, and XVIII in the ninth volume of Pierre Duhem's *Le Système Du Monde* (1913) remain the most comprehensive study available. See also: Dante Alighieri, "Questio de Aqua et Terra," in *La Letteratura Italiana - Storia e Testi*, ed. Francesco Mazzoni, vol. V (Naples: Ricciardi, 1978), 691–880; Giuseppe Boffito, *Intorno alla "Quaestio de aqua et terra" attribuita a Dante* (Carlo Clausen, 1902). Klaus Anselm Vogel, "Sphaera Terrae - Das Mittelalterliche Bild der Erde und die Kosmographische Revolution" (Göttingen: University of Göttingen, 1995).

80. Among them one can list William of Auvergne (ca. 1180/90–1249), bishop of Paris in the last 20 years of his life; philosopher Michael Scot; and astronomer Campanus of Novara. Duhem, *Le Système Du Monde; Histoire des Doctrines Cosmologiques de Platon à Copernic*, IX:109–30.

81. From: Duhem, IX:133.

82. From: Duhem, IX:133.

83. Alighieri, "Questio de Aqua et Terra," 699.

84. Alighieri, 713.

85. John Freccero, "Satan's Fall and the 'Quaestio de Aqua et Terra,'" *Italica* 38, no. 2 (1961): 99–115. For a more recent discussion of realism in Dante's *Divine Comedy* (of which *Inferno* is the first part) see: Teodolinda Barolini, *The Undivine Comedy: Detheologizing Dante* (Princeton, NJ: Princeton University Press, 1992).

86. Albertus Magnus, *On the Causes of the Properties of the Elements (Liber de Causis Proprietatum Elementorum)*, 87.

87. Since it was considered impossible to pass the tropical and equatorial regions, there was no way that human beings living there could receive the Christian message. Belief in the habitability of the antipodes could even be considered heretical. See: Giuseppe Boffito, *L'Eresia Degli Antipodi* (Florence: Ist. alla Querce, 1905).

88. On the rise of theories of a central fire in early modern Europe, and on their theological implications: Luca Ciancio, "An Amphitheatre Built on Toothpicks: Galileo, Nardi and the Hypothesis of Central Fire," *Galilaeana* XV (2018): 83–113; Luca Ciancio, "'Immoderatus Fervor Ad Intra Coërcendus': Reactions to Athanasius Kircher's Central Fire in Jesuit Science and Imagination," *Nuncius* 33, no. 3 (2018): 464–504; Rienk Vermij, "Subterranean Fire. Changing Theories of the Earth During the Renaissance," *Early Science and Medicine* 3, no. 4 (1998): 323–47.

89. "Che 'l mare v'è assai più alto che la terra, e tiensi sì in fra sè, che non cade né corre sopra la terra." Brunetto Latini, *Il tesoro di Brunetto Latini volgarizzato da Bono Giamboni, nuovamento pubblicato secondo l'edizione del MDXXXIII*, trans. Bono Giamboni (Venice: Co' tipi del Gondoliere, 1839), 175.

90. For the sake of clarity, I am using here a modern terminology. Medieval texts would rather say "the center of its weight" and "the center of its magnitude."

91. C. Philipp E. Nothaft, "Origen, Climate Change, and the Erosion of Mountains in Giles of Lessines's Discussion of the Eternity of the World (c. 1260)," *The Mediaeval Journal* 4, no. 1 (January 1, 2014): 51–52. For a detailed discussion of this theory: C. Philipp E. Nothaft, "Climate, Astrology and the Age of the World in Thirteenth-Century Thought: Giles of Lessines and Roger Bacon on the Precession of the Solar Apogee," *Journal of the Warburg and Courtauld Institutes* LXXVII (2014): 35–60.

92. From: Nothaft, "Origen, Climate Change, and the Erosion of Mountains in Giles of Lessines's Discussion of the Eternity of the World (c. 1260)," 69.

93. "Quod redeuntibus corporibus caelestibus omnibus in idem punctum, quod fit in XXX sex millibus annorum, redibunt idem effectus qui sunt modo." English translation from: Edward Grant, *A Source Book in Medieval Science* (Cambridge, MA: Harvard University Press, 1974), 13. This proposition is n. 6 in Grant's list. Latin text from Roland Hissette, *Enquête sur les 219 articles condamnés à Paris le 7 mars 1277* (Peeters Publishers & Booksellers, 1977), 157.

94. Richard C. Dales and Omar Argerami, *Medieval Latin Texts on the Eternity of the World* (Brill, 1991), 123. See also: Richard C. Dales, *Medieval Discussions of the Eternity of the World* (Leiden: E. J. Brill, 1990), 180–81.

95. Theophrastus (371–287 BC), Aristotle's closest pupil and collaborator, argued that the fire and heat produced within the Earth could inflate its surface, making up in the long run for the erosive action of rains and floods. Theophrastus's writings, however, were not widely divulged in Europe until the fifteenth century. See: Duhem, *Le Système Du Monde; Histoire des Doctrines Cosmologiques de Platon à Copernic*, IX:244. On medieval theories on the formation of

mountains, with particular reference to the French context: Ducos, Joelle, "Entre Terre, Air at Eau: La Formation Des Montagnes," in *La Montagne Dans Le Texte Médiéval: Entre Mythe et Réalité*, ed. Claude Alexandre Thomasset and Danièle James-Raoul (Paris: Presses de l'Université de Paris-Sorbonne, 2000), 19–52.

96. Jack Zupko, *John Buridan: Portrait of a Fourteenth-Century Arts Master* (Notre Dame, IN: University of Notre Dame Press, 2003). Johannes M. M. H. Thijssen and Jack Zupko, eds., *The Metaphysics and Natural Philosophy of John Buridan* (Leiden: Brill, 2001). Gyula Klima, *John Buridan* (Oxford: Oxford University Press, 2008).

97. Edith Dudley Sylla, "Ideo Quasi Mendicare Oportet Intellectum Humanum: The Role of Theology in John Buridan's Natural Philosophy," in *The Metaphysics and Natural Philosophy of John Buridan*, ed. Johannes M. M. H. Thijssen and Jack Zupko (Brill, 2001), 221.

98. Jean Buridan, *Ioannis Buridani Expositio et quæstiones in Aristotelis De cælo*, ed. Benoît Patar (Louvain-La-Neuve: Editions Peeters, 1996), 416. See also Patar's discussion in Buridan, 168–69.

99. For a chronology of Jean Buridan's life and works: Buridan, *Ioannis Buridani Expositio et quæstiones in Aristotelis De cælo*, 13–19, 102–18. For the dating of the *Meteorology* see: Buridan, 19. Buridan's *Meteorology* has never been published, but see Sylvie Bagès 1986 doctoral dissertation: Sylvie Bagès, "Les Questiones super tres libros Metheororum Aristotelis de Jean Buridan étude suivie de l'édition du livre I" (Paris, École nationale des chartes, 1986). Duhem published a partial French translation in: Duhem, *Le Système Du Monde; Histoire des Doctrines Cosmologiques de Platon à Copernic*, IX:293–305. While I have drawn upon Duhem's work, I have checked his translation and analysis on copies of the relevant pages of the original manuscript provided by the BNF: Jean Buridan, "Johannis Buridani questiones in Aristotelis physica et in libros metheororum" (n.d.), BNF, lat. 14.723.

100. Joel Kaye, *A History of Balance, 1250–1375: The Emergence of a New Model of Equilibrium and Its Impact on Thought* (Cambridge: Cambridge University Press, 2014), 442–62; Duhem, *Le Système Du Monde; Histoire des Doctrines Cosmologiques de Platon à Copernic*, IX:196–202.

101. Buridan, *Ioannis Buridani Expositio et quæstiones in Aristotelis De cælo*, 417.

102. See also J. Kaye's criticism of the literature that terms Buridan's model as "mechanical": Kaye, *A History of Balance, 1250–1375*, 451–55.

103. "Centum mille millium annorum". Buridan, "Johannis Buridani questiones in Aristotelis physica et in libros metheororum," 201v, column B. Duhem, *Le Système Du Monde; Histoire des Doctrines Cosmologiques de Platon à Copernic*, IX:295–96.

104. Jean Buridan, *Iohannis Buridani Quaestiones Super Libris Quattuor de Caelo et Mundo*, ed. Ernest A. Moody (Cambridge, MA: The Mediaeval Academy of America, 1942), 160.

105. Duhem, *Le Système Du Monde; Histoire des Doctrines Cosmologiques de Platon à Copernic*, IX:303. One League of Paris corresponded to about 3,250 meters.

106. Ernest A. Moody, "Buridan, Jean," in *Dictionary of Scientific Biography*, ed. Charles Coulston Gillispie, vol. 2 (New York: Scribner, 1970), 603. On the intellectual exchange between d'Autrecourt and Buridan: Zupko, *John Buridan*, 183–202. On d'Autrecourt's thought and his condemnation: Luca Bianchi, *L'inizio dei tempi: antichità e novità del mondo da Bonaventura a Newton* (Florence: Olschki, 1987), 52–55. See also: Luca Bianchi, *Pour une histoire de la "double vérité"* (Paris: Vrin, 2008), 105–8.

107. From: Sylla, "Ideo Quasi Mendicare Oportet Intellectum Humanum," 235.

108. Moody, "Buridan, Jean," 605. Edward Grant, "Jean Buridan and Nicole Oresme on Natural Knowledge," *Vivarium* 31, no. 1 (1993): 88.

109. Lucian Petrescu, "The Threefold Object of the Scientific Knowledge. Pseudo-Scotus and the Literature on the Meteorologica in Fourteenth-Century Paris," *Franciscan Studies* 72, no. 1 (October 22, 2016): 477–78.

110. Petrescu, 479.

111. Joëlle Ducos, *La Météorologie En Français Au Moyen Age (XIIIe-XIVe Siècles)* (Paris: Honoré Champion, 1998), 142–44; 351.

112. Recent scholarship has often insisted on the "realism" of Buridan's natural philosophy: Joelle Biard, "The Natural Order in Jean Buridan," in *The Metaphysics and Natural Philosophy of John Buridan*, ed. Johannes. M. M. H. Thijssen and Jack Zupko (Brill, 2001), 91. Among others, Lucien Petrescu writes that Buridan "goes straight to the solution, without arguing: that there are three objects of science, the demonstrated conclusions, the terms that compose the conclusive sentence and the things signified by the terms. The last were the most important." Lucian Petrescu, "The Threefold Object of the Scientific Knowledge. Pseudo-Scotus and the Literature on the Meteorologica in Fourteenth-Century Paris," *Franciscan Studies* 72, no. 1 (2016): 482, 499–500.

113. On the formation of minerals according to Albertus Magnus's *Book of Minerals* – a source Buridan could not ignore—see: Albertus Magnus, *Book of Minerals*, xxxi–xxxv.

114. The "three layers of the habitable land" correspond to a distinction that was common in the Middle Ages: "there are three layers or parts of the earth; one is the lowest around the center, which is the closest to the simple element [of earth] and therefore it is called pure earth ... Above this layer there is one similar to clay, where waters drain. And above there is a third which is like the face of the earth." Peter of Abano, *Conciliator Controversiarum, Quae Inter Philosophos et Medicos Versantur* (Venice: Junta, 1565), 19v. Translation is mine.

115. Buridan, "Johannis Buridani questiones in Aristotelis physica et in libros metheororum," 204r, column A; emphasis mine. See: Duhem, *Le Système Du Monde; Histoire des Doctrines Cosmologiques de Platon à Copernic*, IX:305.

116. On Buridan's "school": J. M. M. H. Thijssen, "The Buridan School Reassessed. John Buridan and Albert of Saxony," *Vivarium* 42, no. 1 (2004): 18–42. Johannes M. M. H. Thijssen, "The Debate over the Nature of Motion: John Buridan, Nicole Oresme and Albert of Saxony. With an Edition of John Buridan's 'Quaestiones Super Libros Physicorum, Secundum Ultimam Lecturam', Book III, Q. 7," *Early Science and Medicine* 14, no. 1/3 (2009): 186–210.

117. Albert of Saxony, *Quæstiones in Aristotelis De cælo*, ed. Benoît Patar (Louvain-la-neuve: Peeters, 2008), 418. Duhem, *Le Système Du Monde; Histoire des Doctrines Cosmologiques de Platon à Copernic*, IX:215. Albert of Saxony, Themon Judeus, and Jean Buridan, *Quaestiones [et] Decisiones Physicales in Octo Libros Physicorum*, ed. George Lokert (Basel: Conrad Resch, 1516), 46r. See also: Duhem, *Le Système Du Monde; Histoire des Doctrines Cosmologiques de Platon à Copernic*, IX:213. On Albert of Saxony, see in particular: Joël Biard, ed., *Itinéraires D'Albert de Saxe, Paris-Vienne Au XIVe Siècle* (J. Vrin, 1991). For a chronology of his life and works, see: Albert of Saxony, *Quæstiones in Aristotelis De cælo*, 9–16. Still useful: Pierre Duhem, *Etudes sur Léonard de Vinci: ceux qu'il a lus et ceux qui l'ont lu* (Paris: Hermann, 1906), 3–19.

118. Graziella Federici-Vescovini, "Note sur la circulation du commentaire d'Albert de Saxe au De Caelo d'Aristote en Italie," in *Itinéraires D'Albert de Saxe, Paris-Vienne Au XIVe Siècle*, ed. Joël Biard (J. Vrin, 1991), 238.

119. Lynn Thorndike, *A History of Magic and Experimental Science: The First Thirteen Centuries of Our Era*, vol. 1 (Columbia University Press, 1958), 156–57.

120. On the historical importance of Albert of Saxony's work for the spread of Buridan's geology, see also: Duhem, *Le Système Du Monde; Histoire des Doctrines Cosmologiques de Platon à Copernic*, IX:309.

121. Dales, *Medieval Discussions of the Eternity of the World*, 260. There is considerably less literature on fourteenth- and fifteenth-century eternalism than on the previous and following centuries. However, see: Bianchi, *L'inizio dei tempi*.

122. Graziella Federici-Vescovini, *Astrologia e Scienza. La crisi dell'aristotelismo sul cadere del trecento e Biagio Pelacani da Parma* (Florence: Nuovedizioni E. Vallecchi, 1979), 384. See also, by the same author, the edition of Pelacani's writing on the soul: Biagio Pelacani da Parma, *Le "Quaestiones de anima". Di Biagio Pelacani da Parma*, ed. Graziella Federici-Vescovini (Florence: L. S. Olschki, 1974), 78.

123. Federici-Vescovini, "Note sur la circulation du commentaire d'Albert de Saxe au De Caelo d'Aristote en Italie," 239. Federici-Vescovini, *Astrologia e scienza. La crisi dell'aristotelismo sul cadere del Trecento e Biagio Pelacani da Parma*, 29–30.

124. John Monfasani, "Aristotelians, Platonists, and the Missing Ockhamists: Philosophical Liberty in Pre-Reformation Italy," *Renaissance Quarterly* 46, no. 2 (1993): 261–64. On B. Pelacani's astrology: Graziella Federici-Vescovini, "Biagio Pelacani: Filosofia, Astrologia e Scienza agli inizi dell'età moderna," in *Filosofia, scienza e astrologia nel Trecento europeo: Biagio Pelacani Parmense: Atti del ciclo di lezioni "Astrologia, Scienza, Filosofia e Società nel Trecento europeo": Parma, 5-6 Ottobre 1990*, ed.

Graziella Federici-Vescovini and Francesco Barocelli (Padua: Il Poligrafo, 1992), 39–52. On the relation between faith and philosophy in Pelacani: Federici-Vescovini, *Astrologia e Scienza. La crisi dell'aristotelismo sul cadere del Trecento e Biagio Pelacani da Parma*, 369–404.

125. Rodolfo Maiocchi, *Codice diplomatico dell' università di Pavia*, vol. 1 (Pavia: Tipografia successori frat. Fusi, 1905), 334.

126. Graziella Federici-Vescovini has listed no fewer than 14 points on which Pelacani and Albert of Saxony dissented: Federici-Vescovini, "Note sur la circulation du commentaire d'Albert de Saxe au De Caelo d'Aristote en Italie," 243–45.

127. On Paul of Venice: Francesco Bottin, "Logica e filosofia naturale nelle opere di Paolo Veneto," in *Scienza e filosofia all'Università di Padova nel Quattrocento* (Padua: Edizioni LINT, 1983), 85–134. On his *Expositio* as a source for Renaissance vernacularizations: Ivano Dal Prete, "Vernacular Meteorology and the Antiquity of the World in Medieval and Renaissance Italy," in *Vernacular Aristotelianism in Italy from the Fourteenth to the Seventeenth Century*, ed. Luca Bianchi, Jill Kraye, and Simon A. Gilson (London: The Warburg Institute, 2016), 152–55.

128. "Uno collecta volumine mentes scholarium vehementius tangere". Paul of Venice, *Expositio librorum naturalium Aristotelis* (Cologne: Johannis de Colonia sociique eius Johannis Manthen de Gherretzem, 1476), sig. a2r. A popular work, it was reprinted in Venice in 1503: Paul of Venice, *Summa Philosophie Naturalis Magistri Pauli Veneti Nouiter Recognita & a Vitijs Purgata Ac Pristine Integritati Restituta* (Venice: Bonetum de Locatellis, 1503). This book was also known as *Summa librorum naturalium Aristotelis*, or *Summa naturalium Aristotelis*.

129. Paul of Venice, *Expositio librorum naturalium Aristotelis*, sig. k1or.

130. Paul of Venice, sig. k1ov.

131. Nicole Oresme, *Le livre du Ciel et du Monde*, ed. Albert Douglas Menut and Denomy Alexander, trans. Albert Douglas Menut (Madison: University of Wisconsin Press, 1968), 570–71.

132. Oresme, 571. The final outcome is clear; the line of Oresme's reasoning, much less so. The literature I am aware of has not been helpful. Like Buridan's theory, Oresme's does not depend on heavenly motions.

133. "Et combien que cesoit faulz, quar le monde n'a pas tant duré que ce peust avoir esté, toutevoies, se le movement du ciel duroit perpetualment, ce seroit bien possible." Oresme, 96–97.

134. Laura Ackerman Smoller, *History, Prophecy, and the Stars: The Christian Astrology of Pierre d'Ailly, 1350–1420* (Princeton: Princeton University Press, 1994), 4. This is not a prediction of when the world will end, but a term "a quo." There is a considerable corpus of literature on d'Ailly, but Smoller's study remains the most relevant to the subject of this book.

135. Smoller, 66.

136. Smoller, 66.

137. D'Ailly adopted a view that had been argued for and against since Arnold of Villanova proposed it at the end of the thirteenth century. He did not question that a complete revolution might be necessary, yet he affirmed that God has the power to accelerate the motion of the eight sphere at will, so that its current pace may not reflect its total duration. Smoller, 90–91.

138. Smoller, 136 and 206–7, note 2.

139. See for example, his explanation of the disappearance and rebirth of rivers: Pierre d'Ailly, *Tractatvs Petri de Eliaco episcopi Cameracensis, super libros Metheororum de impressionibus aeris* (Vienna: per Hieronymu[m] Vietorem & . Ioannem Singreniu[m], 1514), 8r.

140. d'Ailly, 19r.

CHAPTER 3

1. "Le Vilain" means "the peasant," perhaps signifying the humble origin or the author.

2. M. Le Vilain's *Meteorology* was in fact the second Aristotelian treatise ever translated into a European vernacular, preceded only by a Tuscan version of the *Ethics* by the Florentine physician Taddeo Alderotti (ca. 1210–1295). Rolf Edgren proposed 1270 as the latest possible year of composition (Mahieu le Vilain, *Le Metheores d'Aristote*, XV), but his hypothesis has been convincingly refuted by J. Ducos and J. Monfrin. See: Jacques Monfrin, "Jean de Brienne, comte d'Eu, et la traduction des Météorologiques d'Aristote par Mahieu le Vilain (vers 1290)," *Comptes rendus des séances de l'Académie des Inscriptions et Belles-Lettres* 140, no. 1 (1996): 27–36; Joëlle Ducos, *La Météorologie En Français Au Moyen Age (XIIIe-XIVe Siècles)* (Paris: Honoré Champion, 1998), 185–95.

3. Mahieu le Vilain, *Le Metheores d'Aristote*, 1.

4. J. Ducos, "L'oeuvre de Mahieu Le Vilain: Traduction et Commentaire Des Météorologiques," in *Les Traducteurs Au Travail. Leurs Manuscrits et Leurs Méthodes*, ed. Jacqueline Hamesse (Brepols Publishers, 2001), 292.

5. Mahieu le Vilain, *Le Metheores d'Aristote*, 83.

6. Mahieu le Vilain, 67.

7. Mahieu le Vilain, 67.

8. Mahieu le Vilain, 67.

9. Mahieu le Vilain, 70.

10. Nicole Oresme, *Le Livre Du Ciel Du Monde*, ed. Albert Douglas Menut and Denomy Alexander, trans. Albert Douglas Menut (Madison: University of Wisconsin Press, 1968), 179. Oresme's discussion spans pages 131–141 and 161–179. Aristotle's resolute denial that other worlds can exist (thus limiting God's power to create many, if he wishes so) was one of the theses condemned in 1277.

11. For economic and demographic figures: Gabriella Piccinni, in *Duccio: alle origini della pittura senese*, ed. Alessandro Bagnoli (Milan: Silvana, 2003), 26–35.

12. Paul F. Grendler, "What Piero Learned in School: Fifteenth-Century Vernacular Education," *Studies in the History of Art* 48 (1995): 163.

13. Giulio Vaccaro, "Questo libretto che t'ho volgarizzato e chiosato. La traduzione nel Medioevo," in *I Traduttori Come Mediatori Interculturali*, ed. Sergio Portelli and Bart Van Den Bossche (Florence: Cesati, 2016), 11.

14. Ronald Witt, "What Did Giovannino Read and Write? Literacy in Early Renaissance Florence," *I Tatti Studies in the Italian Renaissance* 6 (1995): 83.

15. On scientific books in medieval Italian libraries and their readers: Rita Librandi, "Il lettore di testi scientifici in volgare," in *Lo spazio letterario del medioevo: la ricezione del testo*, ed. Guglielmo Cavallo, Claudio Leonardi, and Enrico Menestò, vol. 2, III (Rome: Salerno, 2003), 125–54. On a particularly successful vernacular medical text, Aldobrandino of Siena's *Régime du Corps*: Elizabeth W. Mellyn, "Passing on Secrets: Interactions between Latin and Vernacular Medicine in Medieval Europe," *I Tatti Studies in the Italian Renaissance* 16, no. 1/2 (2013): 289–309; Sebastiano Bisson, Lada Hordynsky-Caillat, and Odile Redon, "Le Témoin Gênant. Une Version Latine Du 'Régime Du Corps' D'aldebrandin De Sienne," *Médiévales*, no. 42 (2002): 117–30.

16. On Italian translations of Bartholomew the Englishman's work: Rosa Casapullo, "Le Trattato di scienza universal de Vivaldo Belcazer et la tradition du De Proprietatibus Rerum," in *Encyclopédie médiévale et langues européennes: réception et diffusion du De proprietatibus rerum de Barthélemy l'Anglais dans les langues vernaculaires*, ed. Joëlle Ducos (Paris: Honoré Champion éditeur, 2014), 235–58. See also: Vittorio Cian, "Vivaldo Belcazer e l'enciclopedismo italiano delle origini," *Giornale storico della letteratura italiana* Supplemento n. 5 (1902): 10–31; G. Ghinassi, "Nuovi studi sul dialetto mantovano di Vivaldo Belcazer," *Studi di Filologia Italiana* XXIII (1965): 19–172; Rosa Casapullo and Miriam Rita Policardo, "Tecniche Della Divulgazione Scientifica Nel Volgarizzamento Mantovano Del 'De Proprietatibus Rerum' Di Bartolomeo Anglico," *Lingua e Stile*, no. 2/2003 (2003).

17. Rita Librandi, "Ristoro, Brunetto, Bencivenni e la Metaura: intrecci di glosse e rinvii tra le opere di uno scaffale scientifico," in *Lo scaffale della biblioteca scientifica in volgare, secoli XIII-XVI: atti del Convegno (Matera, 14-15 ottobre 2004)*, ed. Rosa Piro and Rita Librandi (Florence: SISMEL, Edizioni del Galluzzo, 2006), 103.

18. Rita Librandi, "Il lettore di testi scientifici in volgare," in *Lo Spazio letterario del medioevo: la ricezione del testo*, ed. Guglielmo Cavallo, Claudio Leonardi, and Enrico Menestò, vol. 2, III (Rome: Salerno, 2003), 144.

19. For a modern English translation: Brunetto Latini, *The Book of the Treasure (Li Livres Dou Tresor)*, trans. Paul Barrette and Spurgeon (New York: Garland, 1993).

20. In the edition of Giamboni's Florentine translation that I have consulted, the biblical story takes less than one page. There is no indication of its chronology, but "many learned men say, that [the first day] was March 14." Brunetto Latini, *Il tesoro di Brunetto Latini volgarizzato da Bono Giamboni, nuovamente pubblicato secondo l'edizione del MDXXXIII*, trans. Bono Giamboni (Venice: Co' tipi del Gondoliere,

1839), 11–12. See also: Brunetto Latini, *La tradizione dei volgarizzamenti toscani del "Tresor" di Brunetto Latini: Con un'edizione critica della redazione α (I.1-129)*, ed. Marco Giola (Verona: QuiEdit, 2010), 267–68. For a modern Italian translation of the French original: Brunetto Latini, *Tresor*, ed. P. G. Beltrami (Torino: G. Einaudi, 2007).

21. Latini, *Il tesoro di Brunetto Latini volgarizzato da Bono Giamboni*, 11. Latini, *La Tradizione Dei Volgarizzamenti Toscani Del "Tresor" Di Brunetto Latini*, 268.

22. Brunetto Latini, *Il tesoro di Brunetto Latini volgarizzato da Bono Giamboni, nuovamente pubblicato secondo l'edizione del MDXXXIII. ...*, trans. Bono Giamboni (Venice: Co' tipi del Gondoliere, 1839), 20. See also: Ibid., 11; and chapter VIII, "On the operations of Nature": Latini, 13–14. In Giola's edition: latini, *la tradizione dei volgarizzamenti toscani del "tresor" di Brunetto Latini*, 268–69.

23. The analogy of the footprint in the dust springs to mind. Latini, *Il tesoro di Brunetto Latini volgarizzato da Bono Giamboni*, 16.

24. On the size and composition of medieval private libraries, with special reference to vernacular scientific texts: Librandi, "Il lettore di testi scientifici in volgare."

25. John of Holywood's astronomical treatise was composed around 1230. On Bencivenni's translation see the critical edition by Gabriella Ronchi: Johannes de Sacrobosco and Zucchero Bencivenni, *Il trattato de la spera*, ed. Gabriella Ronchi (Florence: Accademia della Crusca, 1999).

26. The author explicitly recorded this date even though, due to the discrepancies and local differences in medieval calendars, it may even refer to the modern year 1281. See: Annibale Mottana, "Oggetti e concetti inerenti le Scienze Mineralogiche nella composizione del mondo con le sue cascioni di Restoro d'Arezzo (anno 1282)," *Rendiconti Lincei* 10, no. 3 (1999): 142.

27. On Restoro as a painter: Ada Labriola, "Ricerche Su Margarito e Ristoro d'Arezzo," *Arte Cristiana*, no. 75 (1987): 145–60. Maria Monica Donato, "Un 'savio depentore' fra 'Scienza de le stelle' e 'Sutilita' dell'antico. Restoro d'Arezzo, le arti e il sarcofago romano di Cortona," *Annali della Scuola Normale Superiore di Pisa. Classe di Lettere e Filosofia*, IV, no. 1–2 (1996): 53, 69. The old thesis that Restoro was a monk has been refuted by his modern editor: Restoro of Arezzo, *La Composizione del Mondo*, ed. Alberto Morino (Lavis: La Finestra Editrice, 2007), IX. Any connection between the author, and the Dominican lay brother of the same name who – according to the tradition – would have been one of the architects of Santa Maria Novella in Florence, has also been ruled out: Ilaria Mariotti, "La creazione di un mito: fra Sisto e Restoro architetti della chiesa di Santa Maria Novella a Firenze," *Annali della Scuola Normale Superiore di Pisa. Classe di lettere e filosofia* 1, no. 1 (1996): 249–78.

28. For some later examples, like Brunelleschi or Manetti: Grendler, "What Piero Learned in School," 169.

29. Restoro of Arezzo, *La Composizione del Mondo* (2007), 99. It is significant that Restoro always qualified the worthy craftsman: he never mentioned generic

"carvers" or "draftsmen," but only those who are "learned," "knowledgeable," "ingenious," or "insightful."

30. Restoro of Arezzo, *La composizione del mondo* (2007), 189. See also: Donato, "Un 'Savio Depentore,'" 66.

31. Restoro of Arezzo, *La Composizione del Mondo*, 50.

32. I refer here to the notion of "trading zone" as used by Pamela Long, rather than in Peter Galison's more common definition. Pamela O. Long, *Artisan/Practitioners and the Rise of the New Sciences, 1400-1600* (Corvallis: Oregon State University Press, 2011), 94.

33. Maria Luisa Altieri Biagi, "Nuclei concettuali e strutture sintattiche nella 'Composizione del mondo' di Restoro d'Arezzo," in *L'Avventura della mente: studi sulla lingua scientifica* (Naples: Morano, 1990), 13. On Restoro's sources: Herbert Douglas Austin, *Accredited Citations in Ristoro d'Arezzo's Composizione Del Mondo: A Study of Sources* (Torino: Guido Momo, 1913).

34. Restoro of Arezzo, *La composizione del mondo*, 3–4.

35. Restoro of Arezzo, *La Composizione del Mondo*, 50–51.

36. Restoro of Arezzo, 135–36.

37. "While water dwells upon the earth, it can make mountains, and move earth from one place to another" ("Remanendose l'acqua per la terra, pò fare li monti, e tòllare la terra da uno loco e pònarla ad un altro, secondo quello che noi avemo detto"). Ibid., 135–36.

38. Ristoro d'Arezzo, *La composizione del mondo*, ed. Alberto Morino (Parma: Fondazione Pietro Bembo, 1997), 197-199. Translation is mine.

39. Literature on Restoro has missed this crucial point. Pierre Duhem remarked that Restoro "considers only one such invasion [of waters]; he calls it *the flood*; obviously, he identifies it with the universal flood of *Genesis*." Pierre Duhem, *Etudes sur Léonard de Vinci: ceux qu'il a lus et ceux qui l'ont lu* (Paris: Hermann, 1909), 321. More recently, Annibale Mottana has reached the same conclusion in his well-documented study of Restoro's mineralogy: Mottana, "Oggetti e concetti inerenti le Scienze Mineralogiche ne La composizione del mondo con le sue cascioni di Restoro d'Arezzo (anno 1282)", 149.

40. "E se alcuno savio sarà in quelle parti, che sapia bene de la scienza delle stelle, provedarassi denanti, e vedarà ssé e tutta la sua famellia, secondo che se dice che fece lo savio [Noè] che se providde denanti per la scienza che gli fue data, e vadò sè e tutta la sua famellia dal pericolo del diluvio e ll'arca; e questo magiure accidente del diluvio, ch'avene per la magiure congiuncione che possa essere e llo mondo, si dice ch'avene per purgare li vizii della terra." Restoro of Arezzo, *La Composizione del Mondo*, 137. Translation is mine

41. See for example: Restoro of Arezzo, 138, paragraph 11.

42. Restoro of Arezzo, 136.

43. Domenico De Robertis, "Un monumento della civiltà aretina. 'La composizione del mondo' di Restoro d'Arezzo," *Atti e Memorie dell'Accademia Petrarca di Lettere*

Arti e Scienze di Arezzo 42 (1978-1976): 124. A. Mottana writes of a "scientific project" that was "almost heretical for his time." Annibale Mottana, "Oggetti e concetti inerenti le Scienze Mineralogiche nella composizione del mondo con le sue cascioni di Restoro d'Arezzo (anno 1282)," *Rendiconti Lincei* 10, no. 3 (1999): 148. See also: Restoro of Arezzo, *La composizione del mondo*, (1976) 258, n. 1.

44. "Secondo che disse il filosofo, nullo puote venire ad sapienza, se al tutto non è rimosso da le faccende umane. . . Non si può imprendere sapienza zappando e vagando e correndo e mangiando; è mestiere riposo e unitade." Restoro of Arezzo, *La composizione del mondo*, 258–59. The gloss was extracted from a sermon delivered in 1306 by the Dominican preacher Giordano of Pisa; the copyist placed it at the beginning of the second book, right before Restoro's enthusiastic eulogy of the "most excellent" painters and carvers of the antiquity, whose art he considered indispensable to the understanding of nature.

45. Librandi, "Ristoro, Brunetto, Bencivenni e la Metaura: intrecci di glosse e rinvii tra le opere di uno scaffale scientifico," 111.

46. Albertus Magnus, *Book of Minerals*, ed. Dorothy Wichoff (Oxford: Clarendon Press, 1967), xvi–xvii.

47. Restoro of Arezzo, *La composizione del mondo* (1976), 189.

48. Only five copies are known. The oldest known copy is a luxury codex, beautifully illuminated and composed soon after the completion of the autograph; later ones were of much cheaper manufacture and probably circulated in bourgeois and mercantile milieus. Oliviero, "La composizione dei cieli in Restoro d'Arezzo e in Dante," 353.

49. Restoro of Arezzo, *La Composizione del Mondo* (1976), XXVII–XXVIII. On Restoro as the likely source of Leonardo's analogy, see: Webster Smith, "Observations on the Mona Lisa Landscape," *The Art Bulletin* 67, no. 2 (June 1, 1985): 188–90.

50. On Paul of Venice's *Composition of the World*: Ivano Dal Prete, "Vernacular Meteorology and the Antiquity of the World in Medieval and Renaissance Italy," in *Vernacular Aristotelianism in Italy from the Fourteenth to the Seventeenth Century*, ed. Luca Bianchi, Jill Kraye, and Simon A. Gilson (The Warburg Institute, 2016), 146–49. Duhem, *Etudes sur Léonard de Vinci*, 1909, 325. The Latin version was first printed as: Paul of Venice, *Expositio librorum naturalium Aristotelis* (Cologne: Johannis de Colonia sociique eius Johannis Manthen de Gherretzem, 1476). For illustrated editions: Paul of Venice, *Expositio Magistri Pauli Veneti Super Libros de Generatione et Corruptione Aristotelis. Eiusdem De Compositione Mundi Cum Figuris* (Venice: Bonetum Locatellum, 1498). Paul of Venice, *Liber de compositione mundi excellētissimi viri Pauli Veneti theologi insignis* (Lugduni: Vincent, 1525). Paul of Venice, *Philosophia Naturalis Compendium Clarissimi Philosophi Pauli Veneti: Una Cum Libro de Cōpositione Mundi Qui Astronomia Janua Inscribitur.* (Paris: apud Petrum Gaudoulum, 1514).

51. Lynn Thorndike, "The *De Constitutione Mundi* of John Michael Albert of Carrara," *Romanic Review* 17 (January 1, 1926): 193–216. The cited paper contains material not included in Thorndike's *History of Magic and Experimental Science*, vol. 4.

52. See on this work: Rita Librandi, ed., *La Metaura d'Aristotile: volgarizzamento fiorentino anonimo del XIV secolo: edizione critica*, 2 vols. (Naples: Liguori, 1995).

53. Rita Librandi, ed., *La Metaura d'Aristotile: volgarizzamento fiorentino anonimo del XIV secolo: edizione critica*, vol. 1 (Naples: Liguori, 1995), 52–54.

54. Librandi, "Ristoro, Brunetto, Bencivenni e la Metaura: intrecci di glosse e rinvii tra le opere di uno scaffale scientifico," 121. Albertus Magnus's commentary was still based on the "old translation" of Aristotle's *Meteorology*, while Thomas Aquinas used William of Moerbeke's far more recent and reliable version. For a thorough discussion: Librandi, *La Metaura d'Aristotile*, 1995, 1:15–16; 35–40.

55. "Aristotele parlò molto brieve e somario, ma frate Alberto dalla Magna de' predicatori. . . lo spuose, come vedere potrete apresso, con tutto parlasse molto ploliso [sic] e troppo retripicando." Librandi, *La Metaura d'Aristotile*, 1995, 1:321.

56. Librandi, 1:323.

57. Librandi, 1:323.

58. Librandi, 1:325. Another source explicitly mentioned is a *Treatise on the Sphere*, almost certainly John of Holywood's book in the Florentine translation by Zucchero Bencivegni.

59. Librandi, 1:326. Other interventions hint at business travels to other areas of economic interest for Florentine companies, like Naples and Greece.

60. "We have learned from trustworthy merchants of Genoa, who were there, that in the country of Mauretania, on the shores of the Ocean, they say that the sea is visibly very high and towers over the shore, and scares those who see it for fear that it overruns the earth." Librandi, 1:324. Translation is mine.

61. Librandi, "Ristoro, Brunetto, Bencivenni e la Metaura: intrecci di glosse e rinvii tra le opere di uno scaffale scientifico," 122.

62. Most of the copies catalogued by Rita Librandi belong to the fifteenth century; the last one is dated 1503. Rita Librandi, ed., *La Metaura d'Aristotile: volgarizzamento fiorentino anonimo del XIV secolo: edizione critica*, vol. 2 (Naples: Liguori, 1995), 125–26. The Florentine *Meteorology* was finally printed in 1554: *Opera nuova la quale tratta della filosofia naturale, chiamata la Metaura d'Aristotile; chiosata da San Tommaso d'Aquino, dell'ordine dei Frati Predicatori* (Venice: Comin da Trino, 1554).

63. Kristian Jensen, "The Humanist Reform of Latin and Latin Teaching," in *The Cambridge Companion to Renaissance Humanism*, ed. Jill Kraye (Cambridge: Cambridge University Press, 1996), 63–64. In fifteenth-century Italian academic jargon, a humanist ("umanista") was a teacher or a student of classical literature and associated "arts" (in particular, rhetoric): Nicholas Mann, "The Origins of Humanism," in *The Cambridge Companion to Renaissance Humanism*, ed. Jill Kraye (Cambridge: Cambridge University Press, 1996), 1.

64. Paolo Galluzzi, *Prima di Leonardo: cultura delle macchine a Siena nel Rinascimento* (Milan: Electa, 1991). Michael P. Kucher, *The Water Supply System of Siena, Italy: The Medieval Roots of the Modern Networked City*, (New York: Routledge, 2004). Pamela O. Long, "Hydraulic Engineering and the Study of Antiquity: Rome, 1557–70," *Renaissance Quarterly* 61, no. 4 (2008): 1098–1138. On the culture of mining in early modern Europe, see the special issue on "The Cultural and Material World of Mining in Early Modern Europe," *Renaissance Studies* 34, no. 1 (2020).

65. Francis Ames-Lewis, *The Intellectual Life of the Early Renaissance Artist* (New Haven: Yale University Press, 2000), 109–40. On the *gusto* for antiquity among Renaissance artists and their patrons: Giovanni Agosti, Vincenzo Farinella, Salvatore Settis, "Passione e gusto per l'antico nei pittori italiani del quattrocento," *Annali della Scuola Normale Superiore di Pisa. Classe di Lettere e Filosofia* 17 (1987):1061–1107.

66. Alberti, *On the Art of Building in Ten Books*, p. 328. I have replaced "strata" in the translation with "skins," which is the actual English term for the original Latin *cutes*. F. Ellenberger argued that the Latin term "strata" was used for the first time with reference to rock layers by Niels Steensen in the seventeenth century. François Ellenberger, *History of Geology*, vol. 1 (Rotterdam: A. A. Balkema, 1996), 194.

67. The only previous mention in literature is in Mario Baratta's one-century old book on Leonardo's geology: Mario Baratta, *Leonardo da Vinci ed i problemi della terra* (Torino: Fratelli Bocca, 1903), p. 107. Alberti's description of layered rocks preceded by more than one century Georg Bauer's far more cited chapter in book V of his *On Metals*: Georg Bauer, *Georgii Agricolae De Re Metallica Libri XII* (Basileae: Apud H. Frobenium et N. Episcopium, 1556) . For an English translation: Georg Agricola, *De Re Metallica.*, trans. Herbert Hoover and Lou Henry Hoover, [New ed.] (New York: Dover Publications, 1950), 126–28 and n. 15.

68. The figure reported by Alberti for the distance between two successive fault lines is, however, entirely plausible for observations conducted in the Tuscan Apennines.

69. Anthony Grafton, *Leon Battista Alberti: Master Builder of the Italian Renaissance* (Cambridge, Mass: Harvard University Press, 2002), 82.

70. Alberti wrote *perscrutatores*, a generic Latin term for "investigators", but it is clear from the context that they must be water prospectors. The editor of a 1546 Italian version also translated "water seekers." ("huomini cercatori d'acque"): Leon Battista Alberti, *I dieci libri de l'architettura*, translated by Pietro Lauro (Vinegia: Vincenzo Valgrisi, 1546), p. 222v.

71. Leon Battista Alberti, *L'architettura. (De re aedificatoria)*, edited by Paolo Portoghesi, translated by Giovanni Orlandi, 2 vols., Vol. 1 (Milan: Edizioni Il Polifilo, 1966), pp. 895–897English translation from: Alberti, *On the Art of Building in Ten Books*, p. 328.

72. Alberti, *L'architettura. (De re aedificatoria)*, vol. 2, 895–97. English translation from: Alberti, *On the Art of Building in Ten Books*, 1988, 328.

73. Carmela Baffioni, ed., *Epistles of the Brethren of Purity: On the Natural Sciences: An Arabic Critical Edition and English Translation of Epistles 15-21*, trans. Carmela Baffioni (Oxford: Oxford University Press in association with the Institute of Ismaili Studies, 2013), 233.

74. Holmyard and Mandeville, *Avicennae De Congelatione et Conglutinatione Lapidum*, pp. 30–31.

75. Holmyard and Mandeville, p. 49.

76. Restoro of Arezzo, *La Composizione del Mondo*, 127.

77. Alberti, *On the Art of Building in Ten Books*, 328.

78. For a more extensive discussion: Ivano Dal Prete, "The Ruins of the Earth: Learned Meteorology and Artisan Expertise in Fifteenth-Century Italian Landscapes," *Nuncius* 33, no. 3 (2018): 415–41.

79. "Sunt qui ioco dicant mundi opificem concavitate maris veluti sigillo usum fuisse; cum primo formaret montes"; Leon Battista Alberti, *Libri De re aedificatoria decem* (Paris: Opera magistri Bertholdi Rembolt & Ludouici Hornken, 1512), p. 160v.

80. "Plato affirmed that it was the common opinion of every mathematician, that stars do not return to the same place in the sky before an infinite number of revolutions has taken place in the course of 36,000 years; but it will be impossible to say whether this [present] time, or that [in which the stars will be again in the same places] is closer to the end, and further away from the beginning of the world." Leon Battista Alberti, *Opere Volgari*, edited by Cecil Grayson (Bari: G. Laterza, 1960), book II. Translation is mine.

81. In his book on architecture, Francesco di Giorgio Martini (1439-1501) quoted from Aristotle's *Meteorology, Physics, Methaphysics, Posterior Analitics* and so forth. Francesco di Giorgio Martini, *Trattati di architettura ingegneria e arte militare*, ed. Corrado Maltese (Milan: Il Polifilo, 1967). For a list of the numerous quotations from Aristotle, see p. 575. On Martini, see also: Pari Riahi, *Ars et Ingenium: The Embodiment of Imagination in the Architectural Drawings of Francesco di Giorgio Martini*. Ph.D. Dissertation, McGill University, 2010.

82. On Tuscan engineers of the Renaissance: Paolo Galluzzi, *Prima di Leonardo: cultura delle macchine a Siena nel Rinascimento* (Milan: Electa, 1991). Paolo Galluzzi, *Renaissance Engineers from Brunelleschi to Leonardo Da Vinci* (Florence: Istituto e Museo di Storia della Scienza: Giunti, 2001).

83. In 1495 Leonardo owned no less than 40 books; in 1503, they had become at least 126. Fabio Frosini, "La Biblioteca Di Leonardo," *Biblioteche Dei Filosofi. Biblioteche Filosofiche Private in Età Moderna e Contemporanea*, 2016, 6. No contemporary Florentine private library numbered more than 50 titles. Romain Descendre, "La biblioteca di Leonardo," in *Atlante della letteratura italiana* (Einaudi, 2010), 593. On Leonardo's library, see also: Carlo Vecce, *La biblioteca perduta: i libri di Leonardo* (Rome: Salerno editrice, 2017). Giulia Bologna and Augusto Marinoni, *La biblioteca di Leonardo* (Comune di Milano, 1983). Augusto Marinoni, "La Biblioteca Di Leonardo," *Raccolta Vinciana* 22 (1987): 291–342. And the recent:

Paula Findlen et al., *Leonardo's Library: The World of a Renaissance Reader* (Stanford University Libraries, 2019).

84. Carlo Dionisotti, "Leonardo uomo di lettere," *Italia Medievale e Umanistica* 1962, no. V (n.d.): 183.

85. Carlo Maccagni, *Riconsiderando il problema delle fonti di Leonardo: L'elenco di libri ai fogli 2 verso 3 recto del codice 8936 della Biblioteca nacional di Madrid.* (Florence: G. Barbèra, 1971), 16.

86. Manuscript G, fol. 8 r; Augusto Marinoni, ed., *I Manoscritti Dell'Institut de France*, vol. Manoscritto G (Florence: Giunti Barbèra, 1989), 11. Translation is mine. The theoretical knowledge Leonardo alludes to in this passage is mathematical perspective.

87. Two editions were printed in Venice in 1492 and 1497: Albert of Saxony, *Questiones Subtilissime in Libros De Celo et Mundo*, ed. Hyeronimus Surianus (Venice: Imp. Arte Bonetis locatellis, 1492); Albert of Saxony, *Questiones Subtilissime in Libros De Celo et Mundo*, ed. Hyeronimus Surianus (Venice: per Otinum Papiensem, 1497).

88. The dependence of Leonardo's theory of the Earth from Albert of Saxony's was established by Pierre Duhem more than one century ago: Pierre Duhem, *Etudes sur Léonard de Vinci: ceux qu'il a lus et ceux qui l'ont lu* (Paris: Hermann, 1906). See also: Baratta, *Leonardo da Vinci ed i problemi della terra.* Among recent reappraisals: Stephen Jay Gould, "The Upwardly Mobile Fossils of Leonardo's Living Earth," in *Leonardo's Mountain of Clams and the Diet of Worms: Essays on Natural History* (New York: Harmony Books, 1998), 17–44; Romano Nanni, "Catastrofi e armonie," in *Leonardo da Vinci on Nature: Knowledge and Representation*, ed. Fabio Frosini, Alessandro Nova, and Romano Nanni (Venice: Marsilio, 2015), 95–117; Domenico Laurenza, "Leonardo's Theory of the Earth. Unexplored Issues in Geology from the Codex Leicester," 257–67.

89. Manuscript F, fol. 70 r; Augusto Marinoni, ed., *I Manoscritti Dell'Institut de France*, vol. Manoscritto F (Florence: Giunti Barbèra, 1988), 116. Translation is mine, based on: Edward McCurdy, *The Notebooks of Leonardo Da Vinci*, vol. 2 (New York: Reynal & Hitchcock, 1938), 53.

90. Laurenza, "Leonardo's Theory of the Earth. Unexplored Issues in Geology from the Codex Leicester," 261.

91. Alberti, *On the Art of Building in Ten Books*, 336. On this theory: Dante Alighieri, "Questio de Aqua et Terra," in *La Letteratura Italiana - Storia e Testi*, ed. Francesco Mazzoni, vol. V (Naples: Ricciardi, 1978), 705–06; Pierre Duhem, *Le Système Du Monde; Histoire Des Doctrines Cosmologiques de Platon à Copernic*, vol. IX (Paris: A. Hermann, 1958), 146–48.

92. *Codex Hammer*, f. 35v. Leonardo, *The Leicester Codex (Hammer Codex)*, ed. Francesco Maria Caleca, trans. David Edwards-May (Rome: TREC edizioni pregiati, 2006), 316.

93. *Codex Hammer*, f. 35v. Leonardo, 316.

94. "How the center of gravity of the water and that of the earth, joined together, are one and the same as that of the elements." Carlo Pedretti, ed., *The Codex Hammer of Leonardo Da Vinci*, trans. Carlo Pedretti (Florence: Giunti Barbèra, 1987), 155.

95. "Deepening of certain countries, and the Dead Sea in Syria, that is Sodom and Gomorrah." Codex Hammer, fol. 36 r. Leonardo, *The Leicester Codex (Hammer Codex)*, 319.

96. Stephen Jay Gould, "The Upwardly Mobile Fossils of Leonardo's Living Earth," in *Leonardo's Mountain of Clams and the Diet of Worms: Essays on Natural History* (New York: Harmony Books, 1998), 40–44; Laurenza, "Leonardo's Theory of the Earth," 262–63.

97. Codex Hammer, fol. 36 r. English translation from: Pedretti, *The Codex Hammer of Leonardo Da Vinci*, 6–7. See also: Codex Atlanticus, fol. 784a v. Augusto Marinoni (ed.), *Il Codice Atlantico della Biblioteca Ambrosiana Di Milano*, 20 vols., Vol. 3 (Florence: Giunti, 2000), 1515.

98. Codex Hammer, fol. 36 r. Leonardo, *The Leicester Codex (Hammer Codex)*, 320.

99. Librandi, *La Metaura d'Aristotile*, 1995, 2:268. Leonardo's most explicit reference to this theory is in: Arundel MS 263, fol. 155v; Roberto Marcolongo (ed.), *I manoscritti e i disegni di Leonardo Da Vinci. Il Codice Arundel 263*, 5 vols., Vol. 1 (Rome: Danesi, 1926), 176–77 (I could not consult the 1998 edition by Carlo Pedretti and Carlo Vecce). Leonardo reiterated this notion in a passage dealing with astronomy: Manuscript F, fol. 69 v; Augusto Marinoni (ed.), *I manoscritti dell'Institut de France*, Manoscritto F (Florence: Giunti Barbèra, 1988), 115.

100. Librandi, *La Metaura d'Aristotile*, 1995, 2:268.

101. Codex Hammer, 17v, in Carlo Pedretti (ed.), *The Codex Hammer of Leonardo Da Vinci*, trans. Carlo Pedretti (Firenze: Giunti Barbèra, 1987), 155; Laurenza, "Leonardo's Theory of the Earth. Unexplored Issues in Geology from the Codex Leicester," 267.

102. Alison Brown regards Leonardo's notion that (unlike in the Aristotelian tradition) life on Earth will one day be extinguished as "evocative" of Lucretius and of atomism. See Alison Brown, " 'Natura Idest?' Leonardo, Lucretius and their Views of Nature," in *Leonardo da Vinci on Nature: Knowledge and Representation*, ed. Fabio Frosini and Alessandro Nova, 161.Venice: Marsilio, 2015. A major problem with this hypothesis is that Lucretius considers the Earth itself as a perishable aggregate of atoms, which will one day dissolve. Leonardo never hints at this possibility; he simply discusses how the Earth will one day lose the properties that make terrestrial life possible. The alternatives he considers were already present in the medieval sources available to him, which predated the fifteenth-century rediscovery of Lucretius's work.

103. Frank Fehrenbach, "Leonardo's Point," in *Vision and Its Instruments: Art, Science, and Technology in Early Modern Europe*, ed. Alina Alexandra Payne (University Park: Pennsylvania State University Press, 2015), 84.

104. On Leonardo's reading of Restoro: Smith, "Observations on the Mona Lisa Landscape," 187; Martin Kemp, *Leonardo Da Vinci: The Marvellous Works of Nature and Man* (Oxford University Press, 2006).

105. Stephen Jay Gould, "The Upwardly Mobile Fossils of Leonardo's Living Earth," in *Leonardo's Mountain of Clams and the Diet of Worms: Essays on Natural History* (New York: Harmony Books, 1998), 33–34.

106. Domenico Laurenza, *De figura umana: fisiognomica, anatomia e arte in Leonardo* (Florence: Olschki, 2001), 155–160. On the crisis of the microcosm/macrocosm analogy in Leonardo's late life: Martin Kemp, "La crisi del sapere tradizionale nell'ultimo Leonardo," in Martin Kemp, *Lezioni dell'occhio: Leonardo da Vinci discepolo dell'esperienza* (Milan: Vita e Pensiero, 2004), 133–54.

107. Domenico Laurenza, "Images and Theories. The Study of Fossils in Leonardo, Scilla and Hooke," *Nuncius*, no. 1 (2019): 445–47.

108. Atlantic Codex, fol. 330 v-b. Reproduced and discussed in: Carlo Pedretti, *The Literary Works of Leonardo Da Vinci*, 2 vols., Vol. 2 (Berkeley: University of California Press, 1977), 370. Also in (as fol. 904 v): Marinoni, *Il Codice Atlantico Della Biblioteca Ambrosiana Di Milano*, 2000, III:1676. The strata and "gravel" of the river Adige are briefly mentioned again in: Codex Hammer, fol. 3 r; Pedretti, *The Codex Hammer of Leonardo Da Vinci*, p. 23. I was unable to obtain good reproductions of those sketches.

109. Codex Hammer, fol. 3 r; Pedretti, *The Codex Hammer of Leonardo Da Vinci*, 22. See also: Codex Hammer, fol. 8 v; Pedretti, 72. On the relation between layered rocks, water, and architecture, see also: Arundel MS 263, fol. 157; Marcolongo, *I manoscritti e i disegni di Leonardo Da Vinci*, 178–79.

110. Codex Hammer, fol. 3 r. Pedretti, *The Codex Hammer of Leonardo Da Vinci*, 21–23.

111. Alberti's treatise was universally known, and it is included in a list of books owned by Leonardo. No Italian translations had been printed yet, but Augusto Marinoni considers it "likely" that he possessed a vernacular version: Augusto Marinoni, *Leonardo da Vinci. Scritti letterari* (Milan: Rizzoli, 1974), 255.

112. Gian Battista Vai, "I viaggi di Leonardo lungo le valli romagnole: riflessi di geologia nei quadri, disegni e codici," in *Leonardo, Machiavelli, Cesare Borgia - Arte, Storia e Scienza in Romagna, 1500–1503* (Rome: De Luca, 2003), 44. Using similar arguments, Ann Pizzorusso has proposed that of the two existing versions of the *Virgin of the Rocks* (one at the Louvre and one at the National Gallery) only the Louvre one should be attributed to Leonardo. Ann Pizzorusso, "Leonardo's Geology: The Authenticity of the 'Virgin of the Rocks,'" *Leonardo* 29, no. 3 (1996): 197–200.

113. "How the stratified rocks of the mountains are all in layers of mud deposited one above another by the floods of the rivers. How the different thicknesses of the strata of the rocks are created by the different floods of the rivers, that is the greater and the less floods. How between the various layers of stone are still to be found the tracks of the worms which crawled about upon them when it was not

yet dry." Codex Hammer, fol. 10 r; Pedretti, *The Codex Hammer of Leonardo Da Vinci*, 84.

114. Codex Hammer, f. 35 r. Leonardo, *The Leicester Codex (Hammer Codex)*, 311.

115. Manuscript E, fol. 4 v; Marinoni, *I Manoscritti dell'Institut de France*, Manoscritto E, 10–11. See also: Manuscript F, fol. 79 r; Marinoni, *I Manoscritti Dell'Institut de France*, 1988, Manoscritto F:130–31.

116. Codex Hammer, fol. 9 r; Pedretti, *The Codex Hammer of Leonardo Da Vinci*, 73.

117. Carlo Pedretti, *The Codex Hammer of Leonardo Da Vinci* (Florence: Giunti Barbèra, 1987), 8.

118. Codex Hammer, fol. 8 v, fol. 9 r-v; Pedretti, *The Codex Hammer of Leonardo Da Vinci*, 70–75, 80–82. More references to the formation and/or raising of layered rocks in: Codex Atlanticus, fol. 350 r; Augusto Marinoni, ed., *Il Codice Atlantico della Biblioteca Ambrosiana di Milano*, vol. I (Florence: Giunti, 2000), 608; Manuscript L, fol. 17 r; Augusto Marinoni (ed.), *I Manoscritti dell'Institut de France*, Manoscritto L (Florence: Giunti Barbèra, 1987), 19; Manuscript E, fol. 4 v; Augusto Marinoni (ed.), *I Manoscritti dell'Institut de France*, Manoscritto E (Florence: Giunti Barbèra, 1989), 10–11.

119. Gould, "The Upwardly Mobile Fossils of Leonardo's Living Earth," 22.

120. See: Tiberio Russiliano, *Una reincarnazione di Pico ai tempi di Pomponazzi*, ed. Paola Zambelli (Milan: Il Polifilo, 1994).

121. "At this point natural causes fail us, and therefore in order to resolve such a doubt we must needs [sic] either call in a miracle to our aid or else say that all this water was evaporated by the heat of the Sun." Codex Atlanticus, fol. 155 r b; Augusto Marinoni, ed., *Il Codice Atlantico: Della Biblioteca Ambrosiana Di Milano*, vol. V (Florence: Giunti-Barbèra, 1977), 274–75. English translation from: McCurdy, *The Notebooks of Leonardo Da Vinci*, I:336.

122. Codex Hammer, fol. 8 v; Pedretti, *The Codex Hammer of Leonardo Da Vinci*, 72.

123. Codex Hammer, fol. 9 r; Pedretti, 73.

124. Codex Hammer, fol. 8 v; Pedretti, 71.

125. Codex Hammer, fol. 10 r; Pedretti, 85.

126. Codex Hammer, 10 r; Pedretti, 85. Emphasis mine.

127. See for example: Baratta, *Leonardo da Vinci ed i problemi della terra*, 233.

128. J. M. Jordan, "'Ancient Episteme' and the Nature of Fossils: A Correction of a Modern Scholarly Error," *History and Philosophy of the Life Sciences* 38/1 (2016): 105.

129. Codex Hammer (or Leicester), f. 31r. Pedretti, *The Codex Hammer of Leonardo Da Vinci*, 51. Emphasis mine.

130. For interpretations and the critical fortune of Leonardo's "deluge drawings": Frank Fehrenbach, "Un Nuovo Paradigma: Il Diluvio," in *Leonardo "1952" e la cultura dell'Europa nel dopoguerra*, ed. Romano Nanni and Maurizio Torrini (Florence: Olschki, 2013), 303–20.

131. Leslie Geddes, "'Infinite Slowness and Infinite Velocity': The Representation of Time and Motion in Leonardo's Studies of Geology and Water," in *Leonardo da Vinci on Nature: Knowledge and Representation*, ed. Fabio Frosini and Alessandro Nova (Venice: Marsilio, 2015), 283.

132. Geoff Lehman, "Leonardo, van Eyck, and the Epistemology of Landscape," in *Leonardo in Dialogue: The Artist Amid His Contemporaries*, ed. Francesca Borgo, Alessandro Nova, and Rodolfo Maffeis (Italy: Marsilio, 2019), 97–118.

133. Donald Strong, "The Triumph of Mona Lisa: Science and the Allegory of Time," in *Leonardo e l'età della ragione*, ed. Enrico Bellone, Paolo Rossi (Milan: Scientia, 1982), 255–77.

CHAPTER 4

1. Francesco Beretta, "Orthodoxie philosophique et Inquisition romaine aux 16e-17e siècles," *Historia philosophica* 3 (2005): 72.

2. See also: Eric A. Constant, "A Reinterpretation of the Fifth Lateran Council Decree Apostolici Regiminis (1513)," *The Sixteenth Century Journal* 33, no. 2 (2002): 366–67.

3. Beretta, "Orthodoxie philosophique et Inquisition romaine aux 16e-17e siècles," 73.

4. Beretta, "Orthodoxie philosophique et Inquisition romaine aux 16e-17e siècles," 76.

5. Gaetano of Thiene, *Aristoteles Stagirite Metheororum Liber Primus: Cum Commentariis Caietani de Tienis. Nuper Summa Cum Diligentia Emendatis* (Venice: Jo. de Forolivio, 1491), last page of the first book (pages not numbered).

6. Beretta, "Orthodoxie philosophique et Inquisition romaine aux 16e-17e siècles," 73; John Monfasani, "Aristotelians, Platonists, and the Missing Ockhamists: Philosophical Liberty in Pre-Reformation Italy," *Renaissance Quarterly* 46, no. 2 (1993): 250–51, n. 21. Monfasani found Vernia's declaration "formulistic and reads as if it were prepared to satisfy an inquisitor."

7. Ermolao Barbaro and Hieronymus Wildenberg, *Naturalis scientiae totius compendium, ex Aristotele, & alijs philosophis, Hermolao Barbaro... autore, innumeris, quibus antea scatebat, mendis nunc demum D. Conradi Gesneri*, ed. Konrad Gesner (Basel: Ex officina Ioannis Oporini, 1548), 67–68.

8. On the importance and prestige of natural philosophy in Italian universities: Paul F. Grendler, *The Universities of the Italian Renaissance* (Baltimore: Johns Hopkins University Press, 2002), 267–69. On the institutional weakness of theological faculties in Italy: Monfasani, "Aristotelians, Platonists, and the Missing Ockhamists," 253–56. See also: Constant, "A Reinterpretation of the Fifth Lateran Council Decree Apostolici Regiminis (1513)," 373, and the literature mentioned in Constant, 373, n. 75.

9. "At quod sub Noe fuit, non contingit naturam ut illa: sed divina ultionem nec fuit circa unam regionem sed totam obduxit terram aquis. Et ideo dicitur universale." Jacques Lefèvre d'Etaples and Johannes Cochlaeus, *Meteorologia Aristotelis* (Nuremberg: Peypuß, 1512), 39r.

10. Ermolao Barbaro and Hieronymus Wildenberg, *Naturalis scientiae totius compendium, ex Aristotele, & alijs philosophis, Hermolao Barbaro... autore, innumeris, quibus antea scatebat, mendis nunc demum D. Conradi Gesneri*, ed. Konrad Gesner (Basel: Ex officina Ioannis Oporini, 1548), 199.

11. In: François Ellenberger, *History of Geology*, vol. 1 (Rotterdam: Balkema, 1996), 169.

12. Connell, 7–9, 12–15.

13. William J. Connell, "The Eternity of the World and Renaissance Historical Thought," *California Italian Studies* 2, no. 1 (January 1, 2011): 1.

14. See in particular: Alison Brown, *The Return of Lucretius to Renaissance Florence* (Cambridge, MA: Harvard University Press, 2010).

15. Connell, "The Eternity of the World and Renaissance Historical Thought," 14; Philipp E. Nothaft, "The Early History of Man and the Uses of Diodorus in Renaissance Scholarship: From Annius of Viterbo to Johannes Boemus," in *For the Sake of Learning. Essays in Honor of Anthony Grafton*, ed. Ann Blair and Anja-Silvia Goeing, vol. 2 (Leiden: Brill, 2016), 713.

16. Lodi Nauta, "Philology as Philosophy: Giovanni Pontano on Language, Meaning, and Grammar," *Journal of the History of Ideas* 72, no. 4 (2011): 483.

17. Giovanni Gioviano Pontano, *Pontani Opera. Vrania, siue de stellis libri quinque. Meteororum liber unus. De hortis hesperidum libri duo . . .* (Venice: Aldo Manuzio, 1505), 136r.

18. Nothaft, "The Early History of Man and the Uses of Diodorus in Renaissance Scholarship," 720–23. I consulted a 1535 Lyon edition: Johann Boemus, *Omnium gentium Mores, Leges et Ritus... Ionne Boemo Aubano Teutonico nuper collecti . . .* (Lyon: Apud Haeredes Simonis Vincentii, 1535), 15–20.

19. Niccolò Machiavelli and Francesco Guicciardini, *The Sweetness of Power: Machiavelli's Discourses & Guicciardini's Considerations*, trans. James B. Atkinson and David Sices (Dekalb: Northern Illinois University Press, 2002), 176. I have modified Atkinson and Sices's translation, which reads: ". . . more than five thousand years, *even* if we did not know . . .", thus reversing the meaning of Machiavelli's sentence. See the original: "A quegli filosofi che hanno voluto che il mondo sia stato eterno, credo che si potesse replicare che, se tanta antichità fusse vera, e' sarebbe regionevole che ci fussi memoria di più che cinquemila anni; quando e' non si vedesse come queste memorie de' tempi per diverse cagioni si spengano: delle quali, parte vengono dagli uomini, parte dal cielo." Niccolò Machiavelli, *Discorsi Sopra La Prima Deca Di Tito Livio*, ed. Mario Martelli (Florence: Sansoni, 1971), 171–72.

20. Machiavelli, *Discorsi Sopra La Prima Deca Di Tito Livio*, 172.

21. "Quanto alle cause che vengono dal cielo, sono quelle che spengono la umana generazione, e riducano a pochi gli abitatori *di parte* del mondo." Machiavelli, 173. Emphasis mine.

22. Machiavelli, 173.

23. Machiavelli, 172–73. Machiavelli wrote that Diodorus's chronology spanned over forty or fifty thousand years, instead of ten or twenty thousand.

24. For a recent reassessment of the possible relations between Machiavelli and Leonardo, see: Brown, "'Natura Idest?'. Leonardo, Lucretius and their Views of Nature," 165–68.

25. On Machiavelli as a straight eternalist, see in particular: Gennaro Sasso, *Machiavelli e gli antichi, e altri saggi*, vol. 1 (Milan: Riccardo Ricciardi, 1987), 167–399. For the opposite (and more plausible) reading: Connell, "The Eternity of the World and Renaissance Historical Thought," 16.

26. Connell, "The Eternity of the World and Renaissance Historical Thought," 16.

27. Norman P. Tanner, ed., *Decrees of the Ecumenical Councils*, vol. 1 (Washington: Georgetown University Press, 1990), 606.

28. "Since truth cannot contradict truth, we define that every statement contrary to the enlightened truth of the faith is totally false and we strictly forbid teaching otherwise to be permitted." Tanner, *Decrees of the Ecumenical Councils*, 1:605. See also: Constant, "A Reinterpretation of the Fifth Lateran Council Decree Apostolici Regiminis (1513)."

29. Francesco Beretta, "Orthodoxie philosophique et Inquisition romaine aux 16e-17e siècles," *Historia Philosophica* 3 (2005): 12. The bull, with passages not included in Tanner's edition, is also published in: Constant, "A Reinterpretation of the Fifth Lateran Council Decree Apostolici Regiminis (1513)," 377–79.

30. The bull addressed not only the writings of "philosophers" but even those of certain "poets." Lucretius's *On the Nature of Things* was indeed a scientific poem.

31. Monfasani, "Aristotelians, Platonists, and the Missing Ockhamists," 267–69.

32. Monfasani, "Aristotelians, Platonists, and the Missing Ockhamists," 269.

33. Alessandri, 243r.

34. Alessandro Alessandri, *Alexandri ab Alexandro, Neapolitani I. C. Genialium Dierum Libri sex* (Lyon: Paul Frellon, 1616), 243v.

35. Ellenberger 1988, I, 173.

36. Beretta, "Orthodoxie philosophique et Inquisition romaine aux 16e-17e siècles," 75–76. Constant, "A Reinterpretation of the Fifth Lateran Council Decree Apostolici Regiminis (1513)," 374–75.

37. On Pomponazzi and the mortality of the soul, see also: Lorenzo Casini, "The Renaissance Debate on the Immortality of the Soul: Pietro Pomponazzi and the Plurality of Substantial Forms," in *Mind, Cognition and Representation: The Tradition of Commentaries on Aristotle's De Anima*, ed. Paul J. J. M. Bakker and J. M. M. H. Thijssen (Aldershot: Ashgate Publishing, 2007), 134; Martin L. Pine, *Pietro Pomponazzi: Radical Philosopher of the Renaissance* (Padua: Antenore, 1986); Marco Sgarbi and Maurizio Bertolotti, eds., *Pietro Pomponazzi: Tradizione e Dissenso. Atti del Congresso Internazionale di Studi su Pietro Pomponazzi, Mantova, 23-24 Ottobre 2008* (Florence: L.S. Olschki, 2010). For an Italian translation (with a substantial introduction) of Pomponazzi's treatise on the soul: Pietro Pomponazzi,

Trattato sull'immortalità dell'anima, ed. Vittoria Perrone Compagni (Florence: L. S. Olschki, 1999). For an English translation: Pietro Pomponazzi, *Tractatus de immortalitate animae*, ed. William Henry Hay (Haverford: Haverford College, 1938).

38. Luca Bianchi, "Per una storia dell'aristotelismo 'volgare' nel rinascimento: problemi e prospettive di ricerca," *Bruniana & Campanelliana* 15, no. 2 (2009): 373–74.

39. Bianchi, 375–79.

40. Brian Richardson, *Print Culture in Renaissance Italy: The Editor and the Vernacular Text, 1470–1600* (Cambridge UK: Cambridge University Press, 2003), 90.

41. Paul Grendler estimated the city's male literacy rate at about 33% in the late 1500s. Because of his methodology, this should be considered a conservative figure. Paul F. Grendler, *Schooling in Renaissance Italy: Literacy and Learning, 1300–1600* (Baltimore: Johns Hopkins University Press, 1989), 42–47.

42. On vernacular natural philosophy, and in particular meteorology, in a courtly context and for female readers, see: Craig Martin, "Meteorology for Courtiers and Ladies: Vernacular Aristotelianism in Renaissance Italy," *Philosophical Readings* IV, no. 2 (2012): 3–14.

43. Luca Bianchi, "Volgarizzare Aristotele: per chi?," *Freiburger Zeitschrift Für Philosophie Und Theologie* 59, no. 2 (2012): 495.

44. Marco Sgarbi, "The Instatement of the Vernacular as Language of Culture. A New Aristotelian Paradigm in Sixteenth-Century Italy," *Intersezioni*, no. 3/2016 (2016): 328–34.

45. Angela Piscini, "Gelli, Giovan Battista," *DBI* (Rome: Istituto della Enciclopedia italiana, 2000); Armand L. De Gaetano, *Giambattista Gelli and the Florentine Academy: The Rebellion against Latin* (Florence: L. S. Olschki, 1976); Vittoria Perrone Compagni, "Cose di filosofia si possono dire in volgare. Il programma culturale di Giambattista Gelli," in *Il Volgare Come Lingua Di Cultura Dal Trecento Al Cinquecento: Atti Del Convegno Internazionale, Mantova, 1820 Ottobre 2001*, ed. Arturo Calzona (Florence: L. S. Olschki, 2003), 1000–37; Marco Sgarbi, "Aristotle and the People. Vernacular Philosophy in Renaissance Italy," *Renaissance & Reformation* 39, no. 3 (2016): 102–05.

46. "Quando, le scientie, et l'arti son pervenuti in una provincia a un certo punto di perfettione, che o per diluvii, o per guerre, et tumulti, o per mescolanze di nationi Barbare, et roze, o per mortalità di pestilentie elle manchino et quasi affatto si perdino et di nuovo di poi ricomincionio a racquistare la perfetion loro." In: De Gaetano, *Giambattista Gelli and the Florentine Academy*, 43.

47. On Fausto da Longiano, his translations, and his literary work: F. Pignatti, "Fausto (Fausto da Longiano), Sebastiano," in *DBI*, XLV, Rome, 1995, 394–98. See also, on his *Meteorology*: Martin, "Meteorology for Courtiers and Ladies: Vernacular Aristotelianism in Renaissance Italy." On his activity as a translator: B. Guthmüller, "Fausto da Longiano e il problema del tradurre," *Quaderni Veneti* 12 (1990): 9–152.

48. See Benedetto Varchi's unpublished commentary over the first book of the meteorology: *Comento primo di Benedetto Varchi Fiorentino sopra il primo libro delle Meteore d'Aristotile*, Filze Rinuccini 10, BNCF, particularly ff. 229r-v. See also: Simon A. Gilson, "Vernacularizing Meteorology: Benedetto Varchi's Comento sopra il Primo Libro delle Meteore d'Aristotile," in *Vernacular Aristotelianism in Italy from the Fourteenth to the Seventeenth Century*, ed. Luca Bianchi, Simon A. Gilson, and Jill Kraye (London: The Warburg Institute, 2016), 161–81.

49. Alessandro Piccolomini, *De la sfera del mondo. Libri quattro in lingua Toscana . . . De le stelle fisse. Libro uno con le sue figure, e con le sue tavole* (Venice: al Segno del Pozzo, 1540).

50. On Piccolomini as a vulgarizer of Aristotle: Stefano Caroti, "L''Aristotele Italiano' di Alessandro Piccolomini: Un Progetto Sistematico Di Filosofia Naturale in Volgare a Metà '500," in *Il Volgare Come Lingua Di Cultura Dal Trecento Al Cinquecento: Atti Del Convegno Internazionale, Mantova, 18-20 Ottobre 2001*, ed. Arturo Calzona (Florence: L. S. Olschki, 2003), 361–401.

51. Piccolomini, title page. Sgarbi has pointed out that "only 16% of all vernacular Aristotelian works were proper translations." Sgarbi, "Aristotle and the People," 90.

52. Piccolomini, *De la sfera del mondo. Libri quattro in lingua toscana*, 10r. Pietro d'Abano's *Conciliator* was composed in the early 1300s and reprinted at least eleven times in Venice between 1476 and 1565. See: Laura Turetta, "Bibliografia delle Opere a Stampa di e su Pietro d'Abano," *Medicina nei Secoli* 20, no. 2 (2008): 659–734. I consulted the 1520 Venetian edition: Pietro D'Abano, *Conciliator differentiarum philosophorum [et] medicorum* (Venice: Luceantonii de Giunta, 1520), 28v–29r.

53. Richardson, *Print Culture in Renaissance Italy*, 91.

54. Sebastiano Fausto da Longiano, *Il gentil'huomo del Fausto da Longiano* (Venice: Francesco Bindoni & Maffeo Pasini, 1542); Sebastiano Fausto Longiano, *Il Favsto da Longiano De lo istitvire vn figlio d'un principe da li X. infino à gl'anni de la discretione* (Venice: Francesco Bindoni & Maffeo Pasini, 1542); Sebastiano Fausto Longiano, *Il Fausto da Longiano. De gl'augurij, e de le soperstitioni de gl'antichi* (Venice: Curzio Navò, 1542); Pedanius Dioscorides, *Dioscoride fatto di greco italiano. Al cui fine sono apposte le sue tavole ordinate, con certe avertenze, e trattati necessarii, per la material medesima*, ed. Sebastiano Longiano, trans. Sebastiano Fausto Longiano (Venice: Curzio Navò, 1542). Sebastiano Fausto Longiano, *Meteorologia, cioè Discorso de le impressioni humide & secche generate tanto ne l'aria, quanto ne le cauerne de la terra, non per uia di tradottione, ma di scelta. Trasportata in lingua italiana dal Fausto da Longiano* (Venice: Curzio Navò, 1542).

55. Francesco del Garbo to Benedetto Varchi, November 11, 1540. In: Benedetto Varchi, *La via della dottrina: le lezioni accademiche di Benedetto Varchi*, ed. Annalisa Andreoni (Pisa: ETS, 2012), 47–48. See also: Bianchi, "Volgarizzare Aristotele: Per Chi?," 488. Gilson, "Vernacularizing Meteorology: Benedetto Varchi's *Comento sopra il primo libro delle Meteore d'Aristotile*," 166–67.

56. Vannoccio Biringuccio, *De La Pirotechnia: Libri X* (Venetia: Per Venturino Rossinello ad instantia di Curtio Navò, 1540).

57. Longiano, 2v.

58. Paul of Venice, *Expositio librorum naturalium Aristotelis* (Cologne: Johannis de Colonia sociique eius Johannis Manthen de Gherretzem, 1476). See also: Ivano Dal Prete, "Vernacular Meteorology and the Antiquity of the World in Medieval and Renaissance Italy," in *Vernacular Aristotelianism in Italy from the Fourteenth to the Seventeenth Century*, ed. Luca Bianchi, Jill Kraye, and Simon A. Gilson (London: The Warburg Institute, 2016), 139–59. On the use of medieval vernacular texts by Renaissance translators: Gilson, "Vernacularizing Meteorology: Benedetto Varchi's Comento Sopra Il Primo Libro Delle Meteore d'Aristotile," 168–69.

59. Longiano, *Meteorologia*, 33v.

60. "Da le cose dette si può pruovare, anzi siegue di fatto, che stante la eternità del mondo, è necessario, che 'l mare si trasferisca ad altro loco et dove adesso è mar, habbia ad esser terra, et altre volte vi sia stata, e pe'l contrario dove adesso è terra, è stato, e serà mare." Longiano, *Meteorologia,* 33r.

61. Longiano, 33v.

62. "Non è dunque possibile secondo i naturali, che sia il diluvio universale." Longiano, 33v–34r.

63. Longiano, 44v. Longiano also mentions that "astronomers" and "theologians" disagree on the reason why the earth is not entirely covered by water ("Perch'ella non . . . cuopra tutta la terra, nel modo, che fa l'aere, e 'l fuogo, altrimente senteno gli astronomi et i theologi." Longiano, 34r). However, this has nothing to do with theological issues: "astronomers" are simply the supporters of the theory of the magnetic attraction of the northern stars on the emerged land, as opposed to the uneven density of the sphere of earth; "theologians" are those who invoke direct divine intervention.

64. Martin, *Renaissance Meteorology*, 53–54.

65. Thierry Belleguic and Anouchka Vasak, eds., *Ordre et désordre du monde enquête sur les météores, de la Renaissance à l'âge moderne.* (Paris: Hermann, 2013).

66. Joëlle Ducos, *La Météorologie En Français Au Moyen Age (XIIIe-XIVe Siècles)* (Paris: Honoré Champion, 1998); Claude Thomasset and Joëlle Ducos, *Le temps qu'il fait au Moyen âge: phénomènes atmosphériques dans la littérature, la pensée scientifique et religieuse* (Paris: Presses Paris Sorbonne, 1998).

67. On popular prophecy in early sixteenth-century Italy: Ottavia Niccoli, *Prophecy and People in Renaissance Italy*, trans. Lydia Cochrane (Princeton University Press, 1990).

68. Lorraine Daston and Katharine Park, *Wonders and the Order of Nature, 1150–1750* (New York: Zone Books, 1998).

69. Tanner, *Decrees of the Ecumenical Councils*, 1:634–35.

70. Rienk Vermij, "A Science of Signs. Aristotelian Meteorology in Reformation Germany," *Early Science and Medicine* 15, no. 6 (October 1, 2010): 650–51.

71. Craig Martin, *Renaissance Meteorology: Pomponazzi to Descartes* (Baltimore: Johns Hopkins University Press, 2011), 40.

72. Rienk Vermij, "A Science of Signs. Aristotelian Meteorology in Reformation Germany," *Early Science and Medicine* 15, no. 6 (October 1, 2010): 659–71.

73. Paolo Sarpi, *The History of the Council of Trent* (London: J. Macock, 1676), 1–2.

74. For a general introduction to the Council of Trent: Adriano Prosperi, *Il Concilio Di Trento: Una Introduzione Storica* (Torino: Einaudi, 2001).

75. Girolamo Seripando, *Prediche del reuer.mo mons. Girolamo Seripando, arciuescouo di Salerno . . . sopra il simbolo de gli Apostoli, dichiarato co simboli del concilio Niceno, & di Santo Athanasio...* (Venice: Al segno della Salamandra, 1567), 20v. On Seripando, see: Hubert Jedin, *Papal Legate at the Council of Trent, Cardinal Seripando*, trans. Frederic Clement Eckhoff (St. Louis: B. Herder, 1947); Antonio Cestaro, ed., *Geronimo Seripando e la Chiesa del suo tempo: nel V centenario della nascita: Atti del Convegno di Salerno, 14-16 Ottobre 1994* (Roma: Edizioni di storia e letteratura, 1997).

76. Building on previous German scholarship, Hubert Jedin distinguished between a "Catholic reformation," which from the early sixteenth century called for the spiritual renewal of the Church, and the later "Counter-reformation," namely the defensive struggle to stop and reverse the spread of Protestantism. Erwin Iserloh, Josef Glazik, and Hubert Jedin, *Reformation and Counter Reformation* (New York: Seabury Press, 1986). At least in this case, the distinction seems justified. With particular reference to the Italian context: Prosperi, *Il Concilio Di Trento*, 165–85.

77. Gigliola Fragnito, *Gasparo Contarini: un magistrato veneziano al servizio della cristianità* (Florence: Olschki, 1988), 6.

78. Elisabeth G. Gleason, *Gasparo Contarini: Venice, Rome, and Reform* (Berkeley: University of California Press, 1993), 80–81.

79. Elisabeth G. Gleason, *Gasparo Contarini*, 85.

80. Gasparo Contarini, *De elementis et eorum mixtionibus libri quinque... nunc primum in lucem aediti. Scipionis Capitii de principiis rerum poema* (Paris: Dives Nicolaus, 1548), 39r.

81. Frans Titelmans, *Compendium naturalis philosophiae libri duodecim de consideratione rerum naturalium: earumque ad suum creatorem reductione* (Paris: Francisum Stephanum, 1547). On Titelmans's natural philosophy: Martin, *Renaissance Meteorology*, 28–29; David Lines, "Teaching Physics in Louvain and Bologna. Frans Titelmans and Ulisse Aldrovandi," in *Scholarly Knowledge: Textbooks in Early Modern Europe*, ed. Emidio Campi (Paris: Librairie Droz, 2008), 183–203.

82. Titelmans, *Compendium Naturalis Philosophiae*, 88v–89r.

83. Titelmans's treatise was still reprinted as late as 1658. See: Lines, "Teaching Physics in Louvain and Bologna. Frans Titelmans and Ulisse Adrovandi," 183–84.

84. Francesco Beretta, "Orthodoxie philosophique et Inquisition romaine aux 16e-17e siècles," *Historia Philosophica* 3 (2005): 77–78.

85. Alessandro Piccolomini, *La prima parte della filosofia naturale* (Roma: Valgrisi, 1551), XIIr.

86. Piccolomini, 248–51.

87. "La Terra, o di nuovo (come haviam da credere) fu prodotta al Mondo, ò ver (come stimarono molti) da profondo diluvio d'acque già fatta libera, come rinata, & rinnovata si discoperse." Piccolomini, *La prima parte della filosofia naturale*, iiiv. It should be noted that "di nuovo" ("recently") was traditionally used in the sense of "created".

88. Paul F. Grendler, *L'inquisizione romana e l'editoria a Venezia, 1540–1605* (Rome: Il Veltro, 1983), 35–36, 82. See also the recent English edition: Paul F. Grendler, *The Roman Inquisition and the Venetian Press, 1540–1605* (Princeton: Princeton University Press, 2015) on ecclesiastic censorship in Venice in the late fourteenth–early fifteenth century: Christopher L. C. E. Witcombe, *Copyright in the Renaissance: Prints and the Privilegio in Sixteenth-Century Venice and Rome* (Leiden: Brill, 2004), 62–66.

89. Paul F. Grendler, *L'inquisizione romana e l'editoria a Venezia: 1540–1605* (Rome: Il Veltro, 1983), 72–73.

90. Elisabeth G. Gleason, *Gasparo Contarini*, in particular pp. 99–101.

91. Grendler, *L'inquisizione romana e l'editoria a Venezia*, 38–42.

92. Grendler, 133.

93. Lodovico Nogarola, *Ludouici Nogarolae . . . Dialogus. Qui inscribitur Timotheus, siue de Nilo* (apud Vincentium Valgrysium, 1552). On Nogarola: Paolo Pellegrini, "Nogarola, Ludovico," in *DBI*, vol. 78 (Treccani, 2013), 683–86.

94. On Ramusio, see: Jerome Barnes, *Giovanni Battista Ramusio and the History of Discoveries: An Analysis of Ramusio's Commentary, Cartography, and Imagery in "Delle Navigationi et Viaggi"* (Ph.D. Dissertation, University of Texas at Arlington, 2007).

95. Barnes, *Giovanni Battista Ramusio and the History of Discoveries*, 35. On the political aims and implications of Ramusio's geographic works: Simone Testa, *Italian Academies and Their Networks, 1525-1700: From Local to Global* (New York: Palgrave Macmillan, 2015), 88–91.

96. Giovanni Battista Ramusio, *Primo volume delle nauigationi et viaggi* etc. (Venice: Lucantonio Giunti, 1550), 36. For Ramusio's letter to Fracastoro on the Nile and Fracastoro's reply: Ramusio, 281–88. Many more were published in: Girolamo Fracastoro, *Hieronymi Fracastorii . . . Adami Fumani . . . et Nicolai Archii . . . carminum editio II*, ed. Adamo Fumano, vol. 1 (Padua: J. Cominus, 1739), 110–40.

97. "Ancor che queste cose non sieno eterne, cioè le cose de' fiumi, de' monti, e della terra, sono però vicine all'eterne." Letter from G. Fracastoro to G. B. Ramusio, Verona, Jan. 25, 1548, in: Fracastoro, *Hieronymi Fracastorii . . . Adami Fumani . . . et Nicolai Archii . . . carminum editio II*, 1:92.

98. Ramusio, *Primo volume delle nauigationi et viaggi*, 285–89.

99. "FRA. Contra summos illos viros fateor me dixisse. Sed non temere nec sine causa." Nogarola, *Ludouici Nogarolae... Dialogus*, 21r.

100. Nogarola, 16r-v.

101. Alessandro Scafi, *Mapping Paradise: A History of Heaven on Earth* (London: British Library, 2006). Angelo Cattaneo, *Fra Mauro's Mappamundi and Fifteenth-Century Venetian Culture* (Turnhout: Brepols, 2011), 134–48. On the Renaissance tradition that identified Eden with the whole Earth: Joseph E. Duncan, "Paradise as the Whole Earth," *Journal of the History of Ideas* 30, no. 2 (1969): 171–86. Nogarola's dialogue is briefly mentioned at page 177.

102. "TIM. Quare meo iudicio intelligendum est, hominem posteaquam divina neglexit mandata, non ex alto loco in alium expulsum fuisse atque eiectum, se terrae potius quam incolebat, vim et naturam fuisse mutatam." Nogarola, *Ludouici Nogarolae... Dialogus*, 19r-20v.

103. "FRA. Mihi vero valde placet." Nogarola, *Ludouici Nogarolae... Dialogus*, 21r.

104. Valerio Faenzi, *De montium origine, Valerii Fauenties, ordinis Praedicatorum, dialogus* (Venice: Academia Veneta, 1561). Italian translations have been published in 2001 and 2006: Valerio Faenzi et al., *Sull'origine delle montagne* (Verbania: Tararà, 2006); Valerio Faenzi, *L'origine dei monti*, ed. Ezio Filippi (Verona: Della Scala, 2001). Parts of Faenzi's work have been translated into English: Frank Dawson Adams, *The Birth and Development of the Geological Sciences* (Baltimore: William & Wilkins, 1938), 344–57. On Faenzi's dialogue see, above all: Maria I. Campanale, *Ai confini del Medioevo scientifico: il De montium origine di Valerio Faenzi* (Bari: Edipuglia, 2012); Ivano Dal Prete, "Valerio Faenzi e l'origine dei monti nel Cinquecento Veneto," in *Wissenschaft– Berge – Ideologien. Johann Jakob Scheuchzer (1672-1733) Und Die Frühneuzeitliche Naturforschung*, (Basel: Schwabe, 2010), 199–214. On the short-lived Academy of Fame: Pietro Pagan, "Sulla Accademia 'Veneziana' o Della 'Fama,'" *Atti Dell'Istituto Veneto Di Scienze, Lettere Ed Arti*, Cl. di scienze morali, lettere ed arti, CXXXII (74/ 1973): 359–92. Angela Nuovo, *The Book Trade in the Italian Renaissance*, Library of the Written Word; 26 (Leiden: Brill, 2013). Lina Bolzoni, "L'Accademia Veneziana: splendore e decadenza di un'utopia enciclopedica," in *Università, accademie e società scientifiche in Italia e in Germania dal Cinquecento al Settecento*, ed. Laetitia Boehm and Ezio Raimondi (Bologna: Il mulino, 1981), 117–69. Lina Bolzoni, "Il 'Badoaro' di Francesco Patrizi e l'Accademia veneziana della Fama," *Giornale Storico della Letteratura Italiana* 158, no. 501 (1981): 71–101.

105. "Quod rationi consentaneum fuerit, id tantum veritatis inquirendae proferatur gratia." Faenzi, *De montium origine, Valerii Fauenties, ordinis Praedicatorum, dialogus*, 2r.

106. Camillus was Faenzi's given name before he took his vows. While the preface identifies Camillus with one of Faenzi's brothers, none of them had this name. See: Faenzi, *L'origine dei monti*, 36.

107. Faenzi, *De montium origine, Valerii Fauenties, ordinis Praedicatorum, dialogus*, 14v.

108. Faenzi, 14v.

109. Faenzi, *De Montium Origine*, 16r.

110. Bolzoni, "L'Accademia Veneziana: splendore e decadenza di un'utopia enciclopedica," 150–59.

111. Letter from M. Bonelli to G. A. Facchinetti, May 11, 1566 (n. 4), in Franco Gaeta and Aldo Stella, eds., *Nunziature Di Venezia*, vol. 8 (Roma: Istituto storico italiano per l'età moderna e contemporanea, 1963); Pio Paschini, *Venezia e l'inquisizione Romana da Giulio III a Pio IV* (Padua: Antenore, 1959), 140; Grendler, *L'inquisizione romana e l'editoria a Venezia*, 77.

112. Faenzi remained in charge for more than seven years, against an average of five. He was later granted a minor bishopric, a rather common practice when the inquisitor was (as in his case) a Venetian subject. Grendler, *L'inquisizione romana e l'editoria a Venezia*, 77–78.

113. Letter from G. A. Facchinetti to M. Bonelli, June 10, 1570. In: Gaeta and Stella, *Nunziature Di Venezia*, vol. 8, n. 200.

114. For a reappraisal, see: Neil Tarrant, "Censoring Science in Sixteenth-Century Italy: Recent (and Not-So-Recent) Research," *History of Science* 52, no. 1 (March 1, 2014): 1–27. See also Tarrant's unpublished dissertation: Neil Tarrant, "Disciplining the School of Athens: Censorship, Politics and Philosophy, Italy 1450–1600" (Sussex: University of Sussex, 2009).

115. Scientific titles (of which meteorology was of course just a small part) represented about 11% all of the imprimaturs conceded in Venice to new titles between 1551 and 1607. This figure remained remarkably stable throughout the period. Grendler, *The Roman Inquisition and the Venetian Press, 1540–1605*, 132.

116. Martin, *Renaissance Meteorology*, 60–79.

117. Giasone De Nores, *Breue trattato del mondo, et delle sue parti, semplici, et miste: con molte altre considerationi, che di grado in grado saranno piu notabili, & piu degne di cognitione: di Jason De Nores* (Venice: Andrea Muschio, 1571), 6v.

118. Giovanni Camillo Maffei, *Scala naturale overo fantasia dolcissima, ... intorno alle cose occulte e desiderate nella filosofia* (Venice: Gio. Varisco, 1564), 139.

119. Gozze's *Discourses* are also noteworthy in that the wife of the author, Maria Gondola, took part in their publication. See: Meredith K. Ray, *Daughters of Alchemy: Women and Scientific Culture in Early Modern Italy* (Cambridge, MA: Harvard University Press, 2015), 112–14. Eleonora Carinci and Sandra Plastina, eds., *Corrispondenze scientifiche tra Cinquecento e Seicento. Camilla Erculiani «Lettere di philosophia naturale» (1584). Margherita Sarrocchi «Lettere a Galilei» (1611–1612)* (Lugano: Agorà & Company, 2016), 79–92.

120. Gozze, 10r-v.

121. Gozze, 10r-v.

122. Gozze, 54r.

123. Francesco Patrizi, *Della retorica dieci dialoghi di M. Francesco Patritio: nelli quali si fauella dell'arte oratoria con ragioni repugnanti all'openione, che intorno a quella*

hebbero gli antichi scrittori (Venice: Francesco Senese, 1562). Among the vast literature on Patrizi: Cesare Vasoli, *Francesco Patrizi Da Cherso* (Roma: Bulzoni, 1989); Sandra Plastina, *Gli alunni di Crono: mito, linguaggio e storia in Francesco Patrizi da Cherso (1529–1597)* (Soveria Mannelli: Rubbettino, 1992).

124. On Patrizi and the Academy of Fame: Lina Bolzoni, Il 'Badoaro' di Francesco Patrizi e l'Accademia veneziana della Fama," *Giornale Storico della Letteratura Italiana*, v. 158, n. 501 (1981): 71–101.

125. "Ne primi secoli. . . *dopo l'ultima rinovation del mondo*, questa terra, che noi habitiamo, non fu di questa forma, ne di si picciola grandezza, ch'ella è al presente. Ma di gran lunga maggiore, et di perfetta rotondita." Patrizi, *Della retorica*, 5v. Emphasis mine.

126. Patrizi, 6v. Plastina, *Gli Alunni di Crono*, 38–42. On Patrizi's historical writings in a Renaissance context, see also: Anthony Grafton, *What Was History? The Art of History in Early Modern Europe* (Cambridge; Cambridge University Press, 2007).

127. Girolamo Borro, D*ialogo del flusso e reflusso del mare d'Alseforo Talascopio (G. Borro). Con un ragionamento di Telifilo Filogenio della perfettione delle donne.* (Lucca: V. Busdragho, 1561), 18–20; Girolamo Borro, *Del flusso e reflusso del mare, & dell'inondatione del Nilo* (Florence: Nella stamperia di Giorgio Marescotti., 1583), 22–25.

128. Borro, *Del flusso e reflusso del mare, & dell'inondatione del Nilo*, 175–76.

129. Ugo Baldini and Leen Spruit, eds., *Catholic Church and Modern Science: Documents from the Archives of the Roman Congregations of the Holy Office and the Index* (Roma: Libreria Editrice Vaticana, 2009), 812.

130. Stabile, "Borri, Girolamo," *DBI* (Rome: Treccani, 1971), 16; Baldini and Spruit, *Catholic Church and Modern Science*, 816. The first edition of the *Dialogue* on the tides was examined in 1564, but the inquisitors were looking for references to prohibited books, deterministic astrology, and the mortality of the human soul; Baldini and Spruit, 813–15.

131. Juan Jarava, *I quattro libri della filosofia naturale di Giovan Sarava. Dove... si discopron tutte le principali materie fisiche, le prime cagioni, e gli effetti loro, & i fini... Tradotti di spagnuolo in italiano da Alfonso Ulloa*, trans. Alfonso Ulloa (Venice: Plinio Pietrasanta, 1557), 38–40. Translated from: Juan Jarava, *La philosophia natural brevemente [t]ratada y con mucha diligencia copilada de Aristotilés, Plinio, Platon y otros grades autores* (Martin Nucio, 1546). The same publisher reprinted it in 1565.

132. Eleonora Carinci, "Una 'speziala' padovana: Lettere Di Philosophia Naturale di Camilla Erculiani (1584)," *Italian Studies* 68, no. 2 (July 1, 2013): 211–12.

133. Bartolomeo Arnigio, *Meteoria over discorso intorno alle impressioni imperfette, humide, secche et miste cosi in alto, come nelle viscere della terra generate (etc.)* (Brescia: Fratelli de Marchetti, 1568), 85v.

134. Cesare Rao, *Dell'origine de' monti* (Naples: Orazio Salviani, 1577). On Cesare Rao, see: Donato Verardi, "Rao, Cesare," in *Encyclopedia of Renaissance Philosophy* (Springer, Cham, 2015), 1–3.

135. Rao, *Dell'origine de' monti*, 5v.

136. Rao, 13v.

137. Rao, 20r. The 'mineral virtue' was the power that turned mud and soft materials (including animal tissues) into stone.

138. Cesare Rao, *I Meteori di Cesare Rao di Alessano citta di terra d'Otranto. I quali contengono quanto intorno a tal materia si puo desiderare* (Venice: Giouanni Varisco & Compagni, 1582).

139. Rao, 103.

140. Alessandro Piccolomini, *Della grandezza della terra et dell'acqua* (appresso Giordano Ziletti, all'Insegna della Stella, 1557). A second edition appeared in 1561. Among other reactions elicited by Piccolomini's work, see: Antonio Berga, *Discorso di M. Antonio Berga lettore filosofo nella vniversità di Turino, della grandezza dell'acqua & della terra. Contra l'opionione dil S. Alessandro Picolomini* (Turin: Eredi Bevilaqua, 1579).

141. Agostino Michele, *Trattato della grandezza dell'acqua et della terra. Di Agostino Michele. Nel quale contro l'opinione di molti filosofi, et di molti matematici illustri. Dimostrasi l'acqua esser di maggior quantità della terra* (Venice: Niccolò Moretti, 1583), 5v. On Michele's work: Pasquale Ventrice, *La discussione sulle maree tra astronomia, meccanica e filosofia nella cultura veneto-padovana del Cinquecento* (Venice: Istituto veneto di scienze, lettere ed arti, 1989).

142. Michele, *Trattato della grandezza dell'acqua et della terra*, 6v.

143. Michele, *Trattato della grandezza dell'acqua et della terra*, 5v.

144. Vitale Zuccolo, *Dialogo delle cose meteorologiche. Di D. Vitale Zuccolo padoano... In cui si dichiarano tutte le cose marauigliose, che si generano nell'aere, & alcune mirabili proprietà de' fonti, fiumi, e mari, secondo la dottrina d'Aristotele con le opinioni d'altri illustri scrittori* (Venice: Paolo Megietti, 1590), 162.

145. Grendler, *The Roman Inquisition and the Venetian Press, 1540–1605*, 158.

146. On the disputes between the Jesuits and Cesare Cremonini on the teaching of Aristotelian philosophy, as well as the role of the Venetian government, see: Maurizio Sangalli, "Cesare Cremonini, la Compagnia di Gesù e la Repubblica di Venezia: eterodossia e protezione politica," in Ezio Riondato and Antonino Poppi, eds., *Cesare Cremonini: aspetti del pensiero e scritti: atti del convegno di studio (Padova, 26-27 febbraio 1999)* (Padua: Accademia galileiana di scienze, lettere ed arti in Padova, 2000), 207–18.

147. On Cremonini: *Cesare Cremonini: aspetti del pensiero e scritti: atti del convegno di studio (Padova, 26-27 febbraio 1999)* (Padua: Accademia galileiana di scienze, lettere ed arti in Padova, 2000).

148. Giovanfrancesco Sagredo to Galileo Galilei, Venice, Feb. 7, 1615. Antonio Favaro, ed., *Le Opere di Galileo Galilei: Edizione Nazionale sotto gli auspicii di Sua Maestà*

il Re d'Italia, vol. XII (Florence: Tip. di G. Barbèra, 1902), 139. Translation is mine.

149. Federico Barbierato, *The Inquisitor in the Hat Shop: Inquisition, Forbidden Books and Unbelief in Early Modern Venice* (Leiden: Routledge, 2012), 168.

150. Barbierato, 170.

151. See: Richard J. Blackwell, *Galileo, Bellarmine, and the Bible: Including a Translation of Foscarini's Letter on the Motion of the Earth* (Notre Dame: University of Notre Dame Press, 1991), 11. The Latin term *mores* has actually a broader meaning that includes tradition and custom. For the original text: Norman P. Tanner, ed., *Decrees of the Ecumenical Councils*, vol. 2 (Washington: Sheed & Ward; Georgetown University Press, 1990), 663.

152. Blackwell, *Galileo, Bellarmine, and the Bible*, 12.

153. Blackwell, 29–45. Ugo Baldini, *Legem impone subactis: studi su filosofia e scienza dei Gesuiti in Italia, 1540–1632* (Roma: Bulzoni, 1992), 285–344; Paolo Galluzzi and Ugo Baldini, eds., "L'astronomia del Cardinale Bellarmino," in *Novità celesti e crisi del sapere* (Florence: Giunti Barbèra, 1984), 293–305; Adriana Valerio, "La Verità luogo teologico in Bellarmino," in *Bellarmino e la Controriforma: atti del Simposio Internazionale di Studi, Sora, 15-18 Ottobre, 1986*, ed. Romeo de Maio et al. (Sora, Italy: Centro di Studi Sorani "Vincenzo Patriarca," 1990), 51–87 (especially pp. 60–62).

154. Benito Pereira, *B. Pererii ... Prior Tomus Commentariorum ... Et Disputationum in Genesim; Continens Historiam Mosis Ab Exordio Mundi Usque Ad Noëticum Diluvium, Septem Libris Explanatam. Adjecti Sunt Quatuor Indices, Etc* (Ingolstadt: David Sartorius, 1590).

155. Arnold Williams, *The Common Expositor: An Account of the Commentaries on Genesis, 1527–1633* (Chapel Hill: University of North Carolina Press, 1948), 33.

156. Pereira, *Prior Tomus Commentariorum ... Et Disputationum in Genesim*, 3–4. See also: Blackwell, *Galileo, Bellarmine, and the Bible*, 20–21.

157. Benito Pereira, *De Communibus Omnium Rerum Principiis Libri Quindecim*, 1586. According to William Wallace, Pereira was "more open to Averroist interpretations than other Jesuits at Rome." William Wallace, "Natural Philosophy: Traditional Natural Philosophy," in *The Cambridge History of Renaissance Philosophy*, ed. C. B. Schmitt et al. (Cambridge: Cambridge University Press, 1988), 228.

158. Pereira, *Prior Tomus Commentariorum ... Et Disputationum in Genesim*, 26–27.

159. Pereira, 27. English translation from: Blackwell, *Galileo, Bellarmine, and the Bible*, 22.

160. Pereira, 38–40.

161. Ecclesiastes 1:2

162. Pereira, 40–43.

163. Grafton, "Dating History," 84.

164. Grafton, "From De Die Natali to De Emendatione Temporum," 125. It is impossible here to discuss Renaissance chronology in the details it would deserve. On

Scaliger and his chronological work, see Grafton's monumental study: Anthony Grafton, *Joseph Scaliger: A Study in the History of Classical Scholarship* (Oxford: Oxford University Press, 1983). In the second volume, the introduction and chapter I present an overview of Renaissance chronology before Scaliger.

165. Blackwell, *Galileo, Bellarmine, and the Bible*, 139. Blackwell's work is based on material previously published in: Ugo Baldini, "Galileo, la nuova astronomia e la critica all'aristotelismo nel dialogo epistolare tra Giuseppe Biancani e i revisori romani della Compagnia di Gesù," *Annali dell'Istituto e Museo di Storia della Scienza di Firenze* IX (1984): 13–43. Reprinted in: Ugo Baldini, "Uniformitas et soliditas doctrinae: le censure librorum e opinionum," in *Legem impone subactis: studi su filosofia e scienza dei Gesuiti in Italia, 1540–1632* (Roma: Bulzoni, 1992), 75–111.

166. See in particular, for the cases of Christoph Grienberger and Giuseppe Biancani: Baldini, "Galileo, la nuova astronomia e la critica all'aristotelismo"; Blackwell, *Galileo, Bellarmine, and the Bible*, 158. Later in the century, Jesuit astronomer Giuseppe Ferroni (1628–1709) spent most of the academic year "confuting" the Copernican system so he could teach it. See: Maurizio Torrini, "Giuseppe Ferroni Gesuita e galileiano," *Physis* 15 (1973): 411–23. Even Giambattista Riccioli, the seventeenth-century champion of geocentricism, has often been suspected of being a secret Copernican. Alfredo Dinis, "Was Riccioli a Secret Copernican?," in *Giambattista Riccioli e il merito scientifico dei gesuiti nell'età barocca* (Florence: Olschki, 2002), 1000–29.

167. "Erit quidem pro Aristotele magnum argumentum incorruptibilitatis, qui mundum putavit eternum. At vero in mundo qui nondum duravit 8000 leviusculum est." ARSI, F.G. 655, *Notae in Cosmographiam P. Iosephi Blancani*, f. 114r. Published in: Baldini, "Galileo, la nuova astronomia e la critica all'aristotelismo," 33.

168. Baldini argues that Grienberger followed the values of the Alphonsine Tables, which reckoned 6,984 years between the creation and the year 1252. Baldini, 37, note 7.

169. Giuseppe Biancani, *Aristotelis Loca Mathematica:* . . . *Collecta et Explicata* etc. (Bologna: Bartolomeo Cocchi, 1615). Giuseppe Biancani, *Sphaera mundi, seu Cosmographia* etc. (Bologna: typis Sebastiani Bonomij, 1620). Biancani's *Cosmography* was reprinted in 1635 and 1653 (I also consulted the 1653 edition). On Biancani and his Jesuit censors, see: Blackwell, *Galileo, Bellarmine, and the Bible*, 148–53.

170. On the evolving nature of Renaissance Aristotelianism (or rather Aristotelianisms), see C. Schmitt's classic study: Charles B. Schmitt, *Aristotle and the Renaissance* (Cambridge, MA: Harvard University Press, 1983). On the empirical and experimental turn of Renaissance meteorology: Martin, *Renaissance Meteorology*.

171. Biancani, *Aristotelis Loca Mathematica*, 103.

172. "Quapropter nisi igne illo, quem sacrae literae innuunt cataclysmus ille praeveniatur, aqua mundus interiturus esset." Biancani, *Aristotelis loca*, p. 106.

Biancani's original text read "interiturus est"; one of the censors asked for it to be changed into the more hypothetical "interiturus esset," or even "interiturus videntur". Published in: Baldini, "Galileo, la nuova astronomia e la critica all'aristotelismo," 29.

173. Niccolò Cabeo, *In Quatuor Libros Meteorologicorum Aristotelis Commentaria, et Quaestiones: Quatuor Tomis Compraehensa*, 4 vols. (Rome: Eredi Francesco Corbelletti, 1646). On Cabeo's *Meteorology*: Craig Martin, "With Aristotelians Like These, Who Needs Anti-Aristotelians? Chymical Corpuscular Matter Theory in Niccolò Cabeo's Meteorology," *Early Science and Medicine* 11, no. 2 (April 1, 2006): 135–61. Reprinted in: Craig Martin, *Renaissance Meteorology: Pomponazzi to Descartes* (Baltimore: Johns Hopkins University Press, 2011), 106–24.

174. "Autores tamen etiam Catholici divisi sunt, aliqui enim putant mundum potuisse esse ab aeterno." Cabeo, *In Quatuor Libros Meteorologicorum Aristotelis Commentaria*, 1:410.

175. Cabeo, 1:408–16. See also: Martin, *Renaissance Meteorology*, 122–23.

176. Cabeo, *In Quatuor Libros Meteorologicorum Aristotelis Commentaria*, 1:421–22. The eccentricity of the sphere of water would be proved by the different heights of the surfaces of the Mediterranean and of the Red Sea, reported by a number of classical sources.

177. Cabeo, 1:423.

178. Martin, *Renaissance Meteorology*, 118–19.

179. Athanasius Kircher, *Athanasii Kircheri . . . Mundus Subterraneus: In XII Libros Digestus*, vol. II, 2 vols. (Amstelodami: apud J. Janssonium & E. Weyerstraten, 1665).

180. Michael John Gorman, "The Angel and the Compass: Athanasius Kircher's Magnetic Geography," in *Athanasius Kircher: The Last Man Who Knew Everything*, ed. Paula Findlen (New York: Routledge, 2004), 255–57. Mark A. Waddell, "The World, As It Might Be: Iconography and Probabilism in the Mundus Subterraneus of Athanasius Kircher," *Centaurus* 48, no. 1 (January 1, 2006): 3–4. See also: Mark A. Waddell, *Jesuit Science and the End of Nature's Secrets* (Routledge, 2016), 119–60.

181. On the iconography and symbolism of the *Subterranean World*: Waddell, "The World, As It Might Be." William C. Parcell, "Sign and Symbols in Kircher's Mundus Subterraneus," in *The Revolution in Geology from the Renaissance to the Enlightenment*, ed. Gary D. Rosenberg (Boulder, Colorado: Geological Society of America, 2009), 63–74.

182. On Kircher's world machinery: Tara Nummedal, "Kircher's Subterranean World and the Dignity of the Geocosm," in *Great Art of Knowing: The Baroque Encyclopedia of Athanasius Kircher*, ed. Daniel Stolzenberg (Stanford, CA: Stanford University Press, 2001), 37–47. On theories of a central fire (which replaced the medieval assumption of a central coldness): Rienk Vermij, "Subterranean Fire. Changing

Theories of the Earth During the Renaissance," *Early Science and Medicine* 3, no. 4 (1998): 323–47. Luca Ciancio, "An Amphiteathre Built on Toothpicks": Galileo, Nardi, and the Hypothesis of Subterranean Fire" *Galilaeana*, XV (2018): 83–113.

183. Martin, "With Aristotelians Like These, Who Needs Anti-Aristotelians?" Martin, *Renaissance Meteorology*, 106–24.

184. Hiro Hirai, "Interprétation chymique de la création et origine corpusculaire de la vie chez Athanasius Kircher," *Annals of Science* 64, no. 2 (April 1, 2007): 217–34. The most explicit formulation of his theory is in: Kircher, *Athanasii Kircheri . . . Mundus Subterraneus*, II:327.

185. Stephen Jay Gould, "Father Athanasius on the Isthmus of a Middle State. Understanding Kircher's Palaeontology," in *Athanasius Kircher: The Last Man Who Knew Everything*, ed. Paula Findlen (New York: Routledge, 2004), 235.

186. Anthony Grafton, "Kircher's Chronology," in *Athanasius Kircher: The Last Man Who Knew Everything*, ed. Paula Findlen (New York: Routledge, 2004), 183.

187. See: Baldini, 1984.

188. See: ARSI, Censurae Opinionum I-III (1565–1652).

189. ARSI, FG 663, Censurae Librorum 1662–1663, pp. 319–27ᵃ. The work was reviewed by a number of censors, many praising it for showing the "wonderful" world that the "summum opifex" had hidden from human eyes.

190. On Kircher and the Jesuit censorship: Harald Siebert, "Kircher and His Critics. Censorial Practice and Pragmatic Disregard in the Society of Jesus," in *Athanasius Kircher: The Last Man Who Knew Everything*, ed. Paula Findlen (New York: Routledge, 2004), 79–103. Unfortunately, Siebert's study does not include the censorship of Kircher's last works.

191. Grafton, "Kircher's Chronology," 178. Athanasius Kircher, *Arca Noë* (Amsterdam: apud J. Janssonium & E. Weyerstraten, 1675), 3.

CHAPTER 5

1. Among literature on early modern libertinism: Giorgio Spini, *Ricerca dei libertini: la teoria dell'impostura delle religioni nel Seicento italiano* (Florence: La nuova Italia, 1983); Alain Mothu, ed., *Révolution Scientifique et Libertinage* (Thornout, Belgium: Brepols, 2000).

2. Lydia Barnett, *After the Flood. Imagining the Global Environment in Early Modern Europe* (Baltimore: Johns Hopkins University Press, 2019), 54; Maria Susana Séguin, *Science et religion dans la pensée française du XVIIIe siècle: le mythe du Déluge universel* (Paris: Champion, 2001), 16–17.

3. Jean Buridan, *Ioannis Buridani Expositio et quæstiones in Aristotelis De cælo*, ed. Benoît Patar (Louvain-La-Neuve: Editions Peeters, 1996), 414. Also in: Jean Buridan, *Iohannis Buridani Quaestiones Super Libris Quattuor de Caelo et Mundo*, ed. Ernest A. Moody (Cambridge, Mass: The Mediaeval Academy of America, 1942), 158. See also: Pierre Duhem, *Le Système Du Monde; Histoire Des Doctrines*

Cosmologiques de Platon à Copernic, vol. IX (Paris: A. Hermann, 1958), 197; Ernest A. Moody, "John Buridan on the Habitability of the Earth," *Speculum* 16, no. 4 (1941): 423–24.

4. On the passage from the medieval model of two different spheres of earth and water to the modern conception, see, above all: David Wootton, *The Invention of Science: A New History of the Scientific Revolution* (London: Allen Lane, 2015), 110–43.

5. In: Giuliano Gliozzi, *Adamo e il nuovo mondo* (Milan: Franco Angeli, 1977), 308.

6. Hartmann Schedel, *Liber chronicarum* (Nuremberg: Anton Koberger, 1493), XIv-XIIr.

7. On medieval and early modern "monsters" and the uncertain borders of humanity: Lorraine Daston and Katharine Park, *Wonders and the Order of Nature, 1150–1750* (New York; Cambridge, MA: Zone Books, 1998); Surekha Davies, *Renaissance Ethnography and the Invention of the Human: New Worlds, Maps and Monsters*, (Cambridge: Cambridge University Press, 2016).

8. Andrea Cesalpino, *Andreae CaesalpinI . . . Peripateticarum quaestionum libri quinque* (Venice: apud Iuntas, 1571), 94. See also: Gliozzi, *Adamo e il nuovo mondo*, 318–20. On Pomponazzi and spontaneous generation: Vittoria Perrone Compagni, "Un'ipotesi non impossibile. Pomponazzi sulla generazione spontanea dell'uomo (1518)," *Bruniana & Campanelliana* 13, no. 1 (2007): 99–111.

9. The Swiss physician Theophrastus von Hohenheim (best known as Paracelsus, 1493–1541) placed the Americans among the non-Adamic or pre-Adamic species devoid of immortal soul. Dino Pastine, "Le origini del poligenismo e Isaac Lapeyrère," in *Miscellanea Seicento*, vol. II (Florence: Le Monnier, 1971), 31–38; Gliozzi, *Adamo e il nuovo mondo*, 306–14. On Michel de Montaigne's pre-Adamitism: Gliozzi, 199–219. Both pre-Adamitism and the spontaneous generation of men figured among the charges brought against philosopher Giordano Bruno, burned at the stake in Rome in 1600. Papi, *Antropologia e civiltà nel pensiero di Giordano Bruno*, 79–127; Gliozzi, *Adamo e il nuovo mondo*, 338–47; Pastine, "Le origini del poligenismo e Isaac Lapeyrère," 43; Diego Pirillo, *Filosofia ed eresia nell'Inghilterra del tardo Cinquecento: Bruno, Sidney e i dissidenti religiosi italiani* (Roma: Edizioni di storia e letteratura, 2010), 143–58.

10. Lewis Hanke, *Aristotle and the American Indians: A Study in Race Prejudice in the Modern World* (Bloomington: Indiana University Press, 1970); Lewis Hanke, *All Mankind Is One: A Study of the Disputation between Bartolomé de Las Casas and Juan Ginés de Sepúlveda in 1550 on the Intellectual and Religious Capacity of the American Indians* (DeKalb: Northern Illinois University Press, 1974).

11. Gliozzi, *Adamo e il nuovo mondo*, 291; Barnett, *After the Flood*, 59.

12. Stuart Jenks, "Astrometeorology in the Middle Ages," *Isis* 74, no. 2 (1983): 185–86.

13. Niccoli, 143.

14. McCurdy, *The Notebooks of Leonardo Da Vinci*, 1938, 2:13.

15. On the "great fear" of the year 1524: Lynn Thorndike, *A History of Magic and Experimental Science*, vol. 5 (New York: Columbia University Press, 1941), 178–233; Ottavia Niccoli, "Il diluvio del 1524 fra panico collettivo e irrisione carnevalesca," in *Scienze, credenze occulte, livelli di cultura*, ed. Giancarlo Garfagnini (Florence: Olschki, 1982), 291–368. Reprinted in: Ottavia Niccoli, *Prophecy and People in Renaissance Italy*, trans. Lydia Cochrane (Princeton University Press, 1990), 140–167. For a quantitative analysis of flood-related pamphleteering in early sixteenth-century Germany, see the (impressive) figures provided in: Hans-Joachim Kölher, "The *Flugshriften* and their Importance in Religious Debate: A Quantitative Approach," in *"Astrologi Hallucinati": Stars and the End of the World in Luther's Time*, ed. Paola Zambelli (Berlin: W. de Gruyter, 1986), 153–75.

16. Paola Zambelli convincingly countered Ottavia Niccoli's thesis on the existence of a double apocalyptic literature: reassuring vernacular texts for the lower classes, and much more alarming versions in Latin for the elite. Zambelli, "Fine del mondo o inizio della propaganda," 297, 334–35.

17. Niccoli, *Prophecy and People in Renaissance Italy*, 142, 155.

18. Niccoli, *Prophecy and People in Renaissance Italy*, 150. Norman Cohn, *Noah's Flood: The Genesis Story in Western Thought* (New Haven, CT: Yale University Press, 1996), 26–31.

19. Niccoli, *Prophecy and People in Renaissance Italy*, 167.

20. Dolly MacKinnon, "'Jangled the Belles, and with Fearefull Outcry, Raysed the Secure Inhabitants': Emotion, Memory and Storm Surges in the Early Modern East Anglian Landscape," in *Disaster, Death and the Emotions in the Shadow of the Apocalypse, 1400–1700*, ed. Jennifer Spinks and Charles Zika (London: Palgrave Macmillan, 2016), 155–73.

21. Camilla Erculiani, *Lettere di philosophia naturale di Camilla Herculiana speciala alle tre stelle in Padova, Indrizzate alla Serenissima Regina di Polonia: nella quale si tratta la natural causa delli diluvij et il natural temperamento dell'huomo, et la natural formatione dell'arco celeste* (Krakow: Drukarnia Łazarzowa Kraków, 1584). See also Carinci's recent edition of Erculiani's *Letters*: Eleonora Carinci and Sandra Plastina, eds., *Corrispondenze scientifiche tra Cinquecento e Seicento. Camilla Erculiani «Lettere di philosophia naturale» (1584). Margherita Sarrocchi «Lettere a Galilei» (1611–1612)* (Lugano: Agorà & Company, 2016), 101–46. On Camilla Erculiani: Sandra Plastina, "'Considering the Mutation of the Times and States and Humans': The Letters of Natural Philosophy of Camilla Erculiani," *Bruniana & Campanelliana* 20, no. 1 (2014): 145–56; Carinci and Plastina, *Corrispondenze scientifiche tra Cinquecento e Seicento. Camilla Erculiani «Lettere di philosophia naturale» (1584). Margherita Sarrocchi «Lettere a Galilei» (1611–1612)*, 19–78.

22. Erculiani, *Lettere di philosophia naturale*, fii-v, fiii-r. Erculiani claimed that she elaborated herself this unusual theory (Erculiani, fii-r). I have not found any plausible sources.

23. On the culture and epistolary networks of Renaissance Italian pharmacists: Filippo De Vivo, "La farmacia come luogo di cultura: le spezierie di medicina in Italia," in *Interpretare e curare: medicina e salute nel Rinascimento*, ed. Maria Conforti, Andrea Carlino, and Antonio Clericuzio (Rome: Carocci editore, 2013), 129–42; Valentina Pugliano, "Natural History in the Apothecary's Shop," in *Worlds of Natural History*, ed. Helen Anne Curry et al. (Cambridge: Cambridge University Press, 2018), 44–60.

24. Erculiani, *Lettere di philosophia naturale di Camilla Herculiana speciala alle tre stelle in Padova, Indrizzate alla Serenissima Regina di Polonia*, fiii-r.

25. Erculiani, fiv-r, fiv-v. Giacomo Menochio, *Consiliorum siue Responsorum D. Iacobi Menochii... liber octauus . . .* (Frankfurt: sumptibus Haeredum Andreae Wecheli & Ioan. Gymnici, 1604), 182. Emphasis mine.

26. Menochio, *Consiliorum siue Responsorum D. Iacobi MenochiI . . . liber octauus . . .*, 182.

27. Eleonora Carinci, "Una 'Speziala' Padovana: Lettere Di Philosophia Naturale Di Camilla Erculiani (1584)," *Italian Studies* 68, no. 2 (July 1, 2013): 227–29.

28. Barnett, *After the Flood*. See in particular Ch. 2, pp. 50–88.

29. In 1589, Jesuit censors determined that this opinion was neither erroneous nor dangerous. However, they also resolved that Jesuits should not teach it, as it did not appear sufficiently mainstream. ARSI, Censurae Opinionum I 1565–1627, p. 134. Eighty years later, they still judged inappropriate Simon de Vasconcellos's thesis that Eden was located in Brazil, even though the author only discussed it hypothetically. ARSI, Censurae Opinionum IV 1621–1665, p. 166r (May 20, 1663). See also the broader discussion that follows their opinion, pp. 167r–180ᵃv.

30. Barnett, *After the Flood*, 67.

31. Unnumbered plate between pp. 192 e 193 in: Athanasius Kircher, *Arca Noë* (Amsterdam: apud J. Janssonium & E. Weyerstraten, 1675); Barnett, *After the Flood*, 64–65.

32. Barnett, *After the Flood*, 67–84.

33. Isaac La Peyrère, *Praeadamitae: Sive Exercitatio Super Versibus 12. 13. 14. Cap. V. Epistolae D. Pauli Ad Romanos: Quibus Inducuntur Primi Homines Ante Adamum Conditi* (Amsterdam, 1655). On La Peyrère and his lasting influence: Richard Henry Popkin, *Isaac La Peyrère (1596–1676): His Life, Work, and Influence* (Leiden: Brill, 1987). Anthony Pagden, *The Fall of Natural Man: The American Indian and the Origins of Comparative Ethnology* (Cambridge: Cambridge University Press, 1982). Still essential are the relevant chapters in: Pastine, "Le origini del poligenismo e Isaac Lapeyrère." Gliozzi, *Adamo e il nuovo mondo*. Also: Anthony Grafton, *Defenders of the Text: The Traditions of Scholarship in an Age of Science, 1450-1800* (Cambridge, MA: Harvard University Press, 1991), 204–13; Barnett, *After the Flood*, 58–64.

34. For La Peyrère's enunciation of the local nature of Noah's Flood: Peyrère, *Praeadamitae*, 1655, part II, 202–11.

35. William Poole, *The World Makers: Scientists of the Restoration and the Search for the Origins of the Earth* (Oxford: Peter Lang, 2010), 34–35.

36. Vossius still ended up in the *Index*, but there is little evidence that his ideas on the Deluge were especially targeted. More likely, he was prohibited because of the many Protestant authors he read and cited: Séguin, *Science et religion dans la pensée française du XVIIIe siècle*, 302.

37. For a list of the theses abjured by La Peyrère: Pastine, "Le origini del poligenismo e Isaac Lapeyrère," 193.

38. Pastine, "Le origini del poligenismo e Isaac Lapeyrère," 183–86.

39. Pastine, "Le origini del poligenismo e Isaac Lapeyrère," 173–74. For La Peyrère's demonstration of the extreme antiquity of the world and the doctrine of the "double eternity": La Peyrère, *Praeadamitae*, part II, 129–168.

40. Unnumbered plate in: Kircher, *Arca Noe*.

41. Popkin, *Isaac La Peyrère (1596–1676)*, 16–18.

42. La Peyrère, *Praeadamitae*, 138.

43. Duhem, *Le Système Du Monde; Histoire Des Doctrines Cosmologiques de Platon à Copernic*, IX:288–90. Eleven editions of Peter of Abano's *Conciliator* were printed between 1476 and 1565 in Venice alone: Laura Turetta, "Bibliografia delle Opere a Stampa di e su Pietro d'Abano," *Medicina nei Secoli* 20, no. 2 (2008): 659–734.

44. Albertus Magnus, *Book of Minerals*, ed. Dorothy Wichoff (Oxford: Clarendon Press, 1967), xxxii.

45. Cecco's verses are admittedly obscure, but this seems the most plausible interpretation: "Di fronde vista però vidi impressa / Nel duro marmo, che quando e' si strinse / Nel mezzo delle parti stette oppressa/ Nel molle tempo, come cera al segno / Mostra nel duro sì come dipinse / Natura, che di forma non ha sdegno." See also: Mario Baratta, *Leonardo da Vinci ed i problemi della terra* (Torino: Fratelli Bocca, 1903), 223–24.

46. Cited in Baratta, 224.

47. Leon Battista Alberti, *L'architettura. (De re aedificatoria)*, ed. Paolo Portoghesi, trans. Giovanni Orlandi, (Milan: Edizioni Il Polifilo, 1966), 159.

48. Hanna Rose Shell, "Ceramic Nature," in *Materials and Expertise in Early Modern Europe: Between Market and Laboratory*, ed. Ursula Klein and E. C. Spary (University of Chicago Press, 2010), 50–70.

49. Russiliano, *Una reincarnazione di Pico ai tempi di Pomponazzi*, 166.

50. Luca Ciancio, "Un interlocutore fiammingo di Fracastoro: il medico Iohannes Goropius Becanus (1518–1572) e la teoria dell'origine organica dei fossili," in *Girolamo Fracastoro: fra medicina, filosofia e scienze della natura*, ed. Alessandro Pastore and Enrico Peruzzi (Florence: L.S. Olschki, 2006), 151.

51. Gabriele Falloppio, *Gabrielis Falloppii . . . De medicatis aquis, atque de fossilibus tractatus pulcherrimus, ac maxime utilis: ab Andrea Marcolino . . . collectus. Accessit eiusdem andreae duplex epistola: . . . Cum indice rerum magis obseruandarum*

copiosissimo,ac capitum omnium . . ., ed. Andrea Marcolini (Venice: ex officina Stellae, Iordanis Ziletti, 1564), 109r.

52. Falloppio, 109r.

53. Andrea Cesalpino, *De Metallicis: libri tres* (Rome: Luigi Zannetti, 1596), 5. On his ideas on the formation of rocks and minerals: Hiro Hirai, *Le concept de semence dans les théories de la matière à la Renaissance de Marsile Ficin à Pierre Gassendi* (Turnhout: Brepols, 2005), 155–75 ; François Ellenberger, *History of Geology*, vol. 1 (Rotterdam; Brookfield, VT: A. A. Balkema, 1996), 143–45.

54. On the playfulness of nature in the early modern scientific discourse: Paula Findlen, "Jokes of Nature and Jokes of Knowledge: The Playfulness of Scientific Discourse in Early Modern Europe," *Renaissance Quarterly* 43, no. 2 (July 1, 1990): 292–331.

55. Stephen Jay Gould, "Father Athanasius on the Isthmus of a Middle State. Understanding Kircher's Palaeontology," in *Athanasius Kircher: The Last Man Who Knew Everything*, ed. Paula Findlen (Routledge, 2004), 216. On the influence of Renaissance Platonism on theories of rock and fossil formation: Hirai, *Le concept de semence*. On Cardano: Anthony Grafton, *Cardano's Cosmos: The Worlds and Works of a Renaissance Astrologer* (Cambridge, MA: Harvard University Press, 1999).

56. Ellenberger, *History of Geology*, 1996, 1:137.

57. Giuseppe Biancani, *Aristotelis Loca Mathematica: . . . Collecta et Explicata: Accessere De Natura Mathematicarum Scientiarum Tractatio Atque Clarorum Mathematicorum Chronologia* (Bologna: Bartolomeo Cocchi, 1615), 107.

58. Niccolò Cabeo, *In quatuor libros meteorologicorum aristotelis commentaria, et quaestiones: quatuor tomis compraehensa* (Rome: Eredi Francesco Corbelletti, 1646), 420.

59. Gould, "Father Athanasius on the Isthmus of a Middle State. Understanding Kircher's Palaeontology," 226–34.

60. On early modern Italian museums: Giuseppe Olmi, *l'inventario del mondo: catalogazione della natura e luoghi del sapere nella prima età moderna* (Bologna: Il Mulino, 1992); Paula Findlen, *Possessing Nature: Museums, Collecting, and Scientific Culture in Early Modern Italy* (Berkeley: University of California Press, 1994).

61. Andrea Chiocco and Benedetto Ceruti, *Musaeum Calceolarianum Veronense* (Verona: Apud Angelum Tamum, 1622), 407–9. A landmark in the history of the printed illustration of natural specimens, that famous book reproduced Saraina's entire account of his conversation with Fracastoro. This is the main reason why it did not remain buried in a little-known work of local erudition.

62. Giuliano Gliozzi, "Calzolari, Francesco," in *Dizionario Biografico Degli Italiani* (Rome: Istituto della Enciclopedia Italiana, 1974), 65.

63. Conor Fahy, *Printing a Book at Verona in 1622: The Account Book of Francesco Calzolari Junior* (Paris: Fondation Custodia, 1993), 30–31.

64. Lodovico Moscardo, *Note overo memorie del Museo di Lodovico Moscardo, nobile veronese* (Padua: Paolo Frambotto, 1656), 171. Lodovico Moscardo, *Note overo*

memorie del Museo di Lodovico Moscardo, nobile veronese (Verona: Andrea Rossi, 1672). On Moscardo's museum: Krzysztof Pomian *Collectors and Curiosities: Paris and Venice, 1500–1800,* trans. Elizabeth Wiles-Portier (Cambridge, UK: Polity Press, 1990), 96. K. Pomian misinterpreted the cited sentence, making of Moscardo a supporter of the inorganic origin of fossils. Moscardo's view has been assessed correctly by Alessandro Ottaviani: Alessandro Ottaviani, "Fra diluvio noaico e fuochi sotterranei, Note sulla fortuna sei-settecentesca di Fabio Colonna," *Giornale critico della filosofia italiana,* no. 2 (2017): 280.

65. "Questo rispose il Fracastoro al Saraina con l'ultima sua opinione, la qual veramente è quella, che io stimo degna di un tanto Filosofo; peroché si vede manifestamente, che dove hora sono Monti, già fu il Mare." Moscardo, *Note overo memorie del Museo di Lodovico Moscardo, nobile veronese,* 1656, 173.

66. Moscardo, 173.

67. Moscardo, 174.

68. The biblical catastrophe obtained only a passing mention, unrelated to the formation of marine fossils: Moscardo wrote that according to "some, who saw it," the Flood could have buried the ship allegedly found in a Swiss mine in 1460. Moscardo, 173.

69. Ulisse Aldrovandi, *Vlyssis Aldrouandi . . . Musaeum metallicum in libros IIII distributum,* ed. Matteo Ambrosini (Bologna: typis Io. Baptistae Ferronii, 1648), 820. In: Ottaviani, "Fra diluvio noaico e fuochi sotterranei, Note sulla fortuna sei-settecentesca di Fabio Colonna," 278–79.

70. Francesco Stelluti, *Trattato Del Legno Fossile Minerale Nuovamente Scoperto etc.* (Rome: Mascardi, 1637), 8. For the a catalog of the surviving 199 geological and paleontological drawings of fossils commissioned by Stelluti and Prince Federico Cesi: Andrew C. Scott and David Freedberg, *Fossil Woods and Other Geological Specimens, The Paper Museum of Cassiano Dal Pozzo* (Turnhout, UK: Harvey Miller, 2000); Solinas, Francesco, "Il Trattato Del Legno Fossile di Francesco Stelluti e i quattro volumi della Natural History of Fossils nelle raccolte della Biblioteca Reale Di Windsor," in *Il Museo Cartaceo Di Cassiano Dal Pozzo: Cassiano Naturalista,* eds. Francis Haskell and David Freedberg (Milan: Olivetti [Distribuzione A. Mondadori Arte], 1989), 84–94; Andrew C. Scott, "Federico Cesi and His Field Studies on the Origin of Fossils between 1610 and 1630," *Endeavour* 25, no. 3 (September 1, 2001): 93–103.

71. Fabio Colonna, *Fabii Columnae Lyncei Purpura . . .* (Roma: Apud Jacobum Mascardum, 1616), 37. His chapter on fossil shark teeth is also published, with Italian translation, in: Nicoletta Morello, *La nascita della paleontologia nel seicento: Colonna, Stenone e Scilla* (Milan: FrancoAngeli, 1979), 70–93. See also: Alessandro Ottaviani, *Theatrum Naturae. la ricerca naturalistica tra erudizione e nuova scienza nell'Italia del primo Seicento* (Naples: La Città del Sole, 2007), 15–70.

72. Colonna, *Fabii Columnae Lyncei Purpura . . . ,* 36.

73. Agostino Scilla, *La vana speculazione disingannata dal senso: Lettera risponsiua circa i corpi marini che petrificati si trouano in varij luoghi terrestri* (Naples: Andrea Colicchia, 1670). I also used a recent edition: Agostino Scilla, *La vana speculazione disingannata dal senso*, ed. Paolo Rossi (Florence: Giunti, 1996). Also published in: Morello, *La nascita della paleontologia nel Seicento: Colonna, Stenone e Scilla*, 155–265.

74. Scilla, *La vana speculazione disingannata dal senso*, 1996, 91. The third hypothesis mentioned by Scilla is apparently the Aristotelian doctrine.

75. Paolo Boccone, *Museo di fisica e di esperienze variato, e decorato di osservazioni naturali, note medicinali, e ragionamenti secondo i principij de' moderni etc.* (Venice: Giovan Battista Zuccato, 1697), 285.

76. Niels Steensen, *Elementorum myologiae specimen: seu, Musculi descriptio geometrica. Cui accedunt canis carchariae dissectum caput, et dissectus piscis ex canum genere...* (Florence: Ex Typographia sub signo Stellae, 1667); Niels Steensen, *Nicolai Stenonis De Solido Intra Solidum Naturaliter Contento Dissertationis Prodromus* (Florence: Ex Typographia sub signo Stellae, 1669).

77. For Steno's biography, see: Troels Kardel and Paul Maquet, *Nicolaus Steno: Biography and Original Papers of a 17th-Century Scientist* (Berlin: Springer, 2013).

78. Ellenberger, *History of Geology*, 1996, 1:219. See also the many essays (of uneven quality) dealing with Steno's geological work in: Gary D. Rosenberg, ed., *The Revolution in Geology from the Renaissance to the Enlightenment* (Boulder, CO: Geological Society of America, 2009).

79. Nicolaus Steno, *Geological Papers*, ed. Gustav Scherz (Odense, Denmark: Odense University Press, 1969), 204–05.

80. Rienk Vermij, "Stevin's Physical Geography: The World as a Chemical Furnace," in *Rethinking Stevin, Stevin Rethinking. Constructions of a Dutch Polymath*, ed. Rienk Vermij et al. (Brill, 2020), 113.

81. Moscardo, *Note overo memorie del Museo di Lodovico Moscardo, nobile veronese*, 1656, 173–74. For a well-known contemporary description of the different soils found in excavations of pits: Bernhard Varenius, *Geographia generalis, in qua affectiones generales telluris explicantur. Autore Bernh. Varenio* (Apud Ludovicum Elzevirium, 1650), 69.

82. Steno, *Geological Papers*, 205.

83. Steno, 205.

84. On Steno and the age of the Earth: Nicoletta Morello, "Steno, the Fossils, the Rocks and the Calendar of the Earth," in *Geological Society of America. Special Paper 411* (2006): 81–93; August Ziggelaar, "The Age of Earth in Niels Stensen's Geology," in *The Revolution in Geology from the Renaissance to the Enlightenment*, ed. Gary D. Rosenberg (Boulder, CO: Geological Society of America, 2009), 135–42; Alan H. Cutler, "Nicolaus Steno and the Problem of Deep Time," in *The Revolution in Geology from the Renaissance to the Enlightenment*, ed. Gary D. Rosenberg (Boulder, CO: Geological Society of America, 2009), 143–48.

85. Steno, *Geological Papers*, 203–16. See also: Ellenberger, *History of Geology*, 1996, 1:252–54.

86. Steno, *Geological Papers*, 206–7.

87. See also: Cutler, "Nicolaus Steno and the Problem of Deep Time," 146–47.

88. Stephen Gaukroger, *Descartes' System of Natural Philosophy* (Cambridge: Cambridge University Press, 2002), 161–73; Ellenberger, *History of Geology*, 1996, 1:177–83.

89. Sebastian Olden-Jørgensen, "Nicholas Steno and René Descartes: A Cartesian Perspective on Steno's Scientific Development," in *The Revolution in Geology from the Renaissance to the Enlightenment*, ed. Gary D. Rosenberg (Boulder, CO: Geological Society of America, 2009), 149–57. For Steno's critique of Descartes's anatomy rather than geology: Grigoropoulous Vasiliki, "Steno's Critique of Descartes and Louis de La Forge's Response," in *Steno and the Philosophers*, ed. Raphaële Andrault and Mogens Lærke (Boston: Brill, 2018), 113–37.

90. Giovanni Quirini and Jacopo Grandi, *J. Quirini de Testaceis fossilibus Musæi Septalliani, et J. Grandii de veritate diluvii Universalis, et testaceorum quæ procul a mari reperiuntur generatione, epistolæ* (Venice: Valsalva, 1676). On Jacopo Grandi: Cesare Preti, "Grandi, Jacopo," *Dizionario Biografico Degli Italiani* (Rome: Treccani, 2002). In the context of late seventeenth-century Venetian and Paduan culture: Paolo Ulvioni, *Atene sulle lagune: Bernardo Trevisan e la cultura veneziana tra Sei e Settecento* (Venice: Ateneo veneto, 2000), 87; Ivano Dal Prete, "Francesco Bianchini e la cultura scientifica veronese," in *Unità del sapere, molteplicità dei saperi: Francesco Bianchini (1662-1729), tra natura, storia e religione*, ed. Luca Ciancio and Gian Paolo Romagnani (Verona, Italy: QuiEdit, 2010), 215–17.

91. Quirini and Grandi, *De veritate diluvii*, 6. See also the non-literal interpretations of the Deluge popular in Jansenist literature: Séguin, *Science et religion dans la pensée française du XVIIIe siècle*, particularly pp. 236–45.

92. Quirini and Grandi, *De veritate diluvii*, 5.

93. Quirini and Grandi, 9–10 and following. While Quirini may just have voiced a common opinion, it is unlikely that he did not know Steno's *Prodromus*, and Grandi certainly did. See: Quirini and Grandi, 60.

94. Quirini and Grandi, *De veritate diluvii*, 12 and following.

95. Quirini and Grandi, 60.

96. Quirini and Grandi, 23.

97. The observation of soils found at different depths in deep wells was pioneered by the Dutch mathematician Simon Stevin, in a well-known work first published in 1608. Vermij, "Stevin's Physical Geography: The World as a Chemical Furnace," 113.

98. Quirini and Grandi, *De veritate diluvii*, 52–53. See also: Nicoletta Morello, "Tra diluvio e vulcani. Le concezioni geologiche di Francesco Bianchini e del suo tempo," in *Unità del sapere, molteplicità dei saperi: Francesco Bianchini (1662-1729), tra natura, storia e religione*, ed. Luca Ciancio and Gian Paolo Romagnani (Verona, Italy: QuiEdit, 2010), 201–04.

99. The booklet received a rather extensive review in one of the first Italian literary journals: *Il Giornale de' Letterati. Per tutto l'anno 1676* (Rome: Nicolò Angelo Tinassi, 1676), 12–15.

100. Bernardino Ramazzini, *De fontium Mutinensium scaturigine* (Modena: Eredi Suliani, 1691), 51.

101. Ramazzini, 70. See also: Francesco Luzzini, *Il miracolo inutile: Antonio Vallisneri e le scienze della terra in Europa tra XVII e XVIII secolo* (Florence: Olschki, 2013), 44–45.

102. On Bianchini: Luca Ciancio and Gian Paolo Romagnani, eds., *Unità del sapere, molteplicità dei saperi: Francesco Bianchini (1662–1729), tra natura, storia e religione* (Verona, Italy: QuiEdit, 2010); Valentin Kockel and Brigitte Sölch, eds., *Francesco Bianchini (1662–1729) und die europäische gelehrte Welt um 1700* (Berlin: Walter de Gruyter, 2005). Also: Francesco Uglietti, *Un erudito veronese alle soglie del settecento. Mons. Francesco Bianchini 1662–1729* (Verona: Biblioteca Capitolare di Verona, 1986).

103. Francesco Bianchini, *La Istoria universale provata con monumenti e figurata con simboli de gli antichi . . . da Francesco Bianchini* (Rome: Antonio De Rossi, 1697).

104. Bianchini, 36.

105. Bianchini, 245.

106. "Della qual cosa è difficile allegare altra cagione, che un allagamento generale del globo terrestre, accaduto per il diluvio." Bianchini, 246.

107. On this section of the *Universal History* see: Morello, "Tra diluvio e vulcani. Le concezioni geologiche di Francesco Bianchini e del suo tempo." Morello incorrectly states that Bianchini ignored minor but well documented eruptions. In fact, citing contemporary literature, he argued that only major eruptions needed to be considered because they were the only ones that could produce sizable layers of volcanic rock. Bianchini, *La Istoria Universale*, 250.

108. Bianchini, *La Istoria Universale*, 247–49.

109. M. J. S. Rudwick, *Bursting the Limits of Time: The Reconstruction of Geohistory in the Age of Revolution* (Chicago: University of Chicago Press, 2005), 182.

110. See: Ito Yushi, "Earth Science in the Scientific Revolution 1600–1728" (Melbourne: University of Melbourne, 1985).

111. Ito Yushi, "Hooke's Cyclic Theory of the Earth in the Context of Seventeenth Century England," *The British Journal for the History of Science* 21, no. 3 (1988): 304.

112. In: François Ellenberger, *History of Geology*, trans. Marguerite Carozzi, vol. 2 (Rotterdam: A. A. Balkema, 1999), 91.

113. Poole, *The World Makers*, 34–35; William Poole, "Francis Lodwick's Creation: Theology and Natural Philosophy in the Early Royal Society," *Journal of the History of Ideas* 66, no. 2 (2005): 248.

114. Dmitri Levitin, "Halley and the Eternity of the World Revisited," *Notes and Records of the Royal Society of London* 67, no. 4 (2013): 323. Also: Poole, *The World Makers*, 108–10.

115. Levitin, "Halley and the Eternity of the World Revisited." Levitin's article reassesses previous literature on the topic, in particular: Simon Schaffer, "Halley's Atheism and the End of the World," *Notes Rec. R. Soc. Lond.* 32, no. 1 (July 1, 1977): 17–40. On Halley's cometary theory of the Flood: Sara Schechner, *Comets, Popular Culture, and the Birth of Modern Cosmology* (Princeton, NJ: Princeton University Press, 1999), 156–78.

116. Thomas Burnet, *Telluris theoria sacra: orbis nostri originem & mutationes generales, quas aut jam subiit, aut olim subiturus est, complectens. Libri duo priores de diluvio & paradiso* (London: Roger Norton for Walter Kettilby, 1681). On Burnet: Mirella Pasini, *Thomas Burnet: una storia del mondo tra ragione, mito e rivelazione* (Florence: La Nuova Italia, 1981). Poole, *The World Makers*, 55–64. For a recent re-contextualization of Burnet's work in terms of environmental history: Barnett, *After the Flood*, 89–116.

117. T. Burnet to I. Newton, Jan. 13, 1681. In: Herbert W. Turnbull, ed., *The Correspondence of Isaac Newton*, vol. 3: 1688–1693 (Cambridge: Cambridge University Press, 1961).

118. I. Newton to T. Burnet, Dec. 24, 1680. In: Turnbull, 3: 1688–1693.

119. Poole, *The World Makers*, 59.

120. Poole, 63. Burnet's book was placed in the Index of the forbidden books only in 1739, together with later writings that expounded a highly heterodox doctrine of salvation and hell. The Catholic censors found unacceptable only the last two books of the *Sacred Theory of the Earth*, which dealt with the resurrection and the second coming. Gustavo Costa, *Thomas Burnet e la censura pontificia: con documenti inediti* (Florence: L. S. Olschki, 2006), 78–86.

121. John Woodward, *An Essay toward a Natural History of the Earth: And Terrestrial Bodies, Especially Minerals: As Also of the Sea, Rivers, and Springs: With an Account of the Universal Deluge: And of the Effects That It Had upon the Earth* (London: Printed for R. Wilkin, 1695).

122. John Woodward, *Specimen geographiae physicae Quo agitur de terra, et Corporibus Terrestribus*, trans. Johann Jackob Scheuchzer (Zurich: Typis D. Gessneri, 1704).

123. On Scheuchzer: Simona Boscani Leoni, ed., *Wissenschaft- Berge- Ideologien. Johann Jakob Scheuchzer(1672–1733) und die frühneuzeitliche Naturforschung: Scienza- montagna- ideologie. Johann Jakob Scheuchzer (1672–1733) e la ricerca naturalistica in epoca moderna* (Basel: Schwabe, 2010). On his theory of the Flood: Michael Kempe, *Wissenschaft, Theologie, Aufklärung: Johann Jakob Scheuchzer (1672-1733) und die Sintfluttheorie* (Epfendorf: Bibliotheca Academica, 2003).

124. Bourguet was born in Nîmes, in southern France. His family left for Switzerland in 1685 after the revocation of the Edict of Nantes, which guaranteed the religious freedom of the French Protestants. On Bourguet's scientific career: François Ellenberger, "Bourguet, Louis," in *Dictionary of Scientific Biography* (New York: Scribner, 1970). On his relationship with naturalists in the Venetian area: Ivano Dal Prete, *Scienza e società nel Settecento veneto: il caso veronese, 1680–1796* (Milan:

FrancoAngeli, 2008), 215–29; Lydia Barnett, "Strategies of Toleration: Talking Across Confessions in the Alpine Republic of Letters," *Eighteenth-Century Studies* 48, no. 2 (January 27, 2015): 141–57; Barnett, *After the Flood*, 75–85, 160–87.

125. Francesca Bianca Crucitti Ullrich, *La Bibliothèque italique - cultura "italianisante" e giornalismo letterario* (Milan: Ricciardi, 1974).

126. Vincenzo Ferrone, *Scienza natura religione: mondo newtoniano e cultura italiana nel primo settecento* (Naples: Jovene Editore, 1982), 177–78.

127. Salvatore Rotta, *Il viaggio di Francesco Bianchini in Inghilterra. Contributo alla storia del newtonianesimo in Italia* (Brescia: Paideia, 1966). See also: John Heilbron, "Bianchini and natural philosophy," in *Unità del sapere, molteplicità dei saperi: Francesco Bianchini (1662–1729), tra natura, storia e religione*, ed. Luca Ciancio and Gian Paolo Romagnani (Verona, Italy: QuiEdit, 2010), 33–73. Heilbron, "Francesco Bianchini, Historian. In Memory of Amos Funkenstein," in *Thinking Impossibilities: The Intellectual Legacy of Amos Funkenstein*, ed. David Biale and Robert S. Westman (Toronto: University of Toronto Press, 2008), 257–58. On Newton's chronology: Jed Z. Buchwald and Mordechai Feingold, *Newton and the Origin of Civilization* (Princeton: Princeton University Press, 2013).

128. See Bianchini's glowing review of a work that he did not name, but is perhaps identifiable with the second edition of Newton's *Principia* thanks to references to the *General Scholium* (which did not appear in the 1687 edition): Francesco Bianchini, "Censurae Librorum" (n.d.), fol. 1r, cod. CCCCXXX, n. 5, Biblioteca Capitolare, Verona.

129. A considerable corpus of literature and sources has recently become available on Vallisneri. His works and his correspondence are being published by the "Edizione Nazionale delle Opere di Antonio Vallisneri," directed by Dario Generali and Maria Teresa Monti: http://www.vallisneri.it/default.shtml. I will refer to relevant literature when appropriate.

130. Antonio Vallisneri, *De' corpi marini, che su' monti si trovano; della loro origine; e dello stato del mondo avanti'l diluvio, nel diluvio, e dopo il diluvio: lettere critiche di Antonio Vallisneri* (Venice: Domenico Lovisa, 1721), 46. On Vallisneri's history of the Earth, see above all: Luzzini, *Il miracolo inutile*.

131. A. Vallisneri to L. Bourguet, November 23, 1718. BPUN, Ms. 1282, fol. 226–26; Antonio Vallisneri, *Epistolario 1714–1729*, ed. Dario Generali (Florence: Olschki, 2006), 352–53.

132. Antonio Vallisneri, "Specimen geographiae physicae," *Galleria di Minerva* VI (1708): 17. The review was not signed. For the attribution to Vallisneri: Dario Generali, *Bibliografia delle opere di Antonio Vallisneri* (Florence: Olschki, 2004), 54.

133. L. Bourguet to A. Vallisneri, October 12, 1721. BEM, Mss. It. (ms.IX.F.11).

134. A. Vallisneri to J. Riccati, July 13, 1719. Ms. Ric., II, fols. 309–16. Private archive of the Degli Azzoni Avogadro family (Castelfranco Veneto). Published in: Jacopo Riccati and Antonio Vallisneri, *Jacopo Riccati - Antonio Vallisneri. Carteggio*

(1719–1729), ed. Maria Luisa Soppelsa (Florence: Olschki, 1985), 70–75; Vallisneri, *Epistolario 1714–1729*, 428–29.

135. A. Vallisneri to J. Riccati, September 7, 1721. Ms. Ric., II, folio 509–10. Private Archive of the Azzoni-Avogadro family, Castelfranco Veneto. Published in: Riccati and Vallisneri, *Jacopo Riccati - Antonio Vallisneri. Carteggio (1719–1729)*, 123–24.

136. A. Vallisneri to L. Bourguet, August 14/20, 1719. BPUN, Ms. 1282, fol. 231–32. In: Vallisneri, *Epistolario 1714–1729*, 447. On natural history as a vehicle of interconfessional dialogue between Vallisneri, Bourguet, and Scheuchzer: Barnett, "Strategies of Toleration."

137. On Vallisneri's regional network: Ivano Dal Prete, "'Ingenuous Investigators'. Antonio Vallisneri's Regional Network and the Making of Natural Knowledge in Eighteenth-Century Italy," in *Empires of Knowledge: Scientific Networks in the Early Modern World*, ed. Paula Findlen (New York: Routledge, 2018), 181–204.

138. On Zannichelli: Italo Novelli, *La curiosità e l'ingegno: collezionismo scientifico e metodo sperimentale a Padova nel settecento* (Padua: Università degli Studi di Padova, Centro Musei scientifici, 2000), 78–83; Emilio de Tipaldo, *Biografia degli Italiani illustri nelle scienze: lettere ed arti del secolo XVIII. e de'contemporanei compilata da letterati Italiani di ogni provincia*, vol. VIII (Venice: Alvisopoli, 1841), 478–81; Giovanni Girolamo Zannichelli, *Enumeratio rerum naturalium quae in Museo Zannichelliano asservantur* (Venice: Antonio Bortoli, 1736).

139. Giovanni Girolamo Zannichelli, *Lithographia duorum montium veronensium. Unius nempe vulgo dicti di Boniolo; et alterius di Zoppica* (Venice: Giuseppe Corona, 1721), 28–29.

140. See: G. G. Zannichelli to P. A. Micheli, August 20, 1710. BNCF, Targ. Tozz. 177. Transcription retrieved from: "Antonio Vallisneri - Edizione Nazionale Delle Opere," http://www.vallisneri.it/micheli.shtml.

141. A. Vallisneri to L. F. Marsili, February 20, 1705. BUB, ms. Marsili 80, lett. 21b.

142. O. Alecchi to L. Bourguet, February 2, 1710. BPUN, ms. 1265.

143. O. Alecchi to L. Bourguet [1712]. BPUN, ms. 1265.

144. O. Alecchi to L. Bourguet, May 12, 1714. BPUN, ms. 1265.

145. See: Giacinto Tonti, *Augustiniana de rerum creatione sententia* (Padua: Giuseppe Corona, 1714). A. Vallisneri to L. Bourguet, January 2, 1714. BPMN, ms. 1282. In: Vallisneri, *Epistolario 1714–1729*, 3. See: Dario Generali, "Il « Giornale de' Letterati d'Italia » e la cultura veneta del primo settecento," *Rivista di Storia della Filosofia* 39, no. 2 (1984): 253; Antonio Vallisneri, *Istoria della generazione*, ed. Maria Teresa Monti, vol. 1 (Florence: Olschki, 2009), VI–VIII, LX–LXI and following.

146. A. Vallisneri to L. A. Muratori, November 30, 1720. BEM, Archivio Muratori, Filza 81, fasc. 55, lett. 77. In: Vallisneri, *Epistolario 1714–1729*, 577. I opted for a non-literal translation of the original Italian epithet used by Vallisneri.

147. Giuseppe Monti, *De Monumento Diluviano Nuper in Agro Bononiensi detecto Dissertatio: in qua permultae ipsius inundationis vindiciae, a statu terrae antediluvianae & postdiluvianae desumptae, exponuntur* (Bologna: Rossi, 1719), 22, 44–46.

148. Vallisneri, *De' corpi marini*, 123–32. See also: Luzzini, *Il miracolo inutile*, 214–16. On Vallisneri's discussion of the physical and moral decadence of human bodies as a consequence of the Flood: Lydia Barnett, "The Theology of Climate Change: Sin as Agency in the Enlightenment's Anthropocene," *Environmental History* 20, no. 2 (2015): 217–37; A. Vallisneri to L. Bourguet, January 1, 1722. BPMN, ms. 1282, cc. 251–52. In: Vallisneri, *Epistolario 1714–1729*, 900. Vallisneri treated the topic as a purely hypothetical case, since he made it clear that the size and longevity of the antediluvian men could only be miraculous.

149. A. Vallisneri to U. Landi, January 12, 1714. In: Vallisneri, *Epistolario 1714–1729*, 15.

150. A. Vallisneri to U. Landi, January 12, 1714. In: Vallisneri, 15, and A. Vallisneri to J. Riccati, August 22, 1722. Archivio privato della famiglia Degli Azzoni Avogadro, Ms. Ric., II, cc. 487–89. In Vallisneri, 860.

151. A. Vallisneri to A. Conti, April 18, 1727. BLL, Autografoteca Bastogi, Cass. 30 ins. 2185. In: Vallisneri, *Epistolario 1714–1729*, 1468–69.

152. A. Vallisneri to J. Riccati, August 26, 1721. Private archive of the Degli Azzoni Avogadro family, Castelfranco Veneto, Ms. Ric., II, ff. 497-98. In: Vallisneri, 663.

153. "Naturalmente diceva non poter esser vero tutto ciò che c'insegnano per infallibile verità." A. Vallisneri to A. Conti, August 29, 1727. BLL, Carte Conti. In: Vallisneri, *Epistolario 1714–1729*, 1501. In modern Italian, the adverb *naturalmente* means "obviously" or "of course"; in early modern sources though, it is normally closer to its etymological root of "according to nature."

154. A. Vallisneri to A. Conti, August 29, 1727. BLL, Carte Conti. In: Vallisneri, *Epistolario 1714–1729*, 1501.

CHAPTER 6

1. A. Vallisneri to J. Riccati, August 26, 1721. Private archive of the Degli Azzoni Avogadro family, Castelfranco Veneto, Ms. Ric., II, ff. 497-98. In: Antonio Vallisneri, *Epistolario 1714–1729*, ed. Dario Generali (Florence: Olschki, 2006), 663.

2. A. Vallisneri to L. Bourguet, August 22, 1721. BPUN, Ms. 1282, ff. 249–50. In: Vallisneri, 670.

3. William Buckland, *Geology and Mineralogy Considered with Reference to Natural Theology* (London: W. Pickering, 1836), 14.

4. Charles Lyell, *Principles of Geology: Being an Attempt to Explain the Former Changes of the Earth's Surface, by Reference to Causes Now in Operation*, vol. 1 (London: J. Murray, 1832), 27.

278 *Notes to pages 158–161*

5. Darrin M. McMahon, *Enemies of the Enlightenment: The French Counter-Enlightenment and the Making of Modernity* (Oxford: Oxford University Press, 2001), 6.

6. Bernardino Ramazzini, *De fontium Mutinensium scaturigine* (Modena: Eredi Suliani, 1691), 41–47. On the diffusion of Patrizi's dialogue among British authors: Lydia Barnett, "The Theology of Climate Change: Sin as Agency in the Enlightenment's Anthropocene," *Environmental History* 20, no. 2 (2015): 224.

7. Giuseppe Biancani, *Aristotelis Loca Mathematica: ... Collecta et Explicata : Accessere De Natura Mathematicarum Scientiarum Tractatio Atque Clarorum Mathematicorum Chronologia* (Bologna: Bartolomeo Cocchi, 1615), 103.

8. Gottfried Wilhelm Leibniz, *Protogaea*, eds. Claudine Cohen and Andre Wakefield (Chicago: University of Chicago Press, 2008). *Protogaea* was written between 1691 and 1693. This edition is based on a manuscript that might have been reviewed by Leibniz himself; the first edition (1749) presents some differences. See also: Rhoda Rappaport, "Leibniz on Geology: A Newly Discovered Text," *Studia Leibnitiana* 29, no. 1 (1997): 6–11. The 1693 draft has been published in: D. R. Oldroyd and J. B. Howes, "The First Published Version of Leibniz's Protogaea," *Journal of the Society for the Bibliography of Natural History* 9, no. 1 (November 1, 1978): 56–60. See also the English translation reproduced in: François Ellenberger, *History of Geology*, trans. Marguerite Carozzi, vol. 2 (Rotterdam; Brookfield, VT: A. A. Balkema, 1999), 145–46.

9. Leibniz, *Protogaea*, 5.

10. Leibniz, 15.

11. In: Leibniz, xxiv.

12. Leibniz, 19.

13. A. Conti to A. Vallisneri, January 21, 1727. BLL, Autografoteca Bastogi, Cass. 30 ins. 2185. Published in: Antonio Conti, *Scritti filosofici*, ed. Antonio Badaloni (Naples: Fulvio Rossi, 1972), 402–4. Leibniz was particularly influential in those years at Padua University. See: S. Mazzone and Clara Silvia Roero, *Jacob Hermann and the Diffusion of the Leibnizian Calculus in Italy*, vol. 1 (Florence: Olschki, 1997); Dario Generali, "Antonio Vallisneri 'corrispondente' leibniziano," *Studi e Memorie per la Storia dell'Università di Bologna*, Nuova Serie, 6 (1987): 125–40. On Conti's denunciation: John Lindon, "La 'denonzia' di Antonio Conti per ateismo," in *Antonio Conti: uno scienziato nella République des lettres*, ed. Guido Baldassarri, Silvia Contarini, and Francesca Fedi (Padua: Il poligrafo, 2009), 45–58.

14. Leibniz, *Protogaea*, 14–15.

15. "Et credibile est per magnas illas conversiones etiam animalium species plurimum immutatas". Leibniz, 68. Translation is mine. C. Cohen and A. Wakefield translate instead: "It is also conceivable that many animal species were transformed by these great upheavals." Leibniz, 69.

16. Ellenberger, *History of Geology*, 2:153.

17. See: Gordon Lindsay Campbell, *Lucretius on Creation and Evolution: A Commentary on De Rerum Natura, Book 5, Lines 772–1104* (Oxford: Oxford University Press, 2003).

18. Genesis, 6:2–4.

19. See in particular: Joyce E. Chaplin, *Subject Matter: Technology, the Body, and Science on the Anglo-American Frontier, 1500–1676* (Cambridge, MA: Harvard University Press, 2001).

20. The bulk of the manuscript was probably completed by 1714. Claudine Cohen, *Science, libertinage et clandestinité à l'aube des lumières: le transformisme de Telliamed* (Presses universitaires de France, 2011), 25–26. "Telliamed" is a transparent anagram of the author's last name. On the relation between Leibniz and de Maillet: Claudine Cohen, "Leibniz et Benoît de Maillet: De la Protogée au Telliamed," *Corpus: Revue de philosophie* 59 (2010): 55–78.

21. Pascal Charbonnat, "Usages et réceptions du Telliamed chez les naturalistes durant la seconde moitié du 18e siècle," *Corpus: revue de la philosophie* 59 (2010): 125–52; Benoît de Maillet, *Telliamed: Or, Conversations between an Indian Philosopher and a French Missionary on the Diminution of the Sea*, ed. and trans. Albert V. Carozzi (Urbana: University of Illinois Press, 1968), 4.

22. Benoît de Maillet, *Telliamed: Ou entretiens d'un philosophe indien avec un missionnaire français sur la diminution de la mer* (Paris: Fayard, 1984). For an English version: Maillet, *Telliamed*, 1968. Among recent studies, see in particular: Claudine Cohen, *Science, libertinage et clandestinité à l'aube des lumières: le transformisme de Telliamed* (Presses universitaires de France, 2011); and the essays collected in: *Corpus: Revue de la Philosophie*, 59 (2010).

23. Miguel Benítez, "Fixisme et evolutionnisme au temps des lumieres: le 'Telliamed' de Benoit de Maillet," *Rivista di Storia della Filosofia (1984–)* 45, no. 2 (1990): 247–68.

24. Benítez, 265.

25. De Maillet was the French consul in Egypt between 1692 and 1708, and in Leghorn in 1708–1714. Maillet, *Telliamed*, 1968, 5. On Leghorn as a commercial and naturalistic hub: Adriano Prosperi, ed., *Livorno, 1606–1806: luogo di incontro tra popoli e culture* (Torino: U. Allemandi, 2009); Francesca Trivellato, *The Familiarity of Strangers: The Sephardic Diaspora, Livorno, and Cross-Cultural Trade in the Early Modern Period* (New Haven: Yale University Press, 2009); Corey Tazzara, *The Free Port of Livorno and the Transformation of the Mediterranean World* (Oxford: Oxford University Press, 2017).

26. For a detailed discussion of de Maillet's geological system and observations: Maillet, *Telliamed*, 1968, 34–47.

27. Claudine Cohen, "Leibniz et Benoît de Maillet: de la Protogée au Telliamed," *Corpus: Revue de philosophie*, 59 (2010): 75–77; Cohen, *Science, libertinage et clandestinité à l'aube des lumières*, 59–64.

28. Rhoda Rappaport, "Fontenelle Interprets the Earth's History," *Revue d'histoire Des Sciences* 44, no. 3/4 (1991): 298.

29. Bernard Fontenelle, *Histoire de l'Academie Royale des Sciences. Avec les memoires de mathematique & de physique, pour la meme année, tirés de registres de cette Academie: Année 1703* (chez Gerard Kuyper, 1707), 28.

30. Rappaport, "Fontenelle Interprets the Earth's History," 286–87.

31. "Pourquoi cette grande difference entre le corps d'une meme nature? Ce seroit-la une difficulté considerable de Physique, si elle n'etoit resolue par les grandes & anciennes revolutions arrivées sur notre Globe, que nous indiquent des Coquillages et de Poissons ensevelis dans les terres & sur les Montagnes en toutes les parties du monde." Bernard Fontenelle, "Sur le taches de Mars," *Histoire de l'Academie Royale des Sciences, avec les Memoires de Mathematiques et de Phisique, pour la même Année* [1720], 1722, 95. Translation is mine. See: Jacques Maraldi, "Observations sur le Taches de Mars. Par M. Maraldi," *Histoire de l'Academie Royale des Sciences, avec les Memoires de Mathematiques et de Phisique, pour la même Année* [1720], 1722, 144–53.

32. Fontenelle, "Sur le taches de Mars," 8–9. On the overlaps of Earth history and planetary astronomy in the early eighteenth century: Ivano Dal Prete, "Francesco Bianchini e il pianeta Venere. Astronomia, cronologia e storia della Terra tra Roma e Parigi all'inizio del XVIII secolo," *Nuncius* 20, no. 1 (2005): 95–152. On Maraldi's observations: Dal Prete, 100–08.

33. Claudine Cohen, "L'« Anthropologie » de Telliamed," *Bulletins et Mémoires de la Société d'Anthropologie de Paris* 1, no. 3 (1989): 52. See: Johann Jackob Scheuchzer, *Homo diluvii testis et Theoskopos* (Zurich: Joh. & Byrgkli, 1726).

34. Ellenberger, *History of Geology*, 2:163–70.

35. Jacques Eugéne Louville, "Eugenii De Louville, Equitis, Regia Scientiarum Academiae Socii, nec non Regalis Societatis Londinensis Sodalis, de mutabilitate Eclipticae Dissertatio, Aureliis d. 4 Nov. A. 1718 ad Collectores Actor. Erud. Data, sed nunc demum exhibita," *Acta Eruditorum Lipsiae*, 1719, 292–93. On the seventeenth-century controversy on the obliquity of the ecliptic: John Heilbron, *The Sun in the Church: Cathedrals as Solar Observatories* (Cambridge, MA: Harvard University Press, 1999). On Fracastoro and Louville: Ivano Dal Prete, "Echi fracastoriani nelle cosmologie e nelle teorie della terra del Settecento," in *Girolamo Fracastoro: fra medicina, filosofia e scienze della natura*, ed. Alessandro Pastore and Enrico Peruzzi (Florence: Olschki, 2006), 289–91.

36. A less common variant postulated that the rotation axis of the Earth remained fixed, and what revolved was the plane of the Earth's orbit. Laplace demonstrated at the end of the century that both parameters are in fact stable. Dal Prete, "Echi fracastoriani nelle cosmologie e nelle teorie della terra del Settecento," 290–93 and n. 39.

37. Bernard Fontenelle, "Sur l'equinoxe du printemps du MDCCXIV," *Histoire de l'Academie Royale des Sciences. Pour l'anne 1714*, n.d., 68–69; Bernard Fontenelle, "Sur l'obliquite' de l'ecliptique," *Histoire de l'Académie Royale des Sciences. Pour l'anne 1716*, n.d., 53–54.

38. Maria Susana Séguin, *Science et religion dans la pensée française du XVIIIe siècle: le mythe du Déluge universel* (Paris: Champion, 2001), 91–92.

39. For a general overview on Catholic Enlightenment in Europe: Ulrich L. Lehner, "The Many Faces of the Catholic Enlightenment," in *A Companion to the Catholic Enlightenment in Europe*, ed. Ulrich L. Lehner and Michael O'Neill Printy (Leiden: Brill, 2010), 1–61. Also: Jeffrey D. Burson and Ulrich L. Lehner, eds., *Enlightenment and Catholicism in Europe: A Transnational History* (Notre Dame, Indiana: University of Notre Dame Press, 2014).

40. See for example: A. Vallisneri to A. Conti, September 10, 1727. Vallisneri, *Epistolario 1714–1729*, n. 1506.

41. Séguin, *Science et religion dans la pensée française du XVIIIe siècle*, 92–93.

42. Johann Gottlob Kruger, *Histoire des anciennes revolutions du globe terrestre avec une relation chronologique et historique des tremblemens de terre, arrivés sur notre globe, depuis le commencement de l'ère chrétienne jusqu'a présent* (Amsterdam: Damonneville, 1753), 113–14. Translation is mine. Séguin, *Science et religion dans la pensée française du XVIIIe siècle*, 154.

43. I adopt here the definition of "radical Enlightenment" given in Jonathan Israel, *A Revolution of the Mind: Radical Enlightenment and the Intellectual Origins of Modern Democracy* (Princeton, NJ: Princeton University Press, 2009), vii–viii. For a broader discussion of its intellectual origins: Jonathan Israel, *Radical Enlightenment: Philosophy and the Making of Modernity 1650–1750* (Oxford: Oxford University Press, 2002).

44. On different conceptions of progress between "radical" and mainstream Enlightenment: Israel, *A Revolution of the Mind*, 1–36.

45. See in particular: Anthony Grafton, *What Was History? The Art of History in Early Modern Europe* (Cambridge: Cambridge University Press, 2007).

46. Giovanni Maria Lancisi, "Del modo di filosofar nell'arte medica," *Galleria di Minerva* IV, no. 2 (1700): 33–37. Empahis mine..

47. "I have been looking for the *Dialogues on Rhetoric* etc. by Francesco Patrizi for a long time, but I cannot find them. . . . This is an exceedingly rare book". Vallisneri, *Epistolario 1714–1729*, 448. Translation is mine.

48. On Vallisneri's rethorical strategies and his careless use of second-hand quotations: Maria Teresa Monti, "La scrittura e i gesti dell'Istoria," in *Antonio Vallisneri. Istoria della generazione.*, ed. Maria Teresa Monti, vol. 1, 2 vols. (Florence: Olschki, 2009), XV–CIV.

49. On the categories of *philosophes* and Enlightenment itself as a construction of the conservative propaganda: McMahon, *Enemies of the Enlightenment*, 28–42.

50. Carozzi denied that Voltaire's Earth history might have been influenced by his deism and metaphysical ideas. I remain unconvinced. Marguerite Carozzi, *Voltaire's Attitude toward Geology* (Geneva: Société de Physique et d'Histoire naturelle, 1983), 118.

51. The model had been revisited a few years earlier by the French astronomer Louis Godin (who like Louville, did not mention the geological implications): Louis Godin, "Que l'obliquité de l'Ecliptique diminue, et de quelle manière; e que le noeuds des planètes sont immobiles," *Memoires de l'Academie Royale de Sciences. De l'Annee 1734*, n.d.

52. Voltaire, *Eléments De La Philosophie De Newton*, ed. Robert Walters and W.H. Barber (Oxford: The Voltaire Foundation, 1992), 483. Walters and Barber's text is based on the 1741 edition, but the chapter was already included in the first 1738 Amsterdam edition. To my knowledge, Carozzi's remains the only organic study on Voltaire's geology and Earth history.

53. Voltaire, 477. Translation is mine. On Louville, Godin, and Voltaire see also: Séguin, *Science et religion dans la pensée française du XVIIIe siècle*, 121–25.

54. Carozzi, *Voltaire's Attitude toward Geology*, 16–17. The reference to Louville's time scale was omitted in the 1748 edition, possibly (according to Carozzi) because of Louis Bourguet's criticism.

55. Voltaire, "Dissertation envoyée par l'auteur en italien, à l'Académie de Bologne et traduite par lui-même en français sur le changemens arrivés dans notre globe et sur les pétrifications qu'on prétend en être encore les témoignages," *Oeuvres complètes de Voltaire* (Basle: Jean-Jacques Tourneisen, 1786), 375–89. The Italian and the French versions present some differences. See: Carozzi, *Voltaire's Attitude toward Geology*, 19–24.

56. Voltaire, *Candide or, Optimism: The Robert M. Adams Translation, Backgrounds, Criticism*, ed. Nicholas Cronk, trans. Robert M. Adams (New York: W. W. Norton, 2016), 50. The "large book" may be *Telliamed*, rather than Leibniz's much shorter *Protogaea* (or the Bible, as suggested by the editor).

57. Voltaire, *A Pocket Philosophical Dictionary*, trans. John Fletcher (Oxford: Oxford University Press, 2011), 169. In 1770, however, he referred again to Louville's sources as to "ancient absurdities": Maria Susana Séguin, *Science et religion dans la pensée française du XVIIIe siècle: le mythe du Déluge universel* (Paris: Champion, 2001), 125.

58. Voltaire, *A Pocket Philosophical Dictionary*, 170.

59. From: Thierry Hoquet, *Buffon: histoire naturelle et philosophie* (Champion, 2005), 456. Translation is mine. Hoquet describes Buffon's position as a "new method" deftly employed to "definitively emancipate physics from theological readings," but of course, he was actually deploying an extremely conventional argument.

60. M. J. S. Rudwick, *Bursting the Limits of Time: The Reconstruction of Geohistory in the Age of Revolution* (Chicago: University of Chicago Press, 2005), 139–41.

61. Jacques Roger, *Buffon: A Life in Natural History*, ed. Pearce Williams L., trans. Bonnefoi Sarah Lucille (Ithaca: Cornell University Press, 1997), 187–88.

62. Roger, 409–12.

63. Rudwick, *Bursting the Limits of Time*, 142–49.

64. In: Roger, *Buffon*, 422–23.

65. Roger, 421.

66. Georges-Louis Leclerc comte de Buffon, *Histoire naturelle générale et particuliére: avec la description du Cabinet du Roy: tome septième* (Paris: Imprimerie Royale, 1758), 34–35.

67. Jean-Sylvain Bailly, *Histoire de l'astronomie ancienne, depuis son origine jusq'a l'etablissement de l'école d'Alexandrie* (Paris: chez les freres Debure, 1775). On Bailly as a historian of astronomy: Mirella Pasini, "L'astronomie antédiluvienne. Storia della scienza e origini della civiltà in J. S. Bailly," *Studi Settecenteschi*, no. 11 (1988): 197–236. Another well-connected "notable" who supported top-down, firmly controlled reforms, Bailly might have reconsidered his optimism on the recent course of humankind as he walked to the guillotine in November 1793.

68. On Boulanger's influence on Buffon's *Epochs of Nature*: Roger, *Buffon*, 419–22.

69. On the origins and preservation of the arts in the French Enlightenment: Paola Bertucci, *Artisanal Enlightenment: Science and the Mechanical Arts in Old Regime France* (New Haven: Yale University Press, 2017).

70. Paul Sadrin, *Nicolas-Antoine Boulanger (1722–1759), ou, Avant nous le déluge* (Oxford: Voltaire Foundation, 1986), 3.

71. Nicolas Antoine Boulanger, *Anecdotes physiques de l'histoire de la nature; avec la Nouvelle Mappemonde et le Memoire sur une nouvelle mappemonde*, ed. Pierre Boutin (Paris: Champion, 2006), 23. See Boutin's substantial introduction for an analysis of Boulanger's Earth history and its sources.

72. Boulanger, 239.

73. Boulanger, 333.

74. Boulanger, 241.

75. Boulanger, 242.

76. Boulanger, 234.

77. Boulanger, 244.

78. Boulanger, 243.

79. Boulanger, 249–50.

80. Boulanger, 235.

81. In: Sadrin, *Boulanger*, 63.

82. Boulanger, *Anecdotes physiques de l'histoire de la nature; avec la Nouvelle Mappemonde et le Memoire sur une nouvelle mappemonde*, 234.

83. Boulanger, 245.

84. Boulanger's closest source was Abbé Nöel-Maire Pluche's *Spectacle of Nature* (1732), even though Pluche was a diluvialist and his intentions were apologetic: Séguin, *Science et religion dans la pensée française du XVIIIe siècle*, 414–15. Giovanni Cristani, *D'Holbach e le rivoluzioni del globo: scienze della terra e filosofie della natura nell'età dell'Encyclopédie* (Florence: Olschki, 2003), 122–23.

85. "E quindi sono le scienze, insegnate in enimmi, in favole, in figure, in numeri, in sacrarii, sotto silenzio, et in mille altri nascosti modi". Francesco Patrizi, *Della retorica dieci dialoghi di M. Francesco Patritio: nelli quali si fauella dell'arte oratoria con ragioni repugnanti all'openione, che intorno a quella hebbero gli antichi scrittori* (Venice: Francesco Senese, 1562), 7r.

86. Sadrin, *Boulanger*, 58.

87. Apparently, it could be recognized even by Christian diluvialists who were quick to denounce Buffon's borrowings from Boulanger. Roger, *Buffon*, 422.

88. Sadrin, *Boulanger*, 50.

89. Sadrin, 33–37.

90. Sadrin, 21.

91. English translation from: Nicolas Antoine Boulanger, "Flood," in *Encyclopedia of Diderot & d'Alembert - Collaborative Translation Project* (Ann Arbor: Michigan Publishing, University of Michigan Library, 2009). Originally published as "Déluge," *Encyclopédie ou Dictionnaire raisonné des sciences, des arts et des métiers*, 4:795–803 (Paris, 1754).

92. Cristani, *D'Holbach e le rivoluzioni del globo*, 120.

93. On d'Holbach as editor of Boulangers' *Antiquity Unveiled*: Sadrin, *Boulanger*, 136–38.

94. Boulanger is a central figure in every intellectual biography of d'Holbach. With particular focus on Earth history and the origin of religions, see in particular: Cristani, *D'Holbach e le rivoluzioni del globo*. Anna Minerbi Belgrado, *Paura e ignoranza: studio sulla teoria della religione in d'Holbach* (Florence: Olschki, 1983).

95. Israel, *A Revolution of the Mind*, 15.

96. Paul Henri Dietrich d'Holbach, *Système de la nature ou Des loix du monde physique & du monde moral* (London: M.-M. Rey, 1770).

97. Paul Henri Dietrich d'Holbach, "Fossile," in *Encyclopédie, Ou Dictionnaire Raisonné Des Sciences, Des Arts et Des Métiers*, ed. Denis Diderot and Jean le Rond d'Alembert (Paris: Briasson, 1757), 210. Translation is mine.

98. Cristani, *D'Holbach e le rivoluzioni del globo*, 119–38.

99. Cristani, 121–22.

100. Cristani, 122–24.

101. Jonathan Israel, *Democratic Enlightenment: Philosophy, Revolution, and Human Rights 1750–1790* (Oxford: Oxford University Press, 2013), 140–41.

102. Israel, 145–47.

103. Israel, 146.

104. McMahon, *Enemies of the Enlightenment*, 42.

105. Israel, *Democratic Enlightenment*, 170.

106. Israel, 141–42.

107. Israel, 157.

108. Israel, 159.

CHAPTER 7

1. Letters by G. Rizzetti to A. Vallisneri, October 28, 1721; December 13, 1721, in: BPUN, ms. 1289.4. November 30, 1722; February 8, 1723, in: BEM, Mss. It. (mss. IX.F.11). December 9, 1724, in: ASRE, Archivio Vallisneri, 4/1, fasc. II. On Rizzetti: Lionello Puppi, Ruggero Maschio, and Giacinto Cecchetto, eds., *Giovanni Rizzetti: scienziato e architetto (Castelfranco Veneto 1675-1751)* (Castelfranco Veneto: Banca popolare di Castelfranco Veneto, 1996).

2. A. Vallisneri to L. Bourguet, January 1, 1722. BPUN, ms. 1282, ff. 251-52.

3. A. Vallisneri to L. Bourguet, November 10, 1722. BPUN, ms. 1282, ff. 256-57. In: Antonio Vallisneri, *Epistolario 1714-1729*, ed. Dario Generali (Florence: Olschki, 2006), 971. See: Antonio Conti, *Scritti filosofici*, ed. Antonio Badaloni (Naples: Fulvio Rossi, 1972), 387–88; Luca Ciancio, *Autopsie della terra: illuminismo e geologia in Alberto Fortis (1741-1803)* (Florence: Olschki, 1995), 103. Among recent literature on Conti: Guido Baldassarri, Silvia Contarini, and Francesca Fedi, eds., *Antonio Conti: uno scienziato nella République des lettres* (Padua: Il poligrafo, 2009).

4. Antonio Vallisneri, "Lezione Accademica intorno all'ordine della progressione, e della connessione, che hanno insieme tutte le cose create, etc.," in *Istoria della generazione*, ed. Maria Teresa Monti, vol. 2 (Florence: Olschki, 2009), 524.

5. Girolamo Cesare Fantasti, *La scimia non è specie d'uomo* (Verona: Dionigi Ramanzini, 1732).

6. Girolamo Cesare Fantasti, "Il cervo impietrito, ritrovato in Grezzana," in *La scimia non è specie d'uomo* (Verona: Dionigi Ramanzini, 1732); Girolamo Cesare Fantasti, *Contrascritta di Girolamo Cesare Fantasti filosofo medico che annulla l'opinione del Molto Reverendo Don Gianiacopo Spada Rettore di Grezzana* (Verona: Dionigi Ramanzini, 1737). On the author, see: Ivano Dal Prete, *Scienza e società nel Settecento veneto: il caso veronese, 1680-1796* (Milan, Italy: FrancoAngeli, 2008), 249–56.

7. Simone Contardi, *La rivincita dei "filosofi di carta": saggio sulla filosofia naturale di Antonio Vallisneri junior* (Florence: Olschki, 1994), 5–6, 19.

8. In the late 1750s, Vallisneri *junior* resoundingly disavowed the embryological theories upheld by his father. The about-face spurred a bitter controversy with the most famous of his pupils, biologist Lazzaro Spallanzani: Contardi, 41–85. On Vallisneri *junior*'s Earth history: Contardi, 87–120.

9. See Maffei's controversial works on the ethicality of interest rates: Paola Vismara Chiappa, *Oltre l'usura: la Chiesa moderna e il prestito a interesse* (Catanzaro: Rubbettino, 2004); Giorgio Borelli, "Scipione Maffei e il problema del prestito ad interesse," in *Scipione Maffei nell'Europa del Settecento: atti del convegno, Verona, 23-25 settembre 1996*, ed. Gian Paolo Romagnani (Verona: Cierre Edizioni, 1998), 121–37. On Maffei's political and civic thought: Paolo Ulvioni, *Riformar il mondo: il pensiero civile di Scipione Maffei. Con una nuova edizione del "Consiglio Politico"* (Alessandria: Edizioni dell'Orso, 2008). On the relation between Christian and

pagan morality: Francesca Maria Crasta, *L'eloquenza dei fatti. Filosofia, erudizione e scienze della natura nel Settecento veneto* (Naples: Bibliopolis, 2007), 135–229; Paolo Ulvioni, "La filosofia morale di Scipione Maffei," in *Scipione Maffei nell'Europa del Settecento: atti del convegno, Verona, 23-25 settembre 1996*, ed. Gian Paolo Romagnani (Verona: Cierre Edizioni, 1998), 147–63.

10. On Maffei's projects for the reformation of Padua (and Turin) university: Ulvioni, *Riformar il mondo*, 130–75.

11. Gino Benzoni, "Scipione Maffei e il Mondo delle Accademie," in *Scipione Maffei nell'Europa del Settecento*, ed. Gian Paolo Romagnani (Verona: Cierre, 1998), 252–53.

12. Most scientific articles were written by other contributors, usually prominent astronomers and mathematicians. Francesca Bianca Crucitti Ullrich, "Le 'Osservazioni Letterarie' di Scipione Maffei, o la continuazione del 'Giornale di Venezia,'" in *Scipione Maffei nell'Europa del Settecento: atti del convegno, Verona, 23-25 settembre 1996*, ed. Gian Paolo Romagnani (Verona: Cierre Edizioni, 1998), 321–48; Genziana Panziera, "Le «Osservazioni Letterarie» (1737-1740) di Scipione Maffei" (M.A. Thesis, Padua University, 1992).

13. J.F. Séguier to C. Allioni, August 20th, 1752. BMN, ms. 309, f. 111v. On Maffei's observatory: Dal Prete, *Scienza e società nel Settecento veneto*, 301–21.

14. BMN, ms. 92, p. 116v.

15. "Je puis dire hardiment que son cabinet doit [être] en ce genre infiniment audessus de toutes ceux de l'Europe." J. Séguier, «Pétrifications du Véronois». BMN, ms. 90, c. 44r. For the catalogue of Spada's collection: Giovanni Giacomo Spada, *Catalogus lapidum Veronensium . . . qui apud Joannem Jacobum Spadam Gretianam archibresbyterum asservantur* (Verona: Dionigi Ramanzini, 1739). Giovanni Giacomo Spada, *Catalogi Lapidum Veronensium Mantissa* (Verona: Dionigi Ramanzini, 1740); Giovanni Giacomo Spada, *Corporum lapidefactorum agri Veronensis catalogus quae apud Joan. Jacobum Spadam..., asservantur. Editio altera...* (Verona: Dionigi Ramanzini, 1744).

16. Marble extraction and trade is still an important economic activity in the area where Spada spent most of his life.

17. The argument was mentioned by Vallisneri too, as well as in older literature: Antonio Vallisneri, "Specimen geographiae physicae," *Galleria di Minerva* VI (1708): 17. The quarry of Bolca, where most of Spada's "fishes" were excavated, lies at an elevation of about 600 meters.

18. Giovanni Giacomo Spada, *Dissertazione ove si prova, che li petrificati corpi marini, che nei monti adiacenti a Verona si trovano, non sono scherzi di natura, ne diluviani; ma antediluviani . . .* (Verona: per Dionigi Ramanzini librajo a S. Tomio, 1737), 15–17. The booklet was dedicated to Maffei. Giovanni Giacomo Spada, *Nuova invenzione di Ossa di Cervo petrificate* (Verona, 1737).

19. Giovanni Giacomo Spada, *Giunta alla dissertazione de' corpi marini petrificati ove si prova che sono antediluviani* (Verona: Dionigi Ramanzini, 1737).

20. Spada, 16. See: Antonio Vallisneri, *De' corpi marini, che su' monti si trovano; della loro origine; e dello stato del mondo avanti'l diluvio, nel diluvio, e dopo il diluvio: lettere critiche di Antonio Vallisneri* (Venice: Domenico Lovisa, 1721), 42.

21. Calogero Farinella, "Moro, Anton Lazzaro," in *DBI*, vol. 77 (Rome: Treccani, 2002).

22. Antonio Lazzaro Moro, *De' crostacei: e degli altri marini corpi che si truovano su' monti; libri due* (Venice: Stefano Monti, 1740). Moro's work had two German editions in 1751 and 1765: Farinella, "Moro, Anton Lazzaro." See also: Massimo Baldini, ed., *Anton Lazzaro Moro, 1687-1987 - Atti del Convegno di Studi* (Pordenone: Grafiche editoriali artistiche pordenonesi, 1988). Part of his correspondence has been published in: Anton Lazzaro Moro, *Anton Lazzaro Moro. Carteggio (1735-1764)*, ed. Massimo Baldini et al. (Florence: Olschki, 1993).

23. Moro, *De' crostacei*, 214–18. Moro republished, with some integrations, the account of the phenomenon that was sent to Vallisneri *senior* by one of his former students: Vallisneri, *De' corpi marini*, 117–22.

24. Moro, *De' crostacei*, 313.

25. Moro, 317.

26. Moro was especially influenced by physician Giuseppe Antonio Pujati, who studied at Padua during Vallisneri's tenure as professor of medicine. Marica Roda, "Pujati, Giuseppe Antonio," in *DBI*, vol. 86 (Rome: Treccani, 2016).

27. A. L. Moro to G. Bianchi, April 7, 1740. In: Moro, *Anton Lazzaro Moro. Carteggio (1735-1764)*, 44.

28. Scipione Maffei, *Della formazione de' fulmini tratto del sig. marchese Scipione Maffei, raccolto da varie sue lettere...* (Verona: G. Tumermani, 1747), 119.

29. Voltaire, *Elementi della filosofia del Neuton esposti dal signor di Voltaire tradotti dal francese* (Venice: Giam-Maria Lazzaroni, 1741), 198–214.

30. Voltaire, II.

31. A.L. Moro to L. Alberti, November 5, 1753. BAIS, Autografi Porri, 20.21. This letter is not included among those published in his correspondence. Around ten years earlier, Moro still professed his belief in the fixity of species which received "their specific being . . . from the hand of the Almighty, according to the views expressed in the holy *Genesis*." A. L. Moro to L. Alberti, February 5, 1742. In: Moro, *Anton Lazzaro Moro. Carteggio (1735-1764)*, 70.

32. See: BMN, ms. 285, *Catalogue des libres* [sic] *de J. François Séguier en 1760*.

33. Voltaire, *Principj fisici tratti dagli Elementi di fisica Nevtoniana dell'insigne m.r de Voltaire, e combinati a dovere dal C.A.G.D.C* (Lucca: Filippo Maria Benedini, 1754), 4–5.

34. Voltaire, 7.

35. Madaleine F. Morris, "The Tuscan Editions of the 'Encyclopédie,'" in *Notable Encyclopedias of the Late Eighteenth Century: Eleven Successors of the Encyclopédie*, ed. Frank A. Kafker (Oxford: Voltaire Foundation, 1994), 51–84.

36. John Woodward, *Geografia fisica, ovvero saggio intorno alla storia naturale della terra, del sig. Woodward… con la giunta dell'apologia del saggio contra le osservazioni del dottor Camerario; e d'un trattato de' fossili d'ogni spezie, divisi metodicamente in varie classi* (Venice: Giambatista Pasquali, 1739).

37. Moro, *De' crostacei*, "Avviso" (page not numbered).

38. Gregorio Piccoli, *La scienza dei cieli e dei corpi celesti e loro meravigliosa Posizione, Moto e Grandezza* (Verona: Vallarsi, 1741), 19. Compare with: Gregorio Piccoli, *Ragguaglio di una grotta, ove sono molte ossa di Belve diluviane nei Monti Veronesi* (Verona: Fratelli Merlo, 1739).

39. Woodward, *Geografia fisica*, 232.

40. Paola Bertucci, *Viaggio nel paese delle meraviglie: scienza e curiosità nell'Italia del Settecento* (Turin: Bollati Boringhieri, 2007), 138–42. On Maffei's skepticism of Pivati's electrical researches: Dal Prete, *Scienza e società nel Settecento veneto*, 293–97.

41. Gianfrancesco Pivati, *Nuovo dizionario scientifico e curioso sacro-profano*, 10 vols. (Venice: Milocco, 1746-1751). On Pivati's encyclopedias: Silvano Garofalo, *L'enciclopedismo italiano: Gianfrancesco Pivati* (Ravenna: Longo, 1980).

42. Gianfrancesco Pivati, *Nuovo dizionario scientifico e curioso sacro-profano*, vol. 3 (Venice: Milocco, 1746), 216–29.

43. Giuseppe Antonio Costantini, *Lettere critiche, giocose, morali… del conte Agostino Santi Pupieni* (Venice: Pasinelli, 1743). Giuseppe Antonio Costantini, *Lettere critiche, giocose, morali… del conte Agostino Santi Pupieni* (Venice: Pietro Bassaglia, 1744-). Ten tomes were published starting in 1743.

44. For an example of an almost diluvialist reading of Vallisneri: Piccoli, *Ragguaglio*, 12.

45. Giuseppe Antonio Costantini, *La verità del diluvio universale vindicata dai dubbi, e dimostrata nelle sue testimonianze. Esame critico dell'Avvocato Giuseppe Antonio Costantini autore delle Lettere Critiche* (Venice: Pietro Bassaglia, 1747), X.

46. Costantini, XII.

47. Costantini, X.

48. Costantini, 425.

49. Giuseppe Antonio Costantini, *Nuova raccolta di lettere critiche giocose, morali, scientifiche, ed erudite*, vol. 4 (Venice: Pietro Bassaglia, 1747), 116. See Moro's reply: Anton Lazzaro Moro, *Lettera di Anton Lazzaro Moro all'illustrissimo signor co. Agostino Santi Pupieni autore delle Lettere critiche, giocose, morali, scientifiche ed erudite* [1747].

50. S. Maffei to unknown [July, 1749]. Scipione Maffei, *Epitolario*, ed. Celestino Garibotto, vol. 2 (Milan: Giuffre, 1955), 1248; S. Maffei to G. Lami, January 15, 1750; Maffei, 2:1198.

51. Among the literature on eighteenth-century Republic of Venice, see: Franco Venturi, *Settecento Riformatore*, vol. V:2, VII vols. (Torino: G. Einaudi, 1990). Maffei's extremely moderate *Political Advice*, which proposed the aggregation of 20

representatives of the "subject" cities to the Major Council (an assembly of 1500), circulated secretly and could be printed only after the fall of the Republic. See: Ulvioni, *Riformar il mondo*. Also: Eluggero Pii, "Il pensiero politico di Scipione Maffei: dalla Repubblica di Roma alla Repubblica di Venezia," in *Scipione Maffei nell'Europa del Settecento: atti del convegno, Verona, 23-25 settembre 1996*, ed. Gian Paolo Romagnani (Verona: Cierre Edizioni, 1998), 93–117.

52. Contardi, *La Rivincita Dei "Filosofi Di Carta*,*"* 100–101.

53. Mario Infelise, *L'editoria veneziana nel '700* (Milan: F. Angeli, 2000), 63–67. On the relation between Lodoli and Maffei: Dal Prete, *Scienza e società nel Settecento veneto*, 189–91.

54. Contardi, *La Rivincita Dei "Filosofi Di Carta*,*"* XV.

55. Contardi, 114–20.

56. C. Sibiliato to A.L. Moro, October 7, 1747. In: Moro, *Anton Lazzaro Moro. Carteggio (1735-1764)*, 102.

57. On occasions, natural "curiosities" entered aristocratic houses as part of wholesale purchases. This is how the fossils gathered by physician Antonio Bianchi became part of Count Gomberto Giusti's museum, even though the latter was actually interested in the paintings. O. Alecchi to L. Bourguet, November 17, 1710. BPUN, ms. 1265. On Gomberto Giusti's gallery: Scipione Maffei, *Verona Illustrata*, vol. 3 (Verona: Jacopo Vallarsi, 1732), 334 and following. The same pages offer important information on the composition of Verona's collections around 1730.

58. Krzysztof Pomian, *Collectors and Curiosities: Paris and Venice, 1500-1800*, trans. Elizabeth Wiles-Portier (Cambridge, UK: Polity Press, 1990), 124. See also: Rhoda Rappaport, "Fontenelle Interprets the Earth's History," *Revue d'histoire Des Sciences* 44, no. 3/4 (1991): 281–300. Rhoda Rappaport, *When Geologists Were Historians, 1665-1750* (Ithaca: Cornell University Press, 1997), 211–26.

59. Darrin M. McMahon, *Enemies of the Enlightenment: The French Counter-Enlightenment and the Making of Modernity* (Oxford; New York: Oxford University Press, 2001), 24.

60. Jonathan Israel, *Democratic Enlightenment: Philosophy, Revolution, and Human Rights 1750-1790* (Oxford: Oxford University Press, 2013), 161.

61. "'Quella verità che par menzogna' non si deve mai dir dall'uomo saggio 'perché poi non creduta fa vergogna.'" G. Arduino to G. Vio, April 22, 1764. In: Ezio Vaccari, *Giovanni Arduino (1714-1795): il contributo di uno scienziato veneto al dibattito settecentesco sulle scienze della terra* (Florence: Olschki, 1993), 197.

62. Vaccari, 197–98.

63. Zaccaria Betti, *Descrizione di un meraviglioso ponte naturale nei monti Veronesi* (Verona: Marco Moroni, 1766), 18–19.

64. Alberto Fortis, "Descrizione Orittografica Del Ponte Naturale Di Veja," *Giornale d'Italia, Spettante Alla Scienza Naturale* VI (1770): 249. See also: Ciancio, *Autopsie della terra*, 127–28.

65. Alberto Fortis, "De' cataclismi sofferti dal nostro Pianeta, Saggio Poetico per servire di prodromo ad un Poema Geologico d'Ischiro Euganeo," *L'Europa Letteraria*, 1768, 70 and following.

66. Jacopo Odoardi, "De' corpi marini, che nel Feltrese Distretto si trovano," in *Nuova raccolta d'opuscoli scientifici e filologici*, Vol. 8 (Venice: Simone Occhi, 1761), 162–63.

67. On agricultural academies in late eighteenth-century Veneto: Pietro Del Negro, "La politica di Venezia e le Accademie di Agricoltura," in *La Politica della scienza: Toscana e stati italiani nel tardo Settecento: atti del convegno di Firenze, 27-29 gennaio 1994*, ed. Giulio Barsanti, Vieri Becagli, and Renato Pasta (Florence: Olschki, 1996), 451–89.

68. See: Dal Prete, *Scienza e società nel Settecento veneto*, 415–19.

69. Ciancio, *Autopsie della terra*, 197.

70. Luca Ciancio, "Geologia e ortodossia. L'eredità galileiana nella geologia veneta del secondo Settecento," in *La Politica della scienza: Toscana e stati italiani nel tardo Settecento: atti del convegno di Firenze, 27-29 gennaio 1994*, ed. Giulio Barsanti, Vieri Becagli, and Renato Pasta (Florence: Olschki, 1996), 504–05.

71. Ciancio, *Autopsie della terra*, 187.

72. Ciancio, 194.

73. A. Fortis to A. Vivorio, January 14, 1794. BCV, b. 760, IV.e.6.

74. On Lorgna and the military school of Verona: Calogero Farinella, *L'accademia repubblicana: la Società dei Quaranta e Anton Mario Lorgna* (Milan: FrancoAngeli, 1993).

75. Farinella, 93.

76. Renata Targhetta, *La massoneria veneta dalle origini alla chiusura delle logge (1729-1785)* (Del Bianco, 1988), 23. See also: Alessandro Righi, *Una loggia massonica a Verona nel 1792* (Verona: Franchini, 1912).

77. BCV, ms. 231, p. 2. See also: Giambattista Gazola, *Discorso in morte del signor conte Marco Marioni letto nell'Accademia Filarmonica in Verona il giorno 17 settembre 1795* (Verona: Stamperia Giuliari, 1796). On the history of Lorgna's academy (not to be confused with the later "Academy of the Forty"): Dal Prete, *Scienza e società nel Settecento veneto*, 332–50.

78. On the scientific activities conducted among the *Filarmonici* in the 1770s and 1780s: Dal Prete, *Scienza e società nel Settecento veneto*, 399–401 (see footnotes for archival sources).

79. L. Torri, *Saggio Oritologico sulla notabile e generale rivoluzione a cui soggiacque il Globo Terracqueo*, BCV, b. 51/20. The folder contains four redactions, including a final draft arguably intended for publication.

80. O. Rota to L. Torri [not dated]. BCV, ms. 51, f. 4r.

81. G. Torelli to J.F. Séguier, May 12, 1779. BMN, ms. 148, f. 232v.

82. Dal Prete, *Scienza e società nel Settecento veneto*, 394–95. See footnotes for primary sources.

83. Ivano Dal Prete, "The Gazola Family's Scientific Cabinet. Politics, Society and Scientific Collecting in the Twilight of the Republic of Venice," in *Cabinets of Experimental Philosophy in Eighteenth-Century Europe*, ed. Jim Bennett and Sofia Talas (Leiden: Brill, 2013), 155–72; Ivano Dal Prete, "Giovan Battista Gazola," in *Storia della Società Letteraria di Verona tra otto e novecento*, ed. Gian Paolo Romagnani and Emilio Zangarini, vol. 2, 2 vols. (Verona: Società Letteraria di Verona, 2007), 61–77.

84. Dal Prete, *Scienza e società nel Settecento veneto*, 393. Giovanni Serafino Volta, *Ittiolitologia veronese del Museo Bozziano ora annesso a quello del Conte Giovambattista Gazola e di altri gabinetti di fossili veronesi: Con la versione latina*, vol. 1 (Verona: Stamperia Giuliari, 1796), LVII–LXI.

85. J. Dionisi to G. Toaldo, Sep. 26, 1790; BCapV, cod. DCCCLI. See also: Serafino Volta, *Degl'impietrimenti del territorio veronese ed in particolare dei pesci fossili del celebre Monte Bolca*, 1789, 6. The prospectus of the courses for the year 1790 is also discussed in: Pomian, *Collectors and Curiosities*, 221. Dal Prete, *Scienza e società nel Settecento veneto*, 391.

86. For a nuanced perspective: G. N. Cantor, "What Shall We Do with the 'Conflict Thesis'?," in *Science and Religion: New Historical Perspectives*, ed. Thomas Dixon, Geoffrey N. Cantor, and Stephen Pumfrey (New York: Cambridge University Press, 2010), 283–98.

87. See: Vincenzo Bozza, *Della universale rivoluzione sofferta dal globo terracqueo lettera al molto reverendo p. Orazio Rota m.o.p.p. di Lingue Orientali in Mantova* (Verona, 1788); Orazio Rota, *Dissertazione epistolare sopra i sistemi e le teorie de' due globi celeste e terracqueo che si stabiliscono da Mosé nella storia delle sei giornate della creazione del mondo al cap. primo della sacra Genesi* (Vicenza: Turra, 1789).

88. Bozza, *Della Universale Rivoluzione*, 153.

89. On jurisdictionalism in eighteenth-century Republic of Venice: Franco Trentafonte, *Giurisdizionalismo, illuminismo e massoneria nel tramonto della Repubblica Veneta* (Venice: Deputazione Editrice, 1984); Venturi, *Settecento Riformatore* V:2.

90. On social status and scientific credit in early modern Europe: Shapin and Schaffer, *Leviathan and the Air-Pump*; Shapin, *A Social History of Truth*; With particular reference to eighteenth-century Italy: Ivano Dal Prete, "'Ingenuous Investigators'. Antonio Vallisneri's Regional Network and the Making of Natural Knowledge in Eighteenth-Century Italy," in *Empires of Knowledge: Scientific Networks in the Early Modern World*, ed. Paula Findlen (New York: Routledge, 2018), 181–204.

91. On the diffusion and apologetic use of Newtonianism in Italy (including the Veneto): Vincenzo Ferrone, *The Intellectual Roots of the Italian Enlightenment: Newtonian Science, Religion, and Politics in the Early Eighteenth Century*, trans. Sue Brortherton (Atlantic Highlands, NJ: Humanities Press, 1995). On embryological theories in eighteenth-century Italy: Walter Bernardi, *Le metafisiche dell'embrione: scienze della vita e filosofia da Malpighi a Spallanzani (1672-1793)*

(Florence: Olschki, 1986). On the political implications of the epigenesis vs. preformationism controversy in eighteenth-century Veneto: Ivano Dal Prete, "Cultures and Politics of Preformationism in Eighteenth-Century Italy," in *The Secrets of Generation: Reproduction in the Long Eighteenth Century*, ed. Raymond Stephanson and Darren N. Wagner (Toronto; Buffalo: University of Toronto Press, 2015), 71–75.

92. See also: McMahon, *Enemies of the Enlightenment*, 11.

93. A. Cagnoli to A. Cesaris, Modena, March 30th, 1800. OAB, "Corrispondenza Scientifica." On Cagnoli: Dal Prete, *Scienza e società nel Settecento veneto*, 420–50.

94. Franco Piva, "Cultura francese e censura a Venezia nel secondo settecento (ricerche storico-bibliografiche)," *Memorie dell'Istituto Veneto di Scienze, Lettere ed Arti* XXXVI, no. 3 (1973): 59, 74. *Telliamed* joined a long list of d'Holbach's, Helvetius's, Pierre Bayle's, and Voltaire's works whose importation had been prohibited or restricted for a few decades.

95. Farinella, *L'accademia repubblicana*, 98, n. 39; Farinella, 116; Renata Targhetta, "Ancora sulla massoneria veneta settecentesca, con qualche indugio a proposito di Verona," in *Tra conservazione e novità. Il mondo veneto innanzi alla rivoluzione del 1789* (Verona: Accademia di Agricoltura, Scienze e Lettere, 1991), 23. Righi, *Una loggia massonica a Verona nel 1792*, 9.

96. See: Alessandro Righi, *Il Conte Di Lilla e l'emigrazione Francese a Verona (1794-1796)* (Perugia: Vincenzo Bartelli, 1909), 5, 53–54.

97. Volta, *Ittiolitologia veronese*. See: Franco Riva, "Le avventurose vicende dell'«Ittiolitologia Veronese» del mantovano Serafino Volta," *Civiltà Mantovana* I, no. 5 (1966): 71–77. On the print shop that was founded by count Bartolomeo Giuliari for the sole purpose of producing the *Ittiolitologia*: Franco Riva, *La dimestica stamperia del veronese conte Giuliari* (Florence: Sansoni, 1956).

98. Volta, *Ittiolitologia veronese*, 1:283.

99. Volta, 1:35.

100. Volta, 1:V–XV.

101. On the controversy on "animal electrometry" in the Veneto: Luca Ciancio, "La resistibile ascesa della rabdomanzia. Pierre Thouvenel e la Guerra di Dieci Anni," *Intersezioni* 12, no. 2 (1992): 267–90; Dal Prete, *Scienza e società nel Settecento veneto*, 410–14; Ettore Curi, "La comunità scientifica veronese e gli esperimenti di elettrometria animale alla fine del settecento," *Atti dell'Accademia di Agricoltura, Scienze e Lettere di Verona*, VI, XLII (1990): 96–126.

102. *La guerra di dieci anni: Raccolta polemico-fisica sull'elettrometria galvano-organica: Parte italiana, Parte francese* (Verona, 1802), 215.

103. *La guerra di dieci anni*, 12.

104. François-René Chateaubriand, *Génie du Christianisme. Ou Beautés Poétiques et Morales de la Réligion Chrétienne*, vol. 1 (Paris: Migneret, 1802), 155–62. See: Katia Sainson, "Revolutions in Time: Chateaubriand on the Antiquity of the Earth," *French Forum* 30, no. 1 (2005): 47–63.

105. M. J. S. Rudwick, *Bursting the Limits of Time: The Reconstruction of Geohistory in the Age of Revolution* (Chicago: University of Chicago Press, 2005), 118.

106. Rudwick, 126; 333–36.

CONCLUSION

1. M. J. S. Rudwick, *Bursting the Limits of Time: The Reconstruction of Geohistory in the Age of Revolution* (Chicago: University of Chicago Press, 2005).

2. Rudwick, *Bursting the Limits of Time*, 153–54.

3. Rudwick, 176–78.

4. M. J. S. Rudwick, *Worlds before Adam: The Reconstruction of Geohistory in the Age of Reform* (Chicago: University of Chicago Press, 2008), 11–23; Rudwick, *Bursting the Limits of Time*, 447–51.

5. Giambattista Brocchi, "Discorso sui progressi dello studio della conchiologia fossile in Italia," in *Conchiologia fossile subapennina: con osservazioni geologische sugli Apennini e sul suolo adiacente*, vol. 1 (Milan: Dalla Stamperia reale, 1814), 31–50.

6. On early nineteenth-century British amateur naturalists: M. J. S. Rudwick, *The Great Devonian Controversy: The Shaping of Scientific Knowledge among Gentlemanly Specialists* (Chicago: University of Chicago Press, 1985).

7. Rudwick, *Worlds before Adam*, 72–87; 177–207.

8. William Buckland, *Geology and Mineralogy Considered with Reference to Natural Theology* (London: W. Pickering, 1836), 8–34.

9. Buckland, 9.

10. Buckland, 14.

11. I prefer this term to "scriptural geology," which usually denotes the interpretation of geological records in literalist terms. While a Christian geologist, Buckland could hardly be termed a "scriptural" one.

12. Rudwick, *Bursting the Limits of Time*, 158–72. On the construction of Hutton's "legend" in English literature: Stephen Jay Gould, *Time's Arrow, Time's Cycle: Myth and Metaphor in the Discovery of Geological Time* (Cambridge, MA: Harvard University Press, 1987), 60–97.

13. Charles Lyell, *Principles of Geology: Being an Attempt to Explain the Former Changes of the Earth's Surface, by Reference to Causes Now in Operation*, vol. 1 (London: J. Murray, 1832), 27.

14. Gould, *Time's Arrow, Time's Cycle*, 108. On Lyell's geology and Earth history: Rudwick, *Worlds before Adam*, 244–390; Derek J. Blundell and Andrew C. Scott, eds., *Lyell: The Past Is the Key to the Present*, Geological Society Special Publication, no. 143 (London: Geological Society, 1998).

15. Paul J. McCartney, "Charles Lyell and G. B. Brocchi: A Study in Comparative Historiography," *The British Journal for the History of Science* 9, no. 2 (1976): 175–89.

16. Rudwick, *Worlds before Adam*, 236–50.

17. Rudwick, *Bursting the Limits of Time*, 388. On Lamarck: Pietro Corsi, ed., *Lamarck, philosophe de la nature* (Paris: Presses universitaires de France, 2006).

18. Brocchi, "Conchiologia fossile," 235–40. On Brocchi's influence on British "transmutationism," including on Darwin: Stefano Dominici and Niles Eldredge, "Brocchi, Darwin, and Transmutation: Phylogenetics and Paleontology at the Dawn of Evolutionary Biology," *Evolution: Education and Outreach* 3, no. 4 (December 1, 2010): 576–84.

19. Rudwick, *Worlds before Adam*, 327–29.

20. M. J. S. Rudwick, *Earth's Deep History: How It Was Discovered and Why It Matters* (Chicago: The University of Chicago Press, 2014), 196–97.

21. Rudwick, *Worlds before Adam*, 423–35.

22. Rudwick, 407–20.

23. James A. Secord, *Victorian Sensation: The Extraordinary Publication, Reception, and Secret Authorship of Vestiges of the Natural History of Creation* (Chicago: University of Chicago Press, 2000), 104–05. Robert Chambers, *Vestiges of the Natural History of Creation and Other Evolutionary Writings*, ed. James A. Secord (Chicago: University of Chicago Press, 1994).

24. Cited in: Secord, *Victorian Sensation*, 245.

25. Secord, 258.

26. I cannot review here the impact of evolutionism on other religious tradition, but see: Geoffrey N. Cantor and Marc Swetlitz, eds., *Jewish Tradition and the Challenge of Darwinism* (Chicago: University of Chicago Press, 2006); Marwa Elshakri, *Reading Darwin in Arabic, 1860-1950* (Chicago: The University of Chicago Press, 2013).

27. "He devoted 1368 pages of *Beagle* notes to geological topics, compared with only 368 to biological topics." Frank H. T. Rhodes, "Darwin's Search for a Theory of the Earth: Symmetry, Simplicity and Speculation," *The British Journal for the History of Science* 24, no. 2 (June 1991): 194–95.

28. James A. Secord, "The Discovery of a Vocation: Darwin's Early Geology," *The British Journal for the History of Science* 24, no. 2 (June 1991): 161. The library of the *Beagle* included the first volume of Lyell's *Principles of Geology*. On Darwin and the Genesis Flood: Sandra Herbert, *Charles Darwin, Geologist* (Ithaca: Cornell University Press, 2005), 179–96.

29. Rhodes, "Darwin's Search for a Theory of the Earth," 205. On Darwin's reading of Lyell, see also: Secord, "The Discovery of a Vocation," 150–53.

30. Charles Darwin, *The Works of Charles Darwin, Vol. 16: The Origin of Species, 1876*, ed. Paul H. Barrett and R.B. Freeman (London: Routledge, 2017), 447.

31. On the elaboration of early glacial theories: Rudwick, *Worlds before Adam*, 505–52.

32. Ronald L. Numbers, *The Creationists: From Scientific Creationism to Intelligent Design* (Cambridge, MA: Harvard University Press, 2006), 24–26.

33. Edward J. Larson, *Trial and Error: The American Controversy over Creation and Evolution*, 3rd ed. (New York: Oxford University Press, 2003). See also: Ronald

L. Numbers, *Darwinism Comes to America* (Cambridge, MA: Harvard University Press, 1998).

34. Giuliano Pancaldi, *Darwin in Italy: Science across Cultural Frontiers* (Bloomington: Indiana University Press, 1991), 164–65. According to Pancaldi, the 1864 Italian translation of *On the Origin of Species* was prohibited on account of the translator, a physician and politician with radical inclinations.

35. Frank M. Turner, "The Late Victorian Conflict of Science and Religion as an Event in Nineteenth-Century Intellectual and Cultural History," in *Science and Religion: New Historical Perspectives*, ed. Thomas Dixon, Geoffrey N. Cantor, and Stephen Pumfrey (New York: Cambridge University Press, 2010), 87–109. See also: Rudwick, *Earth's Deep History*, 207–12.

36. For a British perspective: Ciaran Toal, "Preaching at the British Association for the Advancement of Science: Sermons, Secularization and the Rhetoric of Conflict in the 1870s," *The British Journal for the History of Science* 45, no. 1 (2012): 75–95.

37. For a general introduction to the "conflict thesis": Davis Mislin, "Roman Catholics," in *The Warfare between Science and Religion: The Idea That Wouldn't Die*, ed. Jeff Hardin, Ronald L. Numbers, and Ronald A. Binzley (Baltimore, Maryland: Johns Hopkins University Press, 2018), 103–22.

38. John William Draper, *History of the Conflict between Religion and Science* (New York: D. Appleton, 1898), 182.

39. Draper, ix.

40. Stephen Jay Gould, *Rocks of Ages: Science and Religion in the Fullness of Life* (New York: Ballantine Publishing Group, 1999), 103; Mislin, "Roman Catholics," 18–19.

41. Draper, *History of the Conflict between Religion and Science*, 182–83.

42. Draper, 200.

43. Draper, 191.

44. L. Principe has also highlighted the more mundane concerns inciting White (who worked at the secular Cornell University) to attack the denominational colleges that competed with Cornell for funding. Mislin, "Roman Catholics," 10–11.

45. Andrew Dickson White, *A History of the Warfare of Science with Theology in Christendom* (New York: D. Appleton, 1901), ix.

46. White, v–vi.

47. Andrew Dickson White, *A History of the Warfare of Science with Theology in Christendom* (New York: D. Appleton, 1896), 250.

48. White, 251.

49. See in particular, Pierre Duhem's stalwart defense of medieval science in the early 1900s; and James J. Walsh, *The Popes and Science; the History of the Papal Relations to Science During the Middle Ages and down to Our Own Time* (New York: Fordham University Press, 1908). Among Duhem's most recent followers: Edward Grant, *God and Reason in the Middle Ages* (Cambridge: Cambridge University Press, 2001). For an overview of the Catholic reaction to the "conflict thesis": Mislin, "Roman Catholics."

50. G. N. Cantor, "What Shall We Do with the 'Conflict Thesis'?," in *Science and Religion: New Historical Perspectives*, ed. Thomas Dixon, Geoffrey N. Cantor, and Stephen Pumfrey (New York: Cambridge University Press, 2010), 283–98.

51. On the "new atheists" and their claims of inevitable conflict between science and religion: Ronald L. Numbers, "The New Atheists," in *The Warfare between Science and Religion: The Idea That Wouldn't Die*, ed. Jeff Hardin, Ronald L. Numbers, and Ronald A. Binzley (Baltimore: Johns Hopkins University Press, 2018), 220–38.

52. For a scholarly reconstruction of the trial and of its aftermath: Edward J. Larson, *Summer for the Gods: The Scopes Trial and America's Continuing Debate over Science and Religion* (Cambridge MA: Harvard University Press, 1998).

53. Jerome Lawrence and Robert Edwin Lee, *The Complete Text of Inherit the Wind* (New York: 1957), 55.

54. Jeffrey P. Moran, *The Scopes Trial: A Brief History with Documents* (New York: Palgrave, 2002), 159.

55. Moran, 155–57.

56. Numbers, *The Creationists*, 58.

57. Numbers, *The Creationists*, 60.

58. Numbers, 90.

59. Secord, *Victorian Sensation*, 286.

60. Numbers, *The Creationists*, 187.

61. On the topic, see especially: Adam R. Shapiro, *Trying Biology: The Scopes Trial, Textbooks, and the Antievolution Movement in American Schools* (Chicago: The University of Chicago Press, 2013).

62. Apart from Numbers, *The Creationists*, see also: Michael B. Berkman and Eric Plutzer, *Evolution, Creationism, and the Battle to Control America's Classrooms* (New York: Cambridge University Press, 2010).

63. On the perception of the relation between science and religion among religious people in the US: John Hyde Evans, *Morals Not Knowledge: Recasting the Contemporary U.S. Conflict between Religion and Science* (Oakland,: University of California Press, 2018).

64. For an overview of the current state of politics concerning the evolution vs. creationism issue: Berkman and Plutzer, *Evolution, Creationism, and the Battle to Control America's Classrooms*.

65. Numbers, *The Creationists*, 279. Emphasis mine.

66. The resolution did not prohibit their teaching but left the option to local school districts. Various forms of the provision were enacted and finally repealed in the following years, as majorities changed in the Kansas House of Representatives.

67. US presidential primaries, televised debate hosted by CNN. June 5th, 2007. Consulted at: https://www.youtube.com/watch?v=n-BFEhkIujA.

68. Terry Mortenson's study of early nineteenth-century "scriptural geologists" underscores the similarity of those theories to their modern counterparts. In spite of the acknowledgment that there seems to be no direct connection between

the two movements, Mortenson places them within a Christian tradition whose uninterrupted lineage was broken only in the course of the nineteenth century. Compromises with non-literalist interpretations of Genesis would be a recent "mistake." Terry Mortenson, *The Great Turning Point* (Green Forest, AR: Master Books, 2004), 12–15.

Bibliography

MANUSCRIPT SOURCES

Archivio di Stato, Reggio Emilia
Archivio Vallisneri, 4/1, fasc. II

Archivum Romanum Societatis Iesu, Rome
F.G. 655, Censurae Librorum, t. IV (1618–1642), *Notae in Cosmographiam P. Iosephi
    Blancani*
F.G. 657, Censurae Opinionum, t. II (1629–1650)
F.G. 658, Censurae Opinionum, t. III (1650–1652)
F.G. 659, Censurae Opinionum, t. IV (1621–1665)
FG 663, Censurae Librorum (1626–1663)

Biblioteca dell'Accademi degli Intronati, Siena
Autografi Porri, 20.21

Biblioteca Capitolare, Verona
Cod. CCCCXXX, n. 5, "Censurae Librorum"

Biblioteca Estense, Modena
Mss. It. (ms.IX.F.11)
fArchivio Muratori, Filza 81

Biblioteca Labronica, Livorno
Autografoteca Bastogi, Cass. 30 ins. 2185
Carte Conti

Bibliothèque Municipale, Nîmes

ms. 309

ms. 92

ms. 90, *Pétrifications du Véronois*

ms. 285, *Catalogue des libres* [sic] *de J. François Séguier en 1760*.

Bibliothèque National de France, Paris

lat. 14.723, *Johannis Buridani questiones in Aristotelis physica et in libros metheororum*

Biblioteca Nazionale Centrale, Florence

Filze Rinuccini 10, *Comento primo di Benedetto Varchi Fiorentino sopra il primo libro delle Meteore d'Aristotile*

Bibliothèque Publique et Universitaire, Neuchâtel

ms. 1282

ms. 1265

ms. 1289.4

PUBLISHED SOURCES AND LITERATURE

[Anonymous] *Opera nuova la quale tratta della filosofia naturale, chiamata la Metaura d'Aristotile; chiosata da San Tommaso d'Aquino, dell'ordine dei Frati Predicatori.* Venice: Comin da Trino, 1554.

Adams, Frank Dawson. *The Birth and Development of the Geological Sciences.* Baltimore: Williams & Wilkins, 1938.

Agosti, Giovanni, Vincenzo Farinella, and Salvatore Settis. "Passione e gusto per l'antico nei pittori italiani del Quattrocento." *Annali della Scuola Normale Superiore di Pisa. Classe di Lettere e Filosofia* 17, no. 4 (1987): 1061–107.

Agricola, Georg. *De Re Metallica.* Translated by Herbert Hoover and Lou Henry Hoover. [New ed.]. New York, Dover Publications, 1950.

Albert of Saxony. *Quæstiones in Aristotelis De cælo.* Edited by Benoît Patar. Louvain-la-neuve: Peeters, 2008.

Albert of Saxony, Themon Judeus, and Jean Buridan. *Quaestiones [et] Decisiones Physicales in Octo Libros Physicorum.* Edited by George Lokert. Basel: Conrad Resch, 1516.

Albert of Saxony, *Questiones Subtilissime in Libros De Celo et Mundo*, ed. Hyeronimus Surianus. Venice: per Otinum Papiensem, 1497.

Albert of Saxony. *Questiones Subtilissime in Libros De Celo et Mundo*, ed. Hyeronimus Surianus.(Venice: Imp. Arte Bonetis locatellis, 1492.)

Alberti, Leon Battista. *I dieci libri de l'architettura.* Translated by Pietro Lauro. Venice: Vincenzo Valgrisi, 1546.

Alberti, Leon Battista. *L'architettura. (De re aedificatoria).* Edited by Paolo Portoghesi. Translated by Giovanni Orlandi. 2 vols. Trattati di architettura, v. 1. Milan: Edizioni Il Polifilo, 1966.

Alberti, Leon Battista. *Libri De re aedificatoria decem.* Paris: Opera magistri Bertholdi Rembolt & Ludouici Hornken, 1512.

Alberti, Leon Battista. *On the Art of Building in Ten Books.* Translated by Joseph Rykwert, Neil Leach, and Robert Tavernor. Cambridge, MA: MIT Press, 1988.

Alberti, Leon Battista. *Opere Volgari.* Edited by Cecil Grayson. Bari: G. Laterza, 1960.

Albertus Magnus. *Book of Minerals.* Edited by Dorothy Wichoff. Oxford: Clarendon Press, 1967.

Albertus Magnus. *Liber Methauroru[m].* Venice: Impressi p[er] Renaldum de Nouimagio theotonicum, 1488.

Albertus Magnus. *On the Causes of the Properties of the Elements (Liber de Causis Proprietatum Elementorum).* Edited and translated by Irven Michael Resnick. Milwaukee: Marquette University Press, 2010.

Albertus Magnus. *Opus Nobile de Causis Proprietatum Elementorum.* Magdeburg: Jacob Winther, 1506.Aldrovandi, Ulisse. *Vlyssis Aldrouandi . . . Musaeum metallicum in libros IIII distributum.* Edited by Matteo Ambrosini. Bologna: typis Io. Baptistae Ferronii, 1648.

Alessandri, Alessandro. *Alexandri ab Alexandro, Neapolitani I. C. Genialium Dierum Libri sex.* Lyon: Paul Frellon, 1616.

Alexander of Aphrodisias. *Alexandre d'Aphrodisias commentaires sur les Météores d'Aristote.* Edited by A. J. Smet. Louvain: de Wulf-Mansion Centrum, 1968.

Alfred of Sareshel. *Alfred of Sareshel's Commentary on the Metheora of Aristotle: Critical Edition, Introduction, and Notes.* Edited by James K. Otte. Leiden: Brill, 1987.

Alighieri, Dante. "Questio de Aqua et Terra." In *La Letteratura Italiana – Storia e Testi,* edited by Francesco Mazzoni, V:691–880. Naples: Ricciardi, 1978.

Altieri Biagi, Maria Luisa. "Nuclei concettuali e strutture sintattiche nella 'Composizione del mondo' di Restoro d'Arezzo." In *L'Avventura della mente: studi sulla lingua scientifica,* 11–33. Naples: Morano, 1990.

Alverny, Maire-Therese d'. "Translations and Translators." In *Renaissance and Renewal in the Twelfth Century,* edited by Robert Louis Benson, Giles Constable, and Carol Dana Lanham, 421–62. Cambridge, MA: Harvard University Press, 1982.

Ames-Lewis, Francis. *The Intellectual Life of the Early Renaissance Artist.* New Haven: Yale University Press, 2000.

"Antonio Vallisneri – Edizione Nazionale Delle Opere." http://www.vallisneri.it/mich eli.shtml.

Aristotle.*Meteorologica.*TranslatedbyErwinWentworthWebster.Oxford:TheClarendon Press, 1923.

Aristotle. *Meteorologica.* Translated by H. D. P. Lee. Cambridge, MA: Harvard University Press, 2014.

Arnigio, Bartolomeo. *Meteoria over discorso intorno alle impressioni imperfette, humide, secche et miste cosi in alto, come nelle viscere della terra generate (etc.)*. Brescia: Fratelli de Marchetti, 1568.

Austin, Herbert Douglas. *Accredited Citations in Ristoro d'Arezzo's Composizione Del Mondo: A Study of Sources*. Turin: Guido Momo, 1913.

Avicenna. *Avicennae De Congelatione et Conglutinatione Lapidum; Being Sections of the Kitâb al-Shifâ'*. Edited by Eric John Holmyard and Desmond Christopher Mandeville. Paris: P. Guethner, 1927.

Baffioni, Carmela, ed. *Epistles of the Brethren of Purity: On the Natural Sciences: An Arabic Critical Edition and English Translation of Epistles 15–21*. Translated by Carmela Baffioni. Oxford: Oxford University Press in association with the Institute of Ismaili Studies, 2013.

Bagès, Sylvie. "Les Questiones super tres libros Metheororum Aristotelis de Jean Buridan étude suivie de l'édition du livre I." École nationale des chartes, 1986.

Bailly, Jean-Sylvain. *Histoire de l'astronomie ancienne, depuis son origine jusqu'a l'etablissement de l'école d'Alexandrie*. Paris: chez les freres Debure, 1775.

Baldassarri, Guido, Silvia Contarini, and Francesca Fedi, eds. *Antonio Conti: uno scienziato nella République des lettres*. Padua: Il Poligrafo, 2009.

Baldini, Massimo, ed. *Anton Lazzaro Moro, 1687–1987 – Atti del Convegno di Studi*. Pordenone: Grafiche editoriali artistiche pordenonesi, 1988.

Baldini, Ugo. "Galileo, la nuova astronomia e la critica all'aristotelismo nel dialogo epistolare tra Giuseppe Biancani e i revisori romani della Compagnia di Gesù." *Annali Dell'Istituto e Museo di Storia della Scienza di Firenze* IX (1984): 13–43.

Baldini, Ugo. *Legem impone subactis: studi su filosofia e scienza dei Gesuiti in Italia, 1540–1632*. Rome: Bulzoni, 1992.

Baldini, Ugo. "Uniformitas et soliditas doctrinae: le censure librorum e opinionum." In *Legem impone subactis: studi su filosofia e scienza dei Gesuiti in Italia, 1540–1632*, 75–111. Rome: Bulzoni, 1992.

Baldini, Ugo, and Leen Spruit, eds. *Catholic Church and Modern Science: Documents from the Archives of the Roman Congregations of the Holy Office and the Index*. Rome: Libreria editrice vaticana, 2009.

Baratta, Mario. *Leonardo da Vinci ed i problemi della terra*. Turin: Fratelli Bocca, 1903.

Barbaro, Ermolao and Hieronymus Wildenberg. *Naturalis scientiae totius compendium, ex Aristotele, & alijs philosophis, Hermolao Barbaro . . . autore, innumeris, quibus antea scatebat, mendis nunc demum D. Conradi Gesneri*. Edited by Konrad Gesner. Basel: Ex officina Ioannis Oporini, 1548.

Barbierato, Federico. *The Inquisitor in the Hat Shop: Inquisition, Forbidden Books and Unbelief in Early Modern Venice*. Burlington, VT: Routledge, 2012.

Barnes, Jerome. *Giovanni Battista Ramusio and the History of Discoveries: An Analysis of Ramusio's Commentary, Cartography, and Imagery in "Delle Navigationi et Viaggi."* Ph.D. Dissertation, University of Texas at Arlington, 2007.

Barnett, Lydia. *After the Flood. Imagining the Global Environment in Early Modern Europe*. Baltimore: Johns Hopkins University Press, 2019.

Barnett, Lydia. "Strategies of Toleration: Talking Across Confessions in the Alpine Republic of Letters." *Eighteenth-Century Studies* 48, no. 2 (January 27, 2015): 141–57.

Barnett, Lydia. "The Theology of Climate Change: Sin as Agency in the Enlightenment's Anthropocene." *Environmental History* 20, no. 2 (2015): 217–37.

Barolini, Teodolinda. *The Undivine Comedy: Detheologizing Dante*. Princeton, NJ: Princeton University Press, 1992.

Bauer, Georg. *Georgii Agricolae De Re Metallica Libri XII*. Basileae: Apud H. Frobenium et N. Episcopium, 1556.

Belgrado, Anna Minerbi. *Paura e ignoranza: studio sulla teoria della religione in d'Holbach*. Florence: Olschki, 1983.

Benítez, Miguel. "Fixisme et evolutionnisme au temps des lumières: le 'Telliamed' de Benoît de Maillet." *Rivista Di Storia Della Filosofia (1984–)* 45, no. 2 (1990): 247–68.

Belleguic, Thierry and Anouchka Vasak, eds.. *Ordre et désordre du monde: enquête sur les météores, de la Renaissance à l'âge moderne*. Paris: Hermann, 2013.

Benzoni, Gino. "Scipione Maffei e il Mondo delle Accademie." In *Scipione Maffei nell'Europa del Settecento*, edited by Gian Paolo Romagnani, 241–57. Verona: Cierre, 1998.

Beretta, Francesco. "Orthodoxie philosophique et Inquisition romaine aux 16e–17e siècles." *Historia Philosophica* 3 (2005): 67–96.

Beretta, Marco, ed. *From Private to Public: Natural Collections and Museums. Uppsala Studies in History of Science* 32. Sagamore Beach, MA: Science History Publications/ USA, 2005.

Berga, Antonio. *Discorso di M. Antonio Berga lettore filosofo nella vniversità di Turino, della grandezza dell'acqua & della terra. Contra l'opionione dil S. Alessandro Piccolomini*. Turin: Eredi Bevilaqua, 1579.

Berkman, Michael B., and Eric Plutzer. *Evolution, Creationism, and the Battle to Control America's Classrooms*. New York: Cambridge University Press, 2010.

Bernardi, Walter. *Le metafisiche dell'embrione: scienze della vita e filosofia da Malpighi a Spallanzani (1672–1793)*. Florence: Olschki, 1986.

Bertucci, Paola. *Artisanal Enlightenment: Science and the Mechanical Arts in Old Regime France*. New Haven: Yale University Press, 2017.

Bertucci, Paola. *Viaggio nel paese delle meraviglie: scienza e curiosità nell'Italia del Settecento*. Turin: Bollati Boringhieri, 2007.

Betti, Zaccaria. *Descrizione di un meraviglioso ponte naturale nei monti Veronesi*. Verona: Marco Moroni, 1766.

Biancani, Giuseppe. *Aristotelis Loca Mathematica: ... Collecta et Explicata : Accessere De Natura Mathematicarum Scientiarum Tractatio Atque Clarorum Mathematicorum Chronologia*. Bologna: Bartolomeo Cocchi, 1615.

Biancani, Giuseppe. *Sphaera mundi, seu Cosmographia, demonstratiua, ac facili methodo tradita: in qua totius mundi fabrica, vnà cum nouis, Tychonis, Kepleri, Galilaei, aliorumque astronomorum adinuentis continetur. Accessere 1. Breuis introductio ad geographiam. 2. Apparatus ad mathematicarum studium. 3. Echometria, idest geometrica traditio de echo. Authore Iosepho Blancano Bononiensi è Societate Iesu . . .* Bologna: typis Sebastiani Bonomij, 1620.

Bianchi, Luca. *Censure et liberté intellectuelle à l'université de Paris: XIIIe–XIVe Siècles.* Paris: Belles Lettres, 1999.

Bianchi, Luca, ed. *Christian Readings of Aristotle from the Middle Ages to the Renaissance.* Turnhout: Brepols, 2011.

Bianchi, Luca. *Il vescovo e i filosofi: la condanna parigina del 1277 e l'evoluzione dell'aristotelismo scolastico.* Bergamo: Lubrina, 1990.

Bianchi, Luca. *L'errore di Aristotele: la polemica contro l'eternità del mondo nel XIII secolo.* Florence: La Nuova Italia, 1984.

Bianchi, Luca. *L'inizio dei tempi: antichità e novità del mondo da Bonaventura a Newton.* Florence: Olschki, 1987.

Bianchi, Luca. *Pour une histoire de la "double vérité."* Paris: Vrin, 2008.

Bianchi, Luca. "Per una storia dell'aristotelismo 'volgare' nel rinascimento: problemi e prospettive di ricerca." *Bruniana & Campanelliana* 15, no. 2 (2009): 367–85.

Bianchi, Luca. "Volgarizzare Aristotele: per chi?" *Freiburger Zeitschrift Für Philosophie Und Theologie* 59, no. 2 (2012): 495

Bianchini, Francesco. "Censurae Librorum," (n.d.) Cod. CCCCXXX, n. 5. Biblioteca Capitolare, Verona.

Bianchini, Francesco. *La Istoria universale provata con monumenti e figurata con simboli de gli antichi . . . da Francesco Bianchini.* Rome: Antonio De Rossi, 1697.

Biard, Joël, ed. *Itinéraires D'Albert de Saxe, Paris-Vienne Au XIVe Siècle.* Paris: J. Vrin, 1991.

Biard, Joelle. "The Natural Order in Jean Buridan." In *The Metaphysics and Natural Philosophy of John Buridan*, edited by Johannes. M. M. H. Thijssen and Jack Zupko, 77–95. Leiden: Brill, 2001.

Biringuccio, Vannoccio. *De La Pirotechnia: Libri X.* Venice: Per Venturino Rossinello ad instantia di Curtio Navò, 1540.

Bisson, Sebastiano, Lada Hordynsky-Caillat, and Odile Redon. "Le Témoin Gênant. Une version latine du 'Régime Du Corps' D'Aldebrandin de Sienne." *Médiévales*, no. 42 (2002): 117–30.

Blackwell, Richard J. *Galileo, Bellarmine, and the Bible: Including a Translation of Foscarini's Letter on the Motion of the Earth.* Notre Dame: University of Notre Dame Press, 1991.

Blair, Ann. "Mosaic Physics and the Search for a Pious Natural Philosophy in the Late Renaissance." *Isis* 91, no. 1 (2000): 32–58.

Blair, Peter Hunter. *The World of Bede.* Cambridge: Cambridge University Press, 1990.

Blundell, Derek J., and Andrew C. Scott, eds. *Lyell: The Past Is the Key to the Present.* Geological Society Special Publication, no. 143. London: Geological Society, 1998.

Boccone, Paolo. *Museo di fisica e di esperienze variato, e decorato di osservazioni naturali, note medicinali, e ragionamenti secondo i principij de' moderni etc.* Venice: Giovan Battista Zuccato, 1697.

Boemus, Johann. *Omnium gentium Mores, Leges et Ritus . . . Ioanne Boemo Aubano Teutonico nuper collecti . . .* (Lyon: Apud Haeredes Simonis Vincentii, 1535), 15–20.

Boethius of Dacia. *Sull'eternità del mondo.* Edited and translated by Luca Bianchi. Milan: UNICOPLI, 2003.

Boffito, Giuseppe. *Intorno alla "Quaestio de aqua et terra" attribuita a Dante.* Turin: Carlo Clausen, 1902.

Boffito, Giuseppe. *L'Eresia Degli Antipodi.* Florence: Ist. alla Querce, 1905.

Bolzoni, Lina. "Il 'Badoaro' di Francesco Patrizi e l'Accademia veneziana della Fama." *Giornale Storico della Letteratura Italiana* 158, no. 501 (1981): 71–101.

Bolzoni, Lina. "L'Accademia Veneziana: splendore e decadenza di un'utopia enciclopedica." In *Università, accademie e società scientifiche in Italia e in Germania dal Cinquecento al Settecento,* edited by Laetitia Boehm and Ezio Raimondi, 117–69. Bologna: Il Mulino, 1981.

Borelli, Giorgio. "Scipione Maffei e il problema del prestito ad interesse." In *Scipione Maffei nell'Europa del Settecento: atti del convegno, Verona, 23–25 settembre 1996,* edited by Gian Paolo Romagnani, 121–37. Verona: Cierre Edizioni, 1998.

Borro, Girolamo. *Del flusso e reflusso del mare, & dell'inondatione del Nilo.* Florence: Giorgio Marescotti, 1583.

Borro, Girolamo. *Dialogo del flusso e reflusso del mare d'Alseforo Talascopio (G. Borro). Con un ragionamento di Telifilo Filogenio della perfettione delle donne.* Lucca: V. Busdragho, 1561.

Boscani Leoni, Simona, ed. *Wissenschaft- Berge- Ideologien. Johann Jakob Scheuchzer (1672–1733) und die frühneuzeitliche Naturforschung: Scienza- montagna- ideologie. Johann Jakob Scheuchzer (1672–1733) e la ricerca naturalistica in epoca moderna.* Basel: Schwabe, 2010.

Bottin, Francesco. "Logica e filosofia naturale nelle opere di Paolo Veneto." In *Scienza e filosofia all'Università di Padova nel Quattrocento,* 85–134. Padua: Edizioni LINT, 1983.

Boulanger, Nicolas Antoine. *Anecdotes physiques de l'histoire de la nature; avec la Nouvelle Mappemonde et le Memoire sur une Nouvelle Mappemonde.* Edited by Pierre Boutin. Paris: Champion, 2006.

Boulanger, Nicolas Antoine. "Flood." In *Encyclopedia of Diderot & d'Alembert – Collaborative Translation Project.* Ann Arbor: Michigan Publishing, University of Michigan Library, 2009. http://hdl.handle.net/2027/spo.did2222.0000.850.

Bozza, Vincenzo. *Della universale rivoluzione sofferta dal globo terracqueo lettera al molto reverendo p. Orazio Rota m.o.p.p. di Lingue Orientali in Mantova.* Verona: 1788.

Brams, J., and W. Vanhamel, eds. *Guillaume de Moerbeke: Recueil d'études à l'occasion du 700e Anniversaire de Sa Mort (1286).* Leuven: Leuven University Press, 1989.

Brocchi, Giambattista. "Discorso sui progressi dello studio della conchiologia fossile in Italia." In *Conchiologia fossile subapennina: con osservazioni geologische sugli Apennini e sul suolo adiacente*, Vol. 1. Milan: Dalla Stamperia Reale, 1814.

Broucek, James. "Thinking About Religion Before 'Religion': A Review of Brent Nongbri's Before Religion: A History of a Modern Concept." *Soundings: An Interdisciplinary Journal* 98, no. 1 (2015): 98–125.

Brown, Alison. "'Natura Idest?'. Leonardo, Lucretius and their Views of Nature." In *Leonardo da Vinci on Nature: Knowledge and Representation*, edited by Fabio Frosini and Alessandro Nova, 153–79. Venice: Marsilio, 2015.

Brown, Alison. *The Return of Lucretius to Renaissance Florence*. Cambridge, MA: Harvard University Press, 2010.

Brugi, Biagio. "Un parere di Scipione Maffei intorno allo Studio di Padova sui principi del Settecento. Edizione del testo originale con introduzione e note." *Atti del R. Istituto Veneto di scienze, lettere e arti* 69 (October 1909): 575–91.

Buchwald, Jed Z., and Mordechai Feingold. *Newton and the Origin of Civilization*. Princeton, NJ: Princeton University Press, 2013.

Buckland, William. *Geology and Mineralogy Considered with Reference to Natural Theology*. London: W. Pickering, 1836.

Buffon, Georges-Louis Leclerc comte de. *Histoire naturelle générale et particuliére: avec la description du Cabinet du Roy: tome septième*. Paris: Imprimerie Royale, 1758.

Burgess, R. W. *Studies in Eusebian and Post-Eusebian Chronography: 1. The "Chronici Canones" of Eusebius of Caesarea: Structure, Content and Chronology, AD 282–325 – 2. The "Continuatio Antiochiensis Eusebii": A Chronicle of Antioch and the Roman Near East during the Reigns of Constantine and Constantius II, AD 325–350*. Stuttgart: Franz Steiner, 1999.

Buridan, Jean. *Ioannis Buridani Expositio et quæstiones in Aristotelis De cælo*. Edited by Benoît Patar. Louvain-La-Neuve: Editions Peeters, 1996.

Buridan, Jean. *Iohannis Buridani Quaestiones Super Libris Quattuor de Caelo et Mundo*. Edited by Ernest A. Moody. Cambridge, MA: The Mediaeval Academy of America, 1942.

Burnet, Thomas. *Telluris theoria sacra: orbis nostri originem & mutationes generales, quas aut jam subiit, aut olim subiturus est, complectens. Libri duo priores de diluvio & paradiso*. London: Roger Norton for Walter Kettilby, 1681.

Burson, Jeffrey D., and Ulrich L. Lehner, eds. *Enlightenment and Catholicism in Europe: A Transnational History*. Notre Dame, Indiana: University of Notre Dame Press, 2014.

Cabeo, Niccolò. *In Quatuor Libros Meteorologicorum Aristotelis Commentaria, et Quaestiones: Quatuor Tomis Compraehensa*. Vol. 1. 4 vols. Rome: Eredi Francesco Corbelletti, 1646.

Calcidius. *On Plato's Timaeus*. Edited and translated by John Magee. Cambridge, MA: Harvard University Press, 2016.

Campanale, Maria I. *Ai confini del Medioevo scientifico: il De montium origine di Valerio Faenzi*. Bari: Edipuglia, 2012.

Campbell, Gordon Lindsay. *Lucretius on Creation and Evolution: A Commentary on De Rerum Natura, Book 5, Lines 772–1104.* Oxford: Oxford University Press, 2003.

Cantor, Geoffrey N. "What Shall We Do with the 'Conflict Thesis'?" In *Science and Religion: New Historical Perspectives*, edited by Thomas Dixon, Geoffrey N. Cantor, and Stephen Pumfrey, 283–98. New York: Cambridge University Press, 2010.

Cantor, Geoffrey N., and Marc Swetlitz, eds. *Jewish Tradition and the Challenge of Darwinism.* Chicago: University of Chicago Press, 2006.

Carinci, Eleonora. "Una 'speziala' padovana: Lettere di Philosophia Naturale di Camilla Erculiani (1584)." *Italian Studies* 68, no. 2 (July 1, 2013): 202–29.

Carinci, Eleonora, and Sandra Plastina, eds. *Corrispondenze scientifiche tra Cinquecento e Seicento. Camilla Erculiani «Lettere di philosophia naturale» (1584). Margherita Sarrocchi «Lettere a Galilei» (1611–1612).* Lugano: Agorà & Company, 2016.

Caroti, Stefano. "'L'Aristotele Italiano' di Alessandro Piccolomini: un progetto sistematico di filosofia naturale in volgare a metà '500." In *Il volgare come lingua di cultura dal trecento al cinquecento: Atti Del Convegno Internazionale, Mantova, 18–20 Ottobre 2001*, edited by Arturo Calzona, 361–401. Florence: L. S. Olschki, 2003.

Carozzi, Marguerite. *Voltaire's Attitude toward Geology.* Geneva: Société de Physique et d'Histoire Naturelle, 1983.

Casapullo, Rosa. "Le Trattato di scienza universal de Vivaldo Belcazer et la tradition du De Proprietatibus Rerum." In *Encyclopédie médiévale et langues européennes: réception et diffusion du De proprietatibus rerum de Barthélemy l'Anglais dans les langues vernaculaires*, edited by Joëlle Ducos, 235–58. Paris: Honoré Champion éditeur, 2014.

Casapullo, Rosa, and Miriam Rita Policardo. "Tecniche della divulgazione scientifica nel volgarizzamento mantovano del 'De Proprietatibus Rerum' di Bartolomeo Anglico." *Lingua e Stile*, no. 2/2003 (2003): 139–76.

Casini, Lorenzo. "The Renaissance Debate on the Immortality of the Soul: Pietro Pomponazzi and the Plurality of Substantial Forms," in *Mind, Cognition and Representation: The Tradition of Commentaries on Aristotle's De Anima*, edited by Paul J. J. M. Bakker and J. M. M. H. Thijssen, 135–58. Aldershot: Ashgate Publishing, 2007..

Cattaneo, Angelo. *Fra Mauro's Mappamundi and Fifteenth-Century Venetian Culture.* Turnhout: Brepols, 2011.

Cesalpino, Andrea. *Andreae Caesalpini ... Peripateticarum quaestionum libri quinque.* Venice: apud Iuntas, 1571.

Cesalpino, Andrea. *De Metallicis: libri tres.* Rome: Luigi Zannetti, 1596.

Cestaro, Antonio, ed. *Geronimo Seripando e la Chiesa del suo tempo: nel V centenario della nascita: Atti del Convegno di Salerno, 14–16 Ottobre 1994.* Thesaurus Ecclesiarum Italiae Recentioris Aevi. XII, Campania e Basilicata 8. Rome: Edizioni di Storia e Letteratura, 1997.

Chakrabarti, Pratik. *Inscriptions of Nature: Geology and the Naturalization of Antiquity.* Baltimore: Johns Hopkins University Press, 2020.

Chambers, Robert. *Vestiges of the Natural History of Creation and Other Evolutionary Writings*. Edited by James A. Secord. Chicago: University of Chicago Press, 1994.

Chaplin, Joyce E. *Subject Matter: Technology, the Body, and Science on the Anglo-American Frontier, 1500–1676*. Cambridge, MA: Harvard University Press, 2001.

Charbonnat, Pascal. "Usages et réceptions du Telliamed chez les naturalistes durant la seconde moitié du 18e siècle." *Corpus: revue de la philosophie* 59 (2010): 125–52.

Chateaubriand, François-René. *Génie du Christianisme. Ou Beautés Poétiques et Morales de la Réligion Chrétienne*. Vol. 1. Paris: Migneret, 1802.

Chiocco, Andrea, and Benedetto Ceruti. *Musaeum Calceolarianum Veronense*. Verona: Apud Angelum Tamum, 1622.

Christian, William A. "Augustine on the Creation of the World." *The Harvard Theological Review* 46, no. 1 (1953): 1–25.

Cian, Vittorio. "Vivaldo Belcazer e l'enciclopedismo italiano delle origini." *Giornale Storico Della Letteratura Italiana* Supplemento n. 5 (1902): 10–31.

Ciancio, Luca. "An Amphitheatre Built on Toothpicks: Galileo, Nardi and the Hypothesis of Central Fire." *Galilaeana* XV (2018): 83–113.

Ciancio, Luca. *Autopsie della terra: illuminismo e geologia in Alberto Fortis (1741–1803)*. Florence: Olschki, 1995.

Ciancio, Luca. "Geologia e ortodossia. L'eredità galileiana nella geologia veneta del secondo Settecento." In *La Politica della scienza: Toscana e stati italiani nel tardo Settecento: atti del convegno di Firenze, 27–29 gennaio 1994*, edited by Giulio Barsanti, Vieri Becagli, and Renato Pasta. Florence: Olschki, 1996.

Ciancio, Luca. "'Immoderatus Fervor Ad Intra Coërcendus': Reactions to Athanasius Kircher's Central Fire in Jesuit Science and Imagination." *Nuncius* 33, no. 3 (2018): 464–504.

Ciancio, Luca. "La resistibile ascesa della rabdomanzia. Pierre Thouvenel e la Guerra di Dieci Anni." *Intersezioni* 12, no. 2 (1992): 267–90.

Ciancio, Luca. "Un interlocutore fiammingo di Fracastoro: il medico Iohannes Goropius Becanus (1518-1572) e la teoria dell'origine organica dei fossili." In *Girolamo Fracastoro: fra medicina, filosofia e scienze della natura*, edited by Alessandro Pastore and Enrico Peruzzi, 141–155. Florence: L. S. Olschki, 2006.

Ciancio, Luca, and Gian Paolo Romagnani, eds. *Unità del sapere, molteplicità dei saperi: Francesco Bianchini (1662–1729), tra natura, storia e religione*. Verona: QuiEdit, 2010.

Cicogna, Emmanuele Antonio. *Delle inscrizioni Veneziane*. Venice: Orlandelli, 1830.

Classen, Peter. "Res Gestae, Universal History, Apocalypse." In *Renaissance and Renewal in the Twelfth Century*, edited by Robert Louis Benson, Giles Constable, and Carol Dana Lanham, 421–62. Cambridge, MA: Harvard University Press, 1982.

Cohen, Claudine. "L'« Anthropologie » de Telliamed." *Bulletins et Mémoires de la Société d'Anthropologie de Paris* 1, no. 3 (1989): 45–55.

Cohen, Claudine. "Leibniz Et Benoît de Maillet: De la Protogée au Telliamed." *Corpus: Revue de Philosophie* 59 (2010): 55–78.

Cohen, Claudine. *Science, libertinage et clandestinité à l'aube des lumières: le transformisme de Telliamed*. Paris: Presses Universitaires de France, 2011.

Cohen, I. Bernard, ed. *Puritanism and the Rise of Modern Science: The Merton Thesis*. London: Rutgers University Press, 1990.

Cohn, Norman. *Noah's Flood: The Genesis Story in Western Thought*. New Haven: Yale University Press, 1996.

Colonna, Fabio. *Fabii Columnae Lyncei Purpura...* Rome: Apud Jacobum Mascardum, 1616.

Connell, William J. "The Eternity of the World and Renaissance Historical Thought." *California Italian Studies* 2, no. 1 (January 1, 2011).

Constant, Eric A. "A Reinterpretation of the Fifth Lateran Council Decree Apostolici Regiminis (1513)," *The Sixteenth Century Journal* 33, no. 2 (2002): 366–67.

Contardi, Simone. *La rivincita dei "filosofi di carta": saggio sulla filosofia naturale di Antonio Vallisneri junior*. Florence: Olschki, 1994.

Contarini, Gasparo. *De elementis et eorum mixtionibus libri quinque ... nunc primum in lucem aediti. Scipionis Capitii de principiis rerum poema*. Paris: Dives Nicolaus, 1548.

Conti, Antonio. *Scritti filosofici*. Edited by Antonio Badaloni. Naples: Fulvio Rossi, 1972.

Corsi, Pietro, ed. *Lamarck, philosophe de la nature*. Paris: Presses Universitaires de France, 2006.

Costa, Gustavo. *Thomas Burnet e la censura pontificia: con documenti inediti*. Florence: L. S. Olschki, 2006.

Costantini, Giuseppe Antonio. *La verità del diluvio universale vindicata dai dubbi, e dimostrata nelle sue testimonianze. Esame critico dell'Avvocato Giuseppe Antonio Costantini autore delle Lettere Critiche*. Venice: Pietro Bassaglia, 1747.

Costantini, Giuseppe Antonio. *Lettere critiche, giocose, morali . . . del conte Agostino Santi Pupieni*. Venice: Pasinelli, 1743.

Costantini, Giuseppe Antonio. *Lettere critiche, giocose, morali . . . del conte Agostino Santi Pupieni*. Venice: Pietro Bassaglia, 1744.

Costantini, Giuseppe Antonio. *Nuova raccolta di lettere critiche giocose, morali, scientifiche, ed erudite*. Vol. 4. Venice: Pietro Bassaglia, 1747.

Crasta, Francesca Maria. *L'eloquenza dei fatti. Filosofia, erudizione e scienze della natura nel Settecento veneto*. Naples: Bibliopolis, 2007.

Cristani, Giovanni. *D'Holbach e le rivoluzioni del globo: scienze della terra e filosofie della natura nell'età dell'Encyclopédie*. Florence: Olschki, 2003.

Crowe, Michael J. *The Extraterrestrial Life Debate, Antiquity to 1915: A Source Book*. Notre Dame: University of Notre Dame, 2008.

Crucitti Ullrich, Francesca Bianca. *La Bibliothèque italique – cultura "italianisante" e giornalismo letterario*. Milan; Naples: Ricciardi, 1974.

Crucitti Ullrich, Francesca Bianca. "Le 'Osservazioni Letterarie' di Scipione Maffei, o la continuazione del 'Giornale di Venezia.'" In *Scipione Maffei nell'Europa del Settecento: atti del convegno, Verona, 23–25 settembre 1996*, edited by Gian Paolo Romagnani, 321–48. Verona: Cierre Edizioni, 1998.

Curi, Ettore. "La comunità scientifica veronese e gli esperimenti di elettrometria animale alla fine del settecento." *Atti dell'Accademia di Agricoltura, Scienze e Lettere di Verona*, VI, XLII (1990): 96–126.

Cutler, Alan H. "Nicolaus Steno and the Problem of Deep Time." In *The Revolution in Geology from the Renaissance to the Enlightenment*, edited by Gary D. Rosenberg, 143–48. Boulder, CO: Geological Society of America, 2009.

D'Ailly, Pierre. *Tractatvs Petri de Eliaco episcopi Cameracensis, super libros Metheororum de impressionibus aeris.* Vienna: per Hieronymu[m] Vietorem & Ioannem Singreniu[m], 1514.

Dal Prete, Ivano. "The Ruins of the Earth: Learned Meteorology and Artisan Expertise in Fifteenth-Century Italian Landscapes." *Nuncius* 33, no. 3 (2018): 415–41.

Dal Prete, Ivano. "Cultures and Politics of Preformationism in Eighteenth-Century Italy." In *The Secrets of Generation: Reproduction in the Long Eighteenth Century*, edited by Raymond Stephanson and Darren N. Wagner, 59–78. Toronto; Buffalo: University of Toronto Press, 2015.

Dal Prete, Ivano. "Echi fracastoriani nelle cosmologie e nelle teorie della terra del Settecento." In *Girolamo Fracastoro: fra medicina, filosofia e scienze della natura*, edited by Alessandro Pastore and Enrico Peruzzi, 279–98. Florence: Olschki, 2006.

Dal Prete, Ivano. "Francesco Bianchini e il pianeta Venere. Astronomia, cronologia e storia della Terra tra Roma e Parigi all'inizio del XVIII secolo." *Nuncius* 20, no. 1 (2005): 95–152.

Dal Prete, Ivano. "Francesco Bianchini e la cultura scientifica veronese." In *Unità del sapere, molteplicità dei saperi: Francesco Bianchini (1662-1729), tra natura, storia e religione*, edited by Luca Ciancio and Gian Paolo Romagnani, 207–41. Verona, Italy: QuiEdit, 2010.

Dal Prete, Ivano. "Giovan Battista Gazola." In *Storia della Società Letteraria di Verona tra otto e novecento*, edited by Gian Paolo Romagnani and Emilio Zangarini, 2:61–77. Verona: Società Letteraria di Verona, 2007.

Dal Prete, Ivano. "'Ingenuous Investigators'. Antonio Vallisneri's Regional Network and the Making of Natural Knowledge in Eighteenth-Century Italy." In *Empires of Knowledge: Scientific Networks in the Early Modern World*, edited by Paula Findlen, 181–204. New York: Routledge, 2018.

Dal Prete, Ivano. *Scienza e società nel Settecento veneto: il caso veronese, 1680–1796.* Milan: FrancoAngeli, 2008.

Dal Prete, Ivano. "The Gazola Family's Scientific Cabinet. Politics, Society and Scientific Collecting in the Twilight of the Republic of Venice." In *Cabinets of Experimental Philosophy in Eighteenth-Century Europe*, edited by Jim Bennett and Sofia Talas, 155–72. Leiden: Brill, 2013.

Dal Prete, Ivano. "Valerio Faenzi e l'origine Dei Monti Nel Cinquecento Veneto." In *Wissenschaft – Berge – Ideologien. Johann Jakob Scheuchzer (1672-1733) Und Die Frühneuzeitliche Naturforschung*, edited by Simona Boscani Leoni, 199–214. Basel: Schwabe, 2010.

Dal Prete, Ivano. "Vernacular Meteorology and the Antiquity of the World in Medieval and Renaissance Italy." In *Vernacular Aristotelianism in Italy from the Fourteenth to the Seventeenth Century*, edited by Luca Bianchi, Jill Kraye, and Simon A. Gilson, 139–59. London: The Warburg Institute, 2016.

Dales, Richard C. "Anonymi De Elementis: From a Twelfth-Century Collection of Scientific Works in British Museum MS Cotton Galba E. IV." *Isis* 56, no. 2 (1965): 174–89.

Dales, Richard C. *Medieval Discussions of the Eternity of the World*. Leiden: E. J. Brill, 1990.

Dales, Richard C., and Omar Argerami. *Medieval Latin Texts on the Eternity of the World*. Leiden: Brill, 1991.

Darby, Peter. *Bede and the End of Time*. Farnham, Surrey; Burlington, VT: Ashgate, 2012.

Darwin, Charles. *The Works of Charles Darwin, Vol. 16: The Origin of Species, 1876*. Edited by Paul H. Barrett and R.B. Freeman. London: Routledge, 2017.

Daston, Lorraine, and Katharine Park. *Wonders and the Order of Nature, 1150–1750*. New York: Cambridge, MA: Zone Books, 1998.

Davies, Surekha. *Renaissance Ethnography and the Invention of the Human: New Worlds, Maps and Monsters*. Cambridge: Cambridge University Press, 2016.

De Gaetano, Armand L. *Giambattista Gelli and the Florentine Academy: The Rebellion against Latin*. Florence: L. S. Olschki, 1976.

De Robertis, Domenico. "Un monumento della civiltà aretina. 'La composizione del mondo' di Restoro d'Arezzo." *Atti e Memorie Dell'Accademia Petrarca di Lettere Arti e Scienze di Arezzo* 42 (1978 1976): 109–28.

De Vivo, Filippo. "La farmacia come luogo di cultura: le spezierie di medicina in Italia." In *Interpretare e curare: medicina e salute nel Rinascimento*, edited by Maria Conforti, Andrea Carlino, and Antonio Clericuzio, 129–42. Rome: Carocci Editore, 2013.

Del Negro, Pietro. "La politica di Venezia e le Accademie di Agricoltura." In *La Politica della scienza: Toscana e stati italiani nel tardo Settecento : atti del convegno di Firenze, 27–29 gennaio 1994*, edited by Giulio Barsanti, Vieri Becagli, and Renato Pasta, 451–89. Florence: Olschki, 1996.

Descendre, Romain. "La biblioteca di Leonardo." In *Atlante della letteratura italiana*, 1:592–95. Turin: Einaudi, 2010.

Dinis, Alfredo. "Was Riccioli a Secret Copernican?" In *Giambattista Riccioli e il merito scientifico dei gesuiti nell'età barocca*, 1000–1029. Florence: Olschki, 2002.

Dionisotti, Carlo. "Leonardo uomo di lettere." *Italia Medievale e Umanistica* 1962, no. V (n.d.): 183–216.

Dioscorides. *Dioscoride fatto di greco italiano. Al cui fine sono apposte le sue tavole ordinate, con certe avertenze, e trattati necessarii, per la material medesima*, ed.and trans. Sebastiano Fausto da Longiano. Venice: Curzio Navò, 1542.

Dominici, Stefano, and Niles Eldredge. "Brocchi, Darwin, and Transmutation: | Phylogenetics and Paleontology at the Dawn of Evolutionary Biology." *Evolution: Education and Outreach* 3, no. 4 (December 1, 2010): 576–84.

Donato, Maria Monica. "Un 'savio depentore' fra 'scienza de le stelle' e 'sutilita' dell'antico. Restoro d'Arezzo, le arti e il sarcofago romano di Cortona." *Annali Della Scuola Normale Superiore Di Pisa. Classe Di Lettere e Filosofia*, IV, no. 1–2 (1996): 52–78.

Draper, John William. *History of the Conflict between Religion and Science*. 2nd ed. New York: D. Appleton, 1875.

Draper, John William. *History of the Conflict between Religion and Science*. New York: D. Appleton, 1898.

Dronke, Peter. "Bernard Silvestris, Natura, and Personification." *Journal of the Warburg and Courtauld Institutes* 43 (1980): 16–31.

Ducos, J. "L'oeuvre de Mahieu Le Vilain: Traduction et Commentaire Des Météorologiques." In *Les Traducteurs Au Travail. Leurs Manuscrits et Leurs Méthodes*, edited by Jacqueline Hamesse, 285–309. Turnhout: Brepols Publishers, 2001.

Ducos, Joelle. "Entre Terre, Air at Eau: La Formation Des Montagnes." In *La Montagne Dans Le Texte Médiéval: Entre Mythe et Réalité*, edited by Claude Alexandre Thomasset and Danièle James-Raoul, 19–52. Paris: Presses de l'Université de Paris-Sorbonne, 2000.

Ducos, Joëlle. *La Météorologie en Français au Moyen Age (XIIIe–XIVe Siècles)*. 2. Paris: Genève: Honoré Champion; Editions Slatkine, 1998.

Duhem, Pierre. *Etudes sur Léonard de Vinci: ceux qu'il a lus et ceux qui l'ont lu*. Paris: Hermann, 1906.

Duhem, Pierre. *Etudes sur Léonard de Vinci: ceux qu'il a lus et ceux qui l'ont lu*. Paris: Hermann, 1909.

Duhem, Pierre. *Le Système Du Monde; Histoire Des Doctrines Cosmologiques de Platon à Copernic*. Vol. IX. 10 vols. Paris: A. Hermann, 1958.

Duncan, Joseph E. "Paradise as the Whole Earth." *Journal of the History of Ideas* 30, no. 2 (1969): 171–86.

Dunham, Scott A. *The Trinity and Creation in Augustine: An Ecological Analysis*. Albany: State University of New York Press, 2008.

Edwards, Mark J. "Origen." In *The Stanford Encyclopedia of Philosophy*, edited by Edward N. Zalta, Spring 2014. Metaphysics Research Lab, Stanford University, 2014. https://plato.stanford.edu/archives/spr2014/entries/origen/.

Ellenberger, François. "Bourguet, Louis." In *Dictionary of Scientific Biography*, 15, Supplement 1:52–59. New York: Scribner, 1970.

Ellenberger, François. *History of Geology*. Translated by Marguerite Carozzi. 2 vols. Rotterdam: Balkema, 1996–99.

Elshakri, Marwa. *Reading Darwin in Arabic, 1860–1950*. Chicago: University of Chicago Press, 2013.

Erculiani, Camilla. *Lettere di philosophia naturale di Camilla Herculiana speciala alle tre stelle in Padova, Indrizzate alla Serenissima Regina di Polonia: nella quale si tratta la natural causa delli diluvij et il natural temperamento dell'huomo, et la natural formatione dell'arco celeste*. Krakow: Drukarnia Łazarzowa Kraków, 1584.

Eusebius Caesariensis. *Chronicon Bipartitum Nunc Primum Ex Armeniaco Textu in Latinum Conversum Opera P. Jo. Baptistae Aucher . . .* Edited and translated by Johannes Baptista Aucher. Venice: Typis Coenobii PP. Armenorum in insula S. Lazari, 1818.

Evans, John Hyde. *Morals Not Knowledge: Recasting the Contemporary U.S. Conflict between Religion and Science.* Oakland: University of California Press, 2018.

Faenzi, Valerio. *De montium origine, Valerii Fauenties, ordinis Praedicatorum, dialogus.* Venice: in Academia Veneta, 1561.

Faenzi, Valerio. *L'origine dei monti.* Edited by Ezio Filippi. Verona: Della Scala, 2001.

Faenzi, Valerio. *Sull'origine delle montagne.* Edited by Paolo Macini, Ezio Mesini, and Maria Campanale. Verbania: Tararà, 2006.

Fahy, Conor. *Printing a Book at Verona in 1622: The Account Book of Francesco Calzolari Junior.* Paris: Fondation Custodia, 1993.

Falloppio, Gabriele, *Gabrielis Falloppii . . . De medicatis aquis, atque de fossilibus tractatus pulcherrimus, ac maxime utilis: ab Andrea Marcolino . . . collectus. Accessit eiusdem andreae duplex epistola: . . . Cum indice rerum magis obseruandarum copiosissimo, ac capitum omnium . . .* Edited by Andrea Marcolini. Venice: ex officina Stellae, Iordanis Ziletti, 1564.

Fantasti, Girolamo Cesare. *Contrascritta di Girolamo Cesare Fantasti filosofo medico che annulla l'opinione del Molto Reverendo Don Gianiacopo Spada Rettore di Grezzana.* Verona: Dionigi Ramanzini, 1737.

Fantasti, Girolamo Cesare. "Il cervo impietrito, ritrovato in Grezzana." In *La scimia non è specie d'uomo*, 3-17. Verona: Dionigi Ramanzini, 1732.

Fantasti, Girolamo Cesare. *La scimia non è specie d'uomo.* Verona: Dionigi Ramanzini, 1732.

Farinella, Calogero. *L'accademia repubblicana: la Società dei Quaranta e Anton Mario Lorgna.* Milan: FrancoAngeli, 1993.

Farinella, Calogero. "Moro, Anton Lazzaro." In *Dizionario Biografico Degli Italiani*, vol. 77. Rome: Treccani, 2002.

Favaro, Antonio, ed. *Le Opere di Galileo Galilei: Edizione Nazionale sotto gli auspicii di Sua Maestà il Re d'Italia.* Vol. XII. 20 vols. Florence: Tip. di G. Barbèra, 1902.

Federici-Vescovini, Graziella. *Astrologia e scienza. La crisi dell'aristotelismo sul cadere del Trecento e Biagio Pelacani da Parma.* Florence: Nuovedizioni E. Vallecchi, 1979.

Federici-Vescovini, Graziella. "Biagio Pelacani: Filosofia, Astrologia e Scienza agli inizi dell'età moderna." In *Filosofia, scienza e astrologia nel trecento europeo: Biagio Pelacani parmense: Atti del ciclo di lezioni "Astrologia, Scienza, Filosofia e Società nel trecento europeo": Parma, 5-6 Ottobre 1990*, edited by Graziella Federici-Vescovini and Francesco Barocelli, 39-52. Padova: Il Poligrafo, 1992.

Federici-Vescovini, Graziella. "Note sur la circulation du commentaire d'Albert de Saxe au De Caelo d'Aristote en Italie." In *Itinéraires D'Albert de Saxe, Paris-Vienne Au XIVe Siècle*, edited by Joël Biard, 240-52. Paris: J. Vrin, 1991.

Fehrenbach, Frank. "Leonardo's Point." In *Vision and Its Instruments: Art, Science, and Technology in Early Modern Europe*, edited by Alina Alexandra Payne, 69–97. University Park, PA: Pennsylvania State University Press, 2015.

Fehrenbach, Frank. "Un nuovo paradigma: il Diluvio." In *Leonardo "1952" e La Cultura Dell'Europa Nel Dopoguerra*, edited by Romano Nanni and Maurizio Torrini, 303–20. Florence: L. S. Olschki, 2013.

Ferrone, Vincenzo. *Scienza natura religione: mondo newtoniano e cultura italiana nel primo settecento*. Naples: Jovene Editore, 1982.

Ferrone, Vincenzo. *The Intellectual Roots of the Italian Enlightenment: Newtonian Science, Religion, and Politics in the Early Eighteenth Century*. Translated by Sue Brortherton. Atlantic Highlands, NJ: Humanities Press, 1995.

Findlen, Paula. "Jokes of Nature and Jokes of Knowledge: The Playfulness of Scientific Discourse in Early Modern Europe." *Renaissance Quarterly* 43, no. 2 (July 1, 1990): 292–331.

Findlen, Paula. *Possessing Nature: Museums, Collecting, and Scientific Culture in Early Modern Italy*. Berkeley: University of California Press, 1994.

Findlen, Paula, J. G. Amato, Veronica S.-R. Shi, Alexandria R. Tsagaris, and Carlo Vecce. *Leonardo's Library: The World of a Renaissance Reader*. Stanford University Libraries, 2019.

Fitzgerald, Allan, and John C. Cavadini. *Augustine Through the Ages: An Encyclopedia*. Grand Rapids MI: W. B. Eerdmans, 1999.

Fontenelle, Bernard. *Histoire de l'Academie Royale des Sciences. Avec les mémoires de mathematique & de physique, pour la mème année, tirés de registres de cette Academie: Année 1703*. chez Gerard Kuyper, 1707.

Fontenelle, Bernard. "Sur le taches de Mars." In *Histoire de l'Academie Royale des Sciences, avec les Mémoires de Mathematiques et de Phisique, pour la mème Année [1720]*, 1722, 93–95.

Fontenelle, Bernard. "Sur l'equinoxe du printemps du MDCCXIV." *Histoire de l'Academie Royale des Sciences. Pour l'anne 1714*, n.d.

Fontenelle, Bernard. "Sur l'obliquite' de l'ecliptique." *Histoire de l'Academie Royale des Sciences. Pour l'anne 1716*, n.d.

Fortis, Alberto. "De' cataclismi sofferti dal nostro Pianeta, Saggio Poetico per servire di prodromo ad un Poema Geologico d'Ischiro Euganeo." *L'Europa Letteraria*, 1768.

Fortis, Alberto. "Descrizione Orittografica Del Ponte Naturale Di Veja." *Giornale d'Italia, Spettante Alla Scienza Naturale* VI (1770): 241–52.

Fracastoro, Girolamo. *Hieronymi Fracastorii . . . Adami Fumani . . . et Nicolai Archii . . . carminum editio II*. Edited by Adamo Fumano. Vol. 1. 3 vols. Padua: J. Cominus, 1739.

Fracastoro, Girolamo. *Hieronymi Fracastorii Homocentrica. Eivsdem De Cavsis Criticorvm Diervm per Ea Qvæ in Nobis Svnt*. Venice, 1538.

Fracastoro, Girolamo. *Hieronymi Fracastorii Syphilis sive morbus gallicus*. Basel: Bebel, 1536.

Fragnito, Gigliola. *Gasparo Contarini: un magistrato veneziano al servizio della cristianità.* Florence: Olschki, 1988.

Freccero, John. "Satan's Fall and the 'Quaestio de Aqua et Terra.'" *Italica* 38, no. 2 (1961): 99–115.

Frosini, Fabio. "La biblioteca di Leonardo." *Biblioteche dei filosofi. Biblioteche filosofiche private in età moderna e contemporanea,* 2016, 1–13.

Frosini, Fabio, and Alessandro Nova, eds. *Leonardo da Vinci on Nature: Knowledge and Representation.* Venice: Marsilio, 2015.

Gaeta, Franco, and Aldo Stella, eds. *Nunziature Di Venezia.* Vol. 8. Rome: Istituto Storico Italiano per l'età Moderna e Contemporanea, 1963.

Gaetano of Thiene. *Aristoteles Stagirite Metheororum Liber Primus: Cum Commentariis Caietani de Tienis. Nuper Summa Cum Diligentia Emendatis.* Venice: Jo. de Forolivio, 1491.

Galluzzi, Paolo. *Prima di Leonardo: cultura delle macchine a Siena nel Rinascimento.* Milan: Electa, 1991.

Galluzzi, Paolo. *Renaissance Engineers from Brunelleschi to Leonardo Da Vinci.* Florence: Istituto e Museo di Storia della Scienza : Giunti, 2001.

Galluzzi, Paolo, and Ugo Baldini, eds. "L'astronomia del Cardinale Bellarmino." In *Novità celesti e crisi del sapere,* 293–305. Florence: Giunti Barbèra, 1984.

Garofalo, Silvano. *L'enciclopedismo italiano: Gianfrancesco Pivati.* Ravenna: Longo, 1980.

Gaukroger, Stephen. *Descartes' System of Natural Philosophy.* Cambridge: Cambridge University Press, 2002.

Gazola, Giambattista. *Discorso in morte del signor conte Marco Marioni letto nell'Accademia Filarmonica in Verona il giorno 17 settembre 1795.* Verona: Stamperia Giuliari, 1796.

Geddes, Leslie. "'Infinite Slowness and Infinite Velocity': The Representation of Time and Motion in Leonardo's Studies of Geology and Water." In *Leonardo da Vinci on Nature: Knowledge and Representation,* edited by Fabio Frosini and Alessandro Nova, 269–83. Venice: Marsilio, 2015.

Generali, Dario. "Antonio Vallisneri 'corrispondente' leibniziano." *Studi e Memorie per la Storia dell'Università di Bologna,* Nuova Serie, 6 (1987): 125–40.

Generali, Dario. *Bibliografia delle opere di Antonio Vallisneri.* Florence: Olschki, 2004.

Generali, Dario. "Il « Giornale de' Letterati d'Italia » e la cultura veneta del primo settecento." *Rivista di Storia della Filosofia* 39, no. 2 (1984): 243–81.

Ghinassi, G. "Nuovi studi sul dialetto mantovano di Vivaldo Belcazer." *Studi Di Filologia Italiana* XXIII (1965): 19–172.

Gilson, Simon A. "Vernacularizing Meteorology: Benedetto Varchi's Comento sopra il Primo Libro delle Meteore d'Aristotile." In *Vernacular Aristotelianism in Italy from the Fourteenth to the Seventeenth Century,* edited by Luca Bianchi, Simon A. Gilson, and Jill Kraye, 161–81. London: The Warburg Institute, 2016.

Gleason, Elisabeth G. *Gasparo Contarini: Venice, Rome, and Reform.* Berkeley: University of California Press, 1993.

Gliozzi, Giuliano. *Adamo e il nuovo mondo*. Milan: Franco Angeli, 1977.

Gliozzi, Giuliano. "Calzolari, Francesco." In *Dizionario Biografico Degli Italiani*. Vol. 17. Rome: Istituto della Enciclopedia Italiana, 1974.

Godin, Louis. "Que l'obliquité de l'Ecliptique diminue, et de quelle manière; e que le noeuds des planètes sont immobiles." *Mémoires de l'Academie Royale de Sciences. De l'Annee 1734*, n.d.

Gohau, Gabriel. *A History of Geology*. Translated by Albert V. Carozzi and Marguerite Carozzi. New Brunswick: Rutgers University Press, 1991.

Gorman, Michael John. "The Angel and the Compass: Athanasius Kircher's Magnetic Geography." In *Athanasius Kircher: The Last Man Who Knew Everything*, edited by Paula Findlen, 239–59. New York: Routledge, 2004.

Gould, Stephen Jay. "Father Athanasius on the Isthmus of a Middle State. Understanding Kircher's Paleontology." In *Athanasius Kircher: The Last Man Who Knew Everything*, edited by Paula Findlen, 207–37. Routledge, 2004.

Gould, Stephen Jay. *Rocks of Ages: Science and Religion in the Fullness of Life*. New York: Ballantine Publishing Group, 1999.

Gould, Stephen Jay. "The Upwardly Mobile Fossils of Leonardo's Living Earth." In *Leonardo's Mountain of Clams and the Diet of Worms: Essays on Natural History*, 17–44. New York: Harmony Books, 1998.

Gould, Stephen Jay. *Time's Arrow, Time's Cycle: Myth and Metaphor in the Discovery of Geological Time*. Cambridge, MA: Harvard University Press, 1987.

Gozze, Niccolò Vito di. *Discorsi di M. Nicolo Vito di Gozze, gentil'huomo ragugeo, dell'Academia de gli Occulti, sopra le metheore d'Aristotele, ridotti in dialogo & diuisi in quattro giornate. Interlocutori esso M. Nicolo di Gozze, e M. Michiele Monaldi*. Venice: Francesco Ziletti, 1584.

Grafton, Anthony. *Cardano's Cosmos: The Worlds and Works of a Renaissance Astrologer*. Cambridge, MA: Harvard University Press, 1999.

Grafton, Anthony. "Dating History: The Renaissance & the Reformation of Chronology." *Daedalus* 132, no. 2 (2003): 74–85.

Grafton, Anthony. *Defenders of the Text: The Traditions of Scholarship in an Age of Science, 1450–1800*. Cambridge, MA: Harvard University Press, 1991.

Grafton, Anthony. "From De Die Natali to De Emendatione Temporum: The Origins and Setting of Scaliger's Chronology." *Journal of the Warburg and Courtauld Institutes* 48 (1985): 100–43.

Grafton, Anthony. *Joseph Scaliger: A Study in the History of Classical Scholarship*. Oxford: New York: Clarendon Press; Oxford University Press, 1983.

Grafton, Anthony. "Kircher's Chronology." In *Athanasius Kircher: The Last Man Who Knew Everything*, edited by Paula Findlen, 171–87. New York: Routledge, 2004.

Grafton, Anthony. *Leon Battista Alberti: Master Builder of the Italian Renaissance*. Cambridge, MA: Harvard University Press, 2002.

Grafton, Anthony. *What Was History? The Art of History in Early Modern Europe*. Cambridge: Cambridge University Press, 2007.

Grafton, Anthony. *What Was History? The Art of History in Early Modern Europe.* Cambridge: Cambridge University Press, 2007.

Grafton, Anthony, and Megan Hale Williams. *Christianity and the Transformation of the Book: Origen, Eusebius, and the Library of Caesarea.* Cambridge, MA: Belknap Press of Harvard University Press, 2006.

Grant, Edward. *A Source Book in Medieval Science.* Cambridge, MA: Harvard University Press, 1974.

Grant, Edward. *God and Reason in the Middle Ages.* Cambridge: Cambridge University Press, 2001.

Grant, Edward. "Jean Buridan and Nicole Oresme on Natural Knowledge." *Vivarium* 31, no. 1 (1993): 84–105.

Grendler, Paul F. *L'inquisizione romana e l'editoria a Venezia, 1540–1605.* Rome: Il Veltro, 1983.

Grendler, Paul F. *The Universities of the Italian Renaissance.* Baltimore: Johns Hopkins University Press, 2002.

Grendler, Paul F. *The Roman Inquisition and the Venetian Press, 1540–1605.* Princeton, NJ: Princeton University Press, 2015.

Grendler, Paul F. "What Piero Learned in School: Fifteenth-Century Vernacular Education." *Studies in the History of Art* 48 (1995): 160–74.

Grendler, Paul F. *Schooling in Renaissance Italy: Literacy and Learning, 1300–1600.* Baltimore: Johns Hopkins University Press, 1989.

Griffel, Frank. "Al-Ghazali." In *The Stanford Encyclopedia of Philosophy*, edited by Edward N. Zalta, Winter 2016. Metaphysics Research Lab, Stanford University, 2016. https://plato.stanford.edu/archives/win2016/entries/al-ghazali/.

Guthmüller, Bodo. "Fausto da Longiano e il problema del tradurre." *Quaderni Veneti*, 12, 1990: 9–152.

Hanke, Lewis. *All Mankind Is One: A Study of the Disputation between Bartolomé de Las Casas and Juan Ginés de Sepúlveda in 1550 on the Intellectual and Religious Capacity of the American Indians.* DeKalb: Northern Illinois University Press, 1974.

Hanke, Lewis. *Aristotle and the American Indians: A Study in Race Prejudice in the Modern World.* Bloomington: Indiana University Press, 1970.

Hardie, Philip. "The Speech of Pythagoras in Ovid Metamorphoses 15: Empedoclean Epos." *The Classical Quarterly* 45, no. 1 (1995): 204–14.

Haring, N. "The Creation and Creator of the World According to Thierry of Chartres and Clarenbaldus of Arras." *Archives d'histoire Doctrinale et Litteraire Du Moyen Age* 22 (1955): 137–216.

Harkness, Deborah E. *The Jewel House: Elizabethan London and the Scientific Revolution.* New Haven: Yale University Press, 2007.

Harrison, Peter. "'Science' and 'Religion': Constructing the Boundaries." In *Science and Religion: New Historical Perspectives*, edited by Thomas Dixon, G. N. Cantor, and Stephen Pumfrey, 23–49. New York: Cambridge University Press, 2010.

Hasse, Dag Nikolaus. "Latin Averroes Translations of the First Half of the Thirteenth Century." In *Universality of Reason, Plurality of Philosophies in the Middle Ages*, edited by A. Musco, 149–78. Palermo: Officina di Studi Medievali 2012.

Heilbron John. "Francesco Bianchini, Historian. In Memory of Amos Funkenstein." In *Thinking Impossibilities: The Intellectual Legacy of Amos Funkenstein*, edited by David Biale and Robert S. Westman, 227–79. Toronto; Buffalo: University of Toronto Press, 2008.

Heilbron John. "Bianchini and natural philosophy." In *Unità del sapere, molteplicità dei saperi: Francesco Bianchini (1662–1729), tra natura, storia e religione*, edited by Luca Ciancio and Gian Paolo Romagnani, 33–73. Verona: QuiEdit, 2010.

Heilbron John. *The Sun in the Church: Cathedrals as Solar Observatories*. Cambridge, MA: Harvard University Press, 1999.

Henryk, Anzulewicz. "The Systematic Theology of Albert the Great." In *A Companion to Albert the Great: Theology, Philosophy, and the Sciences*, edited by Irven Michael Resnick, 15–67. Leiden: Brill, 2013.

Herbert, Sandra. *Charles Darwin, Geologist*. Ithaca: Cornell University Press, 2005.

Hillier. "Ibn Rushd (Averroes) | Internet Encyclopedia of Philosophy." http://www.iep.utm.edu/ibnrushd/#H3.

Hine, Harry M. "Seneca and Anaxagoras in Pseudo-Bede's De Mundi Celestis Terrestrisque Constitutione." *Viator* 19 (January 1, 1988): 111–28.

Hirai, Hiro. "Interprétation chymique de la création et origine corpusculaire de la vie chez Athanasius Kircher." *Annals of Science* 64, no. 2 (April 1, 2007): 217–34.

Hirai, Hiro. *Le concept de semence dans les théories de la matière à la Renaissance de Marsile Ficin à Pierre Gassendi*. Turnhout: Brepols, 2005.

Hissette, Roland. *Enquête sur les 219 articles condamnés à Paris le 7 mars 1277*. Peeters Publishers & Booksellers, 1977.

Holbach, Paul Henri Dietrich d'. "Fossile." In *Encyclopédie, ou Dictionnaire Raisonné des Sciences, des Arts et des Métiers*, edited by Denis Diderot and Jean le Rond d'Alembert, 7:209–11. Paris: Briasson, 1757.

Holbach, Paul Henri Dietrich d'. "Déluge." In *Encyclopédie, ou Dictionnaire Raisonné des Sciences, des Arts et des Métiers*, edited by Denis Diderot and Jean le Rond d'Alembert, 4: 795–803. Paris: 1754.

Holbach, Paul Henri Dietrich d'. *Système de la nature ou Des loix du monde physique & du monde moral*. London: M.-M. Rey, 1770.

Hoquet, Thierry. *Buffon: histoire naturelle et philosophie*. Paris: Champion, 2005.

Ḥunayn ibn Isḥāq al-'Ibādī. "Liber de proprietatibus ælementorum." In *Hoc in uolumine continentur infrascripta opera Aristotelis uidelicet: in principio: Vita eiusdem etc.* Venice: Gregorio de Gregori, 1496.

Infelise, Mario. *L'editoria Veneziana Nel '700*. Milan: F. Angeli, 2000.

Ingham, Patricia Clare. *The Medieval New: Ambivalence in an Age of Innovation*. Philadelphia: University of Pennsylvania Press, 2015.

Iserloh, Erwin, Josef Glazik, and Hubert Jedin. *Reformation and Counter Reformation.* New York: Seabury Press, 1986.

Isidore of Seville. *The Etymologies of Isidore of Seville.* Translated by Stephen A. Barney. Cambridge: Cambridge University Press, 2006.

Israel, Jonathan. *A Revolution of the Mind: Radical Enlightenment and the Intellectual Origins of Modern Democracy.* Princeton, NJ: Princeton University Press, 2009.

Israel, Jonathan. *Democratic Enlightenment: Philosophy, Revolution, and Human Rights 1750–1790.* Oxford: Oxford University Press, 2013.

Israel, Jonathan. *Radical Enlightenment: Philosophy and the Making of Modernity 1650– 1750.* Oxford: Oxford University Press, 2002.

Jarava, Juan. *I quattro libri della filosofia naturale di Giovan Sarava. Dove... si discopron tutte le principali materie fisiche, le prime cagioni, e gli effetti loro, & i fini... Tradotti di spagnuolo in italiano da Alfonso Ulloa.* Translated by Alfonso Ulloa. Venice: Plinio Pietrasanta, 1557.

Jarava, Juan. *La philosophia natural brevemente [t]ratada y con mucha diligencia copilada de Aristotilés, Plinio, Platon y otros grades autores.* Antwerp: Martin Nucio, 1546.

Jedin, Hubert. *Papal Legate at the Council of Trent, Cardinal Seripando.* Translated by Frederic Clement Eckhoff. St. Louis: B. Herder, 1947.

Jenks, Stuart. "Astrometeorology in the Middle Ages." *Isis* 74, no. 2 (1983): 185–210.

Jensen, Kristian. "The Humanist Reform of Latin and Latin Teaching." In *The Cambridge Companion to Renaissance Humanism*, edited by Jill Kraye, 63–81. Cambridge: Cambridge University Press, 1996.

Johannes de Sacrobosco, and Zucchero Bencivenni. *Il trattato de la spera.* Edited by Gabriella Ronchi. Florence: Accademia della Crusca, 1999.

Jordan, J. M. "'Ancient Episteme' and the Nature of Fossils: A Correction of a Modern Scholarly Error." *History and Philosophy of the Life Sciences* 38, no. 1 (2016): 90–116.

Kardel, Troels, and Paul Maquet. *Nicolaus Steno: Biography and Original Papers of a 17th Century Scientist.* Berlin: Springer, 2013.

Kauntze, Mark. *Authority and Imitation: A Study of the Cosmographia of Bernard Silvestris.* Leiden; Boston: Brill, 2014.

Kauntze, Mark. "The Creation Grove in the 'Cosmographia' of Bernard Silvestris." *Medium Ævum* 78, no. 1 (2009): 16–34.

Kaye, Joel. *A History of Balance, 1250–1375: The Emergence of a New Model of Equilibrium and Its Impact on Thought.* Cambridge: Cambridge University Press, 2014.

Kemp, Martin. "La crisi del sapere tradizionale nell'ultimo Leonardo." In *Lezioni dell'occhio: Leonardo da Vinci discepolo dell'esperienza*, 133–54. Milan: Vita e Pensiero, 2004.

Kemp, Martin. *Leonardo Da Vinci: The Marvellous Works of Nature and Man.* Oxford UK: Oxford University Press, 2006.

Kempe, Michael. *Wissenschaft, Theologie, Aufklärung: Johann Jakob Scheuchzer (1672– 1733) und die Sintfluttheorie.* Epfendorf: Bibliotheca Academica, 2003.

Kircher, Athanasius. *Arca Noë.* Amsterdam: apud J. Janssonium & E. Weyerstraten, 1675.

Kircher, Athanasius. *Athanasii Kircheri ... Mundus Subterraneus: In XII Libros Digestus.*2 vols. Amstelodami: apud J. Janssonium & E. Weyerstraten, 1665.

Klima, Gyula. *John Buridan.* Oxford: Oxford University Press, 2008.

Kockel, Valentin, and Brigitte Sölch, eds. *Francesco Bianchini (1662–1729) und die europäische gelehrte Welt um 1700.* Berlin: Walter de Gruyter, 2005.

Kölher, Hans-Joachim. "The *Flugshriften* and their Importance in Religious Debate: A Quantitative Approach." In *"Astrologi Hallucinati": Stars and the End of the World in Luther's Time,* edited by Paola Zambelli, 153–75. Berlin: W. de Gruyter, 1986.

Kruger, Johann Gottlob. *Histoire des anciennes revolutions du globe terrestre avec une relation chronologique et historique des tremblemens de terre, arrivés sur notre globe, depuis le commencement de l'ère chrétienne jusqu'à présent.* Amsterdam: Damonneville, 1753.

[Anonymous] *La guerra di dieci anni: Raccolta polemico-fisica sull'eletrometria galvano-organica : Parte italiana, Parte francese.* Verona, 1802.

Kucher, Michael P. *The Water Supply System of Siena, Italy: The Medieval Roots of the Modern Networked City.* New York: Routledge, 2004

La Peyrère, Isaac. *Praeadamitae: Sive Exercitatio Super Versibus 12. 13. 14. Cap. V. Epistolae D. Pauli Ad Romanos : Quibus Inducuntur Primi Homines Ante Adamum Conditi.* [Amsterdam], 1655.

Labriola, Ada. "Ricerche su Margarito e Ristoro d'Arezzo." *Arte Cristiana,* no. 75 (1987): 145–60.

Lancisi, Giovanni Maria. "Del modo di filosofar nell'arte medica." *Galleria di Minerva* IV, no. 2 (1700): 33–37.

Larsen, Andrew E. *The School of Heretics: Academic Condemnation at the University of Oxford, 1277–1409.* Leiden; Boston: Brill, 2011.

Larson, Edward J. *Summer for the Gods: The Scopes Trial and America's Continuing Debate over Science and Religion.* Cambridge, MA: Harvard University Press, 1998.

Larson, Edward J. *Trial and Error: The American Controversy over Creation and Evolution.* 3rd ed. New York: Oxford University Press, 2003.

Latini, Brunetto. *Il tesoro di Brunetto Latini volgarizzato da Bono Giamboni, nuovamento pubblicato secondo l'edizione del MDXXXIII.* Translated by Bono Giamboni. Venice: Co' tipi del Gondoliere, 1839.

Latini, Brunetto. *La tradizione dei volgarizzamenti toscani del "Tresor" di Brunetto Latini: con un'edizione critica della redazione α (I.1–129).* Edited by Marco Giola. Verona: QuiEdit, 2010.

Latini, Brunetto. *The Book of the Treasure (Li Livres Dou Tresor).* Translated by Paul Barrette and Spurgeon. New York: Garland Pub, 1993.

Latini, Brunetto. *Tresor.* Edited by P. G. Beltrami. Turin: G. Einaudi, 2007.

Laurenza, Domenico. *De figura umana : fisiognomica, anatomia e arte in Leonardo.* Florence: Olschki, 2001.

Laurenza, Domenico. "Images and Theories. The Study of Fossils in Leonardo, Scilla and Hooke." *Nuncius,* no. 1 (2019): 442–463.

Laurenza, Domenico. "Leonardo's Theory of the Earth. Unexplored Issues in Geology from the Codex Leicester." In *Leonardo da Vinci on Nature: knowledge and representation*, edited by Fabio Frosini and Alessandro Nova, 257–67. Venice: Marsilio, 2015.

Lawrence, Jerome, and Robert Edwin Lee. *The Complete Text of Inherit the Wind*. New York: 1957.

Lefèvre d'Etaples, Jacques, and Johannes Cochlaeus. *Meteorologia Aristotelis*, 39r. Nuremberg: Peypuß, 1512.

Leftow, Brian. "Boethius on Eternity." *History of Philosophy Quarterly* 7, no. 2 (1990): 123–42.

Leguay, Jean-Pierre. *Les catastrophes au Moyen Age*. Editions Jean-paul Gisserot, 2005.

Lehman, Geoff. "Leonardo, van Eyck, and the Epistemology of Landscape." In *Leonardo in Dialogue: The Artist Amid His Contemporaries*, edited by Francesca Borgo, Alessandro Nova, and Rodolfo Maffeis, 97–118. Italy: Marsilio, 2019.

Lehner, Ulrich L. "The Many Faces of the Catholic Enlightenment." In *A Companion to the Catholic Enlightenment in Europe*, edited by Ulrich L. Lehner and Michael O'Neill Printy, 1–61. Leiden: Brill, 2010.

Leibniz, Gottfried Wilhelm. *Protogaea*. Edited by Claudine Cohen and Andre Wakefield. Chicago: University of Chicago Press, 2008.

Lemay, Helen Rodnite. "Science and Theology at Chartres: The Case of the Supracelestial Waters." *The British Journal for the History of Science* 10, no. 3 (1977): 226–36.

Leonardo da Vinci. *The Leicester Codex (Hammer Codex)*. Edited by Francesco Maria Caleca. Translated by David Edwards-May. Rome: TREC edizioni pregiati, 2006.

Lettinck, Paul. *Aristotle's Meteorology and Its Reception in the Arab World: With an Edition and Translation of Ibn Suwār's Treatise on Meteorological Phenomena and Ibn Bājja's Commentary on the Meteorology*. Leiden: Brill, 1999.

Levitin, Dmitri. "Halley and the Eternity of the World Revisited." *Notes and Records of the Royal Society of London* 67, no. 4 (2013): 315–29.

Librandi, Rita. "Il lettore di testi scientifici in volgare." In *Lo spazio letterario del medioevo: la ricezione del testo*, edited by Guglielmo Cavallo, Claudio Leonardi, and Enrico Menestò, 2, III:125–54. Rome: Salerno, 2003.

Librandi, Rita, ed. *La Metaura d'Aristotile: volgarizzamento fiorentino anonimo del XIV secolo: edizione critica*. 2 vols. Naples: Liguori, 1995.

Librandi, Rita. "Ristoro, Brunetto, Bencivenni e la Metaura: intrecci di glosse e rinvii tra le opere di uno scaffale scientifico." In *Lo scaffale della biblioteca scientifica in volgare, secoli XIII-XVI: atti del Convegno (Matera, 14–15 ottobre 2004)*, edited by Rosa Piro and Rita Librandi, 101–22. Florence: SISMEL, Edizioni del Galluzzo, 2006.

Lindon, John. "La 'denonzia' di Antonio Conti per ateismo." In *Antonio Conti: uno scienziato nella République des lettres*, edited by Guido Baldassarri, Silvia Contarini, and Francesca Fedi, 45–58. Padua: Il Poligrafo, 2009.

Lines, David. "Teaching Physics in Louvain and Bologna. Frans Titelmans and Ulisse Adrovandi." In *Scholarly Knowledge: Textbooks in Early Modern Europe*, edited by Emidio Campi, 183–203. Paris: Librairie Droz, 2008.

Long, Pamela O. *Artisan/Practitioners and the Rise of the New Sciences, 1400–1600.* Corvallis, OR: Oregon State University Press, 2011.

Long, Pamela O. "Hydraulic Engineering and the Study of Antiquity: Rome, 1557–70." *Renaissance Quarterly* 61, no. 4 (2008): 1098–1138.

Long, Pamela O. *Openness, Secrecy, Authorship: Technical Arts and the Culture of Knowledge from Antiquity to the Renaissance.* Baltimore: Johns Hopkins University Press, 2001.

Long, Pamela O. *Technology, Society, and Culture in Late Medieval and Renaissance Europe, 1300–1600.* Washington, DC: Society for the History of Technology and the American Historical Association, 2000.

Longiano, Sebastiano Fausto da. *Il Favsto da Longiano De lo istitvire vn figlio d'un principe da li X. infino à gl'anni de la discretione.* Venice: Francesco Bindoni & Maffeo Pasini, 1542.

Longiano, Sebastiano Fausto da. *Il Fausto da Longiano. De gl'augurij, e de le soperstitioni de gl'antichi.* Venice: Curzio Navò, 1542.

Longiano, Sebastiano Fausto da. *Meteorologia, cioè Discorso de le impressioni humide & secche generate tanto ne l'aria, quanto ne le cauerne de la terra, non per uia di tradottione, ma di scelta. Trasportata in lingua italiana dal Fausto da Longiano.* Venice: Curzio Navò, 1542.

Louville, Jacques Eugéne. "Eugenii De Louville, Equitis, Regia Scientiarum Academiae Socii, nec non Regalis Societatis Londinensis Sodalis, de mutabilitate Eclipticae Dissertatio, Aureliis d. 4 Nov. A. 1718 ad Collectores Actor. Erud. Data, sed nunc demum exhibita." *Acta Eruditorum Lipsiae,* 1719.

Lucretius Carus, Titus. *On the Nature of Things.* Translated by W. H. D. Rouse and Martin Ferguson Smith. Cambridge, MA: Harvard University Press, 2014.

Luzzini, Francesco. *Il miracolo inutile: Antonio Vallisneri e le scienze della terra in Europa tra XVII e XVIII secolo.* Florence: Olschki, 2013.

Lyell, Sir Charles. *Principles of Geology: Being an Attempt to Explain the Former Changes of the Earth's Surface, by Reference to Causes Now in Operation.* Vol. 1. 3 vols. London: J. Murray, 1832.

Maccagni, Carlo. *Riconsiderando il problema delle fonti di Leonardo: L'elenco di libri ai fogli 2 verso 3 recto del codice 8936 della Biblioteca nacional di Madrid.* Florence: G. Barbèra, 1971.

Machiavelli, Niccolò and Francesco Guicciardini. *The Sweetness of Power: Machiavelli's Discourses & Guicciardini's Considerations,* translated by James B. Atkinson and David Sices. Dekalb: Northern Illinois University Press, 2002.

Machiavelli, Niccolò. *Discorsi Sopra La Prima Deca Di Tito Livio,* ed. Mario Martelli. Florence: Sansoni, 1971.

MacKinnon, Dolly. "'Jangled the Belles, and with Fearefull Outcry, Raysed the Secure Inhabitants': Emotion, Memory and Storm Surges in the Early Modern East Anglian Landscape." In *Disaster, Death and the Emotions in the Shadow of the Apocalypse,*

1400–1700, edited by Jennifer Spinks and Charles Zika, 155–73. London: Palgrave Macmillan UK, 2016.

Maffei, Giovanni Camillo. *Scala naturale overo fantasia dolcissima, . . . intorno alle cose occulte e desiderate nella filosofia*. Venice: Gio. Varisco, 1564.

Maffei, Scipione. *Della formazione de' fulmini tratto del sig. marchese Scipione Maffei, raccolto da varie sue lettere . . .* Verona: G. Tumermani, 1747.

Maffei, Scipione. *Epitolario*. Edited by Celestino Garibotto. 2 vols. Milan: Giuffrè, 1955.

Maffei, Scipione. *Verona Illustrata*. 4 vols. Verona: Jacopo Vallarsi, 1732.

Mahieu le Vilain. *Le Metheores d'Aristote*. Edited by Rolf Edgren. Uppsala: Almqvist & Wiksells Boktryckeri Aktiebolag, 1945.

Maillet, Benoît de. *Telliamed: Or, Conversations between an Indian Philosopher and a French Missionary on the Diminution of the Sea*. Edited and translated by Albert V. Carozzi. Urbana: University of Illinois Press, 1968.

Maillet, Benoît de. *Telliamed: Ou entretiens d'un philosophe indien avec un missionnaire français sur la diminution de la mer*. Edited by Francine Markovits. Paris: Fayard, 1984.

Maiocchi, Rodolfo. *Codice diplomatico dell' università di Pavia*. Vol. 1. Pavia: Tipografia successori frat. Fusi, 1905.

Mann, Nicholas. "The Origins of Humanism." In *The Cambridge Companion to Renaissance Humanism*, edited by Jill Kraye, 1–19. Cambridge: Cambridge University Press, 1996.

Maraldi, Jacques. "Observations sur le Taches de Mars. Par M. Maraldi." *Histoire de l'Academie Royale des Sciences, avec les Mémoires de Mathematiques et de Phisique, pour la mème Année [1720]*, 1722, 144–53.

Marcolongo, Roberto, ed. *I manoscritti e i disegni di Leonardo da Vinci. Il Codice Arundel 263*. Vol. I. 4 vols. Rome: Danesi, 1926.

Marinoni, Augusto. *Leonardo da Vinci. Scritti letterari*. Milan: Rizzoli, 1974, 255.

Marinoni, Augusto. "La Biblioteca di Leonardo," *Raccolta Vinciana* 22 (1987): 291–342.

Marinoni, Augusto, ed. *I Manoscritti Dell'institut de France*. Vol. Manoscritto L. 12 vols. Florence: Giunti Barbèra, 1987.

Marinoni, Augusto, ed. *I Manoscritti Dell'Institut de France*. Vol. Manoscritto F. 12 vols. Florence: Giunti Barbèra, 1988.

Marinoni, Augusto, ed. *I Manoscritti Dell'Institut de France*. Vol. Manoscritto G. 12 vols. Florence: Giunti Barbèra, 1989.

Marinoni, Augusto, ed. *I Manoscritti Dell'Institut de France*. Vol. Manoscritto E. 12 vols. Florence: Giunti Barbèra, 1989.

Marinoni, Augusto, ed. *Il Codice Atlantico: Della Biblioteca Ambrosiana di Milano*. Vol. V. 12 vols. Florence: Giunti Barbèra, 1977.

Marinoni, Augusto, ed. *Il Codice Atlantico della Biblioteca Ambrosiana di Milano*. 3 vols. Florence: Giunti, 2000.

Mariotti, Ilaria. "La creazione di un mito: Fra Sisto e Ristoro architetti della chiesa di Santa Maria Novella a Firenze." *Annali Della Scuola Normale Superiore Di Pisa. Classe Di Lettere e Filosofia* 1, no. 1 (1996): 249–78.

Bibliography

Markus, R. A. *Bede and the Tradition of Ecclesiastical Historiography*. Jarrow Lecture 1975. Jarrow on Tyne [Eng.]: St. Paul's Rectory, 1975.

Martin, Craig. "Meteorology for Courtiers and Ladies: Vernacular Aristotelianism in Renaissance Italy." *Philosophical Readings* IV, no. 2 (2012): 3–14

Martin, Craig. *Renaissance Meteorology: Pomponazzi to Descartes*. Baltimore: Johns Hopkins University Press, 2011.

Martin, Craig. "With Aristotelians Like These, Who Needs Anti-Aristotelians? Chymical Corpuscular Matter Theory in Niccolò Cabeo's Meteorology." *Early Science and Medicine* 11, no. 2 (April 1, 2006): 135–61.

Martini, Giorgio. *Trattati di architettura ingegneria e arte militare*. Edited by Corrado Maltese. Milan: Il Polifilo, 1967.

May, Gerhard. *Creatio Ex Nihilo: The Doctrine of "Creation out of Nothing" in Early Christian Thought*. Translated by A. S. Worrall. Edinburgh: T&T Clark, 1994.

Mazzone, S., and Clara Silvia Roero. *Jacob Hermann and the Diffusion of the Leibnizian Calculus in Italy*. Vol. 1. Florence: Olschki, 1997.

McCartney, Paul J. "Charles Lyell and G. B. Brocchi: A Study in Comparative Historiography." *The British Journal for the History of Science* 9, no. 2 (1976): 175–89.

McCurdy, Edward. *The Notebooks of Leonardo Da Vinci*. Vol. I. 2 vols. New York: Reynal & Hitchcock, 1938.

McGinnis, Jon. "Arabic and Islamic Natural Philosophy and Natural Science." In *The Stanford Encyclopedia of Philosophy*, edited by Edward N. Zalta, Fall 2015. Metaphysics Research Lab, Stanford University, 2015. https://plato.stanford.edu/archives/fall2015/entries/arabic-islamic-natural/.

McGinnis, Jon. *Avicenna*. New York: Oxford University Press, 2009.

McMahon, Darrin M. *Enemies of the Enlightenment: The French Counter-Enlightenment and the Making of Modernity*. Oxford: Oxford University Press, 2001.

Mellyn, Elizabeth W. "Passing on Secrets: Interactions between Latin and Vernacular Medicine in Medieval Europe." *I Tatti Studies in the Italian Renaissance* 16, no. 1/2 (2013): 289–309.

Menochio, Giacomo. *Consiliorum siue Responsorum D. Iacobi Menochii… liber octauus…* Frankfurt: sumptibus Haeredum Andreae Wecheli & Ioan. Gymnici, 1604.

Michele, Agostino. *Trattato della grandezza dell'acqua et della terra. Di Agostino Michele. Nel quale contro l'opinione di molti filosofi, et di molti matematici illustri. Dimostrasi l'acqua esser di maggior quantità della terra*. Venice: Niccolò Moretti, 1583.

Mislin, Davis. "Roman Catholics." In *The Warfare between Science and Religion: The Idea That Wouldn't Die*, edited by Jeff Hardin, Ronald L. Numbers, and Ronald A. Binzley, 103–22. Baltimore: Johns Hopkins University Press, 2018.

Monfasani, John. "Aristotelians, Platonists, and the Missing Ockhamists: Philosophical Liberty in Pre-Reformation Italy." *Renaissance Quarterly* 46, no. 2 (1993): 247–76.

Monfrin, Jacques. "Jean de Brienne, comte d'Eu, et la traduction des Météorologiques d'Aristote par Mahieu le Vilain (vers 1290)." *Comptes rendus des séances de l'Académie des Inscriptions et Belles-Lettres* 140, no. 1 (1996): 27–36.

Monti, Giuseppe. *De Monumento Diluviano Nuper in Agro Bononiensi detecto Dissertatio: in qua permultae ipsius inundationis vindiciae, a statu terrae antediluvianae & postdiluvianae desumptae, exponuntur.* Bologna: Rossi, 1719.

Monti, Maria Teresa. "La scrittura e i gesti dell'Istoria." In *Antonio Vallisneri. Istoria della generazione*, edited by Maria Teresa Monti, 1:XV–CIV. Florence: Olschki, 2009.

Moody, Ernest A. "Buridan, Jean." In *Dictionary of Scientific Biography*, edited by Charles Coulston Gillispie, 2:603–8. New York: Scribner, 1970.

Moody, Ernest A. "John Buridan on the Habitability of the Earth." *Speculum* 16, no. 4 (1941): 415–25.

Moran, Jeffrey P. *The Scopes Trial: A Brief History with Documents.* New York: Palgrave, 2002.

Morello, Nicoletta. *La nascita della paleontologia nel seicento: Colonna, Stenone e Scilla.* Milan: FrancoAngeli, 1979.

Morello, Nicoletta. "Steno, the Fossils, the Rocks and the Calendar of the Earth." In *Geological Society of America. Special Paper 411*, 81–93, 2006.

Morello, Nicoletta. "Tra diluvio e vulcani. Le concezioni geologiche di Francesco Bianchini e del suo tempo." In *Unità del sapere, molteplicità dei saperi: Francesco Bianchini (1662–1729), tra natura, storia e religione*, edited by Luca Ciancio and Gian Paolo Romagnani, 185–206. Verona: QuiEdit, 2010.

Moro, Anton Lazzaro. *Anton Lazzaro Moro. Carteggio (1735–1764).* Edited by Massimo Baldini, Lino Conti, L. Cristante, and Rita Piutti. Florence: Olschki, 1993.

Moro, Anton Lazzaro. *Lettera di Anton Lazzaro Moro all'illustrissimo signor co. Agostino Santi Pupieni autore delle Lettere critiche, giocose, morali, scientifiche ed erudite* [1747].

Moro, Antonio Lazzaro. *De' crostacei: e degli altri marini corpi che si truovano su' monti; libri due.* Venice: Stefano Monti, 1740.

Morris, Madaleine F. "The Tuscan Editions of the 'Encyclopédie.'" In *Notable Encyclopedias of the Late Eighteenth Century: Eleven Successors of the Encyclopédie*, edited by Frank A. Kafker, 51–84. Oxford: Voltaire Foundation, 1994.

Mortenson, Terry. *The Great Turning Point.* Green Forest, AR: Master Books, 2004.

Moscardo, Lodovico. *Note overo memorie del Museo di Lodovico Moscardo, nobile veronese.* Padua: Paolo Frambotto, 1656.

Moscardo, Lodovico. *Note overo memorie del Museo di Lodovico Moscardo, nobile veronese.* Verona: Andrea Rossi, 1672.

Mosshammer, Alden A. *The Chronicle of Eusebius and Greek Chronographic Tradition.* Lewisburg, PA: Bucknell University Press, 1979.

Mothu, Alain, ed. *Révolution Scientifique et Libertinage.* Thornout: Brepols, 2000.

Mottana, Annibale. "Oggetti e concetti inerenti le Scienze Mineralogiche ne La composizione del mondo con le sue cascioni di Restoro d'Arezzo (anno 1282)." *Rendiconti Lincei* 10, no. 3 (1999): 133.

Myers, K. Sara. *Ovid's Causes: Cosmogony and Aetiology in the Metamorphoses.* Ann Arbor: University of Michigan Press, 1994.

Nanni, Romano. "Catastrofi e armonie." In *Leonardo da Vinci on Nature: knowledge and representation*, edited by Fabio Frosini, Alessandro Nova, and Romano Nanni, 95–117. Venice: Marsilio, 2015.

Nardi, Bruno. *La caduta di Lucifero: e l'autenticità della "Quaestio de aqua et terra."* Turin: Società Editrice Internazionale, 1959.

Nauta, Lodi. "Philology as Philosophy: Giovanni Pontano on Language, Meaning, and Grammar," *Journal of the History of Ideas* 72, no. 4 (2011): 483.

Netton, Ian Richard. *Muslim Neoplatonists: An Introduction to the Thought of the Brethren of Purity, Ikhwān al-Ṣafāʾ*. London: G. Allen & Unwin, 1982.

Niccoli, Ottavia. "Il diluvio del 1524 fra panico collettivo e irrisione carnevalesca." In *Scienze, credenze occulte, livelli di cultura*, edited by Giancarlo Garfagnini, 291–368. Florence: Olschki, 1982.

Niccoli, Ottavia. *Prophecy and People in Renaissance Italy*. Translated by Lydia Cochrane. Princeton, NJ: Princeton University Press, 1990.

Nogarola, Lodovico. *Ludouici Nogarolae ... Dialogus. Qui inscribitur Timotheus, siue de Nilo*. Venice : apud Vincentium Valgrysium, 1552.

Nongbri, Brent. *Before Religion: A History of a Modern Concept*. New Haven: Yale University Press, 2013.

Nores, Giasone De. *Breue trattato del mondo, et delle sue parti, semplici, et miste: con molte altre considerationi, che di grado in grado saranno piu notabili, & piu degne di cognitione: di Jason De Nores*. Venice: Andrea Muschio, 1571.

North, J. D. "Chronology and the Age of the World." In *Cosmology, History, and Theology*, edited by Wolfgang Yourgrau and Allen D. Breck, 307–33. New York: Springer US, 1977.

Nothaft, C. Philipp E. "Climate, Astrology and the Age of the World in Thirteenth-Century Thought: Giles of Lessines and Roger Bacon on the Precession of the Solar Apogee." *Journal of the Warburg and Courtauld Institutes* LXXVII (2014): 35–60.

Nothaft, C. Philipp E. *Dating the Passion: The Life of Jesus and the Emergence of Scientific Chronology (200–1600)*. Leiden: Brill, 2011.

Nothaft, C. Philipp E. "Origen, Climate Change, and the Erosion of Mountains in Giles of Lessines's Discussion of the Eternity of the World (c. 1260)." *The Mediaeval Journal* 4, no. 1 (January 1, 2014): 43–69.

Nothaft, C. Philipp E. "The Early History of Man and the Uses of Diodorus in Renaissance Scholarship: From Annius of Viterbo to Johannes Boemus." In *For the Sake of Learning. Essays in Honor of Anthony Grafton*, edited by Ann Blair and Goeing Anja-Silvia, 2:711–28. Leiden: Brill, 2016.

Nothaft, C. Philipp E. "Walter Odington's De Etate Mundi and the Pursuit of a Scientific Chronology in Medieval England." *Journal of the History of Ideas* 77, no. 2 (2016): 183–201.

Novelli, Italo. *La curiosità e l'ingegno: collezionismo scientifico e metodo sperimentale a Padova nel settecento*. Padua: Università degli Studi di Padova, Centro Musei scientifici, 2000.

Numbers, Ronald L. *Darwinism Comes to America.* Cambridge, MA: Harvard University Press, 1998.

Numbers, Ronald L. *The Creationists: From Scientific Creationism to Intelligent Design.* Cambridge, MA: Harvard University Press, 2006.

Numbers, Ronald L. "The New Atheists." In *The Warfare between Science and Religion: The Idea That Wouldn't Die,* edited by Jeff Hardin, Ronald L. Numbers, and Ronald A. Binzley, 220–38. Baltimore: Johns Hopkins University Press, 2018.

Nummedal, Tara. "Kircher's Subterranean World and the Dignity of the Geocosm." In *The Great Art of Knowing: The Baroque Encyclopedia of Athanasius Kircher,* edited by Daniel Stolzenberg, 37–47. Stanford, CA: Stanford University Press, 2001.

Nuovo, Angela. *The Book Trade in the Italian Renaissance.* Library of the Written Word; 26. Leiden: Brill, 2013.

Obrist, Barbara, and Irene Caiazzo, eds. *Guillaume de Conches: philosophie et science au XIIe siècle.* Florence: SISMEL edizioni del Galluzzo, 2011.

Odoardi, Jacopo. "De' corpi marini, che nel Feltrese Distretto si trovano." In *Nuova raccolta d'opuscoli scientifici e filologici,* 8:106–96. Venice: Simone Occhi, 1761.

Olden-Jørgensen, Sebastian. "Nicholas Steno and René Descartes: A Cartesian Perspective on Steno's Scientific Development." In *The Revolution in Geology from the Renaissance to the Enlightenment,* edited by Gary D. Rosenberg, 149–57. Boulder, CO: Geological Society of America, 2009.

Oldroyd, D. R., and J. B. Howes. "The First Published Version of Leibniz's Protogaea." *Journal of the Society for the Bibliography of Natural History* 9, no. 1 (November 1, 1978): 56–60.

Oliviero, Adriana. "La composizione dei cieli in Restoro d'Arezzo e in Dante." In *Dante e la scienza,* edited by Patrick Boyd and Vittorio Russo, 351–62. Ravenna: Longo, 1995.

Olmi, Giuseppe. *L'inventario del mondo: catalogazione della natura e luoghi del sapere nella prima età moderna.* Bologna: Il Mulino, 1992.

Oresme, Nicole. *Le Livre du Ciel et du Monde.* Edited by Albert Douglas Menut and Denomy Alexander. Translated by Albert Douglas Menut. Madison: University of Wisconsin Press, 1968.

Ottaviani, Alessandro. "Fra diluvio noaico e fuochi sotterranei, Note sulla fortuna sei-settecentesca di Fabio Colonna." *Giornale critico della filosofia italiana,* no. 2 (2017): 272–303.

Ottaviani, Alessandro. *Theatrum Naturae. La ricerca naturalistica tra erudizione e nuova scienza nell'Italia del primo Seicento.* Naples: La Città del Sole, 2007.

Otte, James K. "The Life and Writings of Alfredus Anglicus." *Viator* 3 (January 1, 1972): 275–92.

Otte, James K. "The Role of Alfred of Sareshel (Alfredus Anglicus) and His Commentary on the Metheora in the Reacquisition of Aristotle." *Viator* 7 (January 1, 1976): 197–210.

Pagan, Pietro. "Sulla Accademia 'Veneziana' o Della 'Fama.'" *Atti Dell'Istituto Veneto Di Scienze, Lettere Ed Arti*, Cl. di scienze morali, lettere ed arti, CXXXII (74 1973): 359–92.

Pagden, Anthony. *The Fall of Natural Man: The American Indian and the Origins of Comparative Ethnology*. Cambridge: Cambridge University Press, 1982.

Pancaldi, Giuliano. *Darwin in Italy: Science across Cultural Frontiers*. Bloomington: Indiana University Press, 1991.

Panziera, Genziana. "Le «Osservazioni Letterarie» (1737–1740) di Scipione Maffei." M.A. Thesis, Padua University, 1992.

Papi, Fulvio. *Antropologia e civiltà nel pensiero di Giordano Bruno*. 2. ed. Naples: Liguori, 2006.

Parcell, William C. "Sign and Symbols in Kircher's Mundus Subterraneus." In *The Revolution in Geology from the Renaissance to the Enlightenment*, edited by Gary D. Rosenberg. Boulder, CO: Geological Society of America, 2009.

Paschini, Pio. *Venezia e l'inquisizione Romana Da Giulio III a Pio IV*. Padua: Antenore, 1959.

Pasini, Mirella. "L'astronomie antédiluvienne. Storia della scienza e origini della civiltà in J.S. Bailly." *Studi Settecenteschi*, no. 11 (1988): 197–236.

Pasini, Mirella. *Thomas Burnet: Una storia del mondo tra ragione, mito e rivelazione*. Florence: La Nuova Italia, 1981.

Pastine, Dino. "Le origini del poligenismo e Isaac Lapeyrère." In *Miscellanea Seicento*, II:9–234. Florence: Le Monnier, 1971.

Pastore, Alessandro. "Il consulto di Girolamo Fracastoro sul tifo petecchiale (Trento, 1547)." In *Girolamo Fracastoro*, edited by Alessandro Pastore and Enrico Peruzzi, 91–101. Florence: Olschki, 2006.

Patrizi, Francesco. *Della retorica dieci dialoghi di M. Francesco Patritio: nelli quali si fauella dell'arte oratoria con ragioni repugnanti all'openione, che intorno a quella hebbero gli antichi scrittori*. Venice: Francesco Senese, 1562.

Paul of Venice. *Expositio librorum naturalium Aristotelis*. Cologne: Johannis de Colonia sociique eius Johannis Manthen de Gherretzem, 1476.

Paul of Venice. *Expositio Magistri Pauli Veneti Super Libros de Generatione et Corruptione Aristotelis. Eiusdem De Compositione Mundi Cum Figuris*. Venice: Bonetum Locatellum, 1498.

Paul of Venice. *Liber de compositione mundi excellētissimi viri Pauli Veneti theologi insignis*. Lyon: Vincent, 1525.

Paul of Venice. *Philosophia Naturalis Compendium Clarissimi Philosophi Pauli Veneti: Una Cum Libro de Cōpositione Mundi Qui Astronomia Janua Inscribitur*. Paris: apud Petrum Gaudoulum, 1514.

Paul of Venice. *Summa Philosophie Naturalis Magistri Pauli Veneti Nouiter Recognita & a Vitijs Purgata Ac Pristine Integritati Restituta*. Venice: Bonetum de Locatellis, 1503.

Pedretti, Carlo, ed. *The Codex Hammer of Leonardo Da Vinci*. Translated by Carlo Pedretti. Florence, Italy: Giunti Barbèra, 1987.

Pedretti, Carlo. *The Literary Works of Leonardo Da Vinci*. Vol. II. 2 vols. Berkeley: University of California Press, 1977.

Pelacani da Parma, Biagio. *Le "Quaestiones de anima". Di Biagio Pelacani da Parma*. Edited by Graziella Federici-Vescovini. Florence: L. S. Olschki, 1974.

Pellegrini, Paolo. "Nogarola, Ludovico." In *Dizionario Biografico Degli Italiani*, Vol. 78. Rome: Treccani, 2013.

Pereira, Benito. *B. Pererii . . . Prior Tomus Commentariorum . . . Et Disputationum in Genesim; Continens Historiam Mosis Ab Exordio Mundi Usque Ad Noëticum Diluvium, Septem Libris Explanatam. Adjecti Sunt Quatuor Indices, Etc*. Ingolstadt: David Sartorius, 1590.

Pereira, Benito. *De Communibus Omnium Rerum Principiis Libri Quindecim*, Venice: Andrea Muschio, 1586.

Perrone Compagni, Vittoria. "Cose di filosofia si possono dire in volgare. Il programma culturale di Giambattista Gelli." In *Il volgare come lingua di cultura dal trecento al cinquecento: Atti del Convegno Internazionale, Mantova, 18–20 Ottobre 2001*, edited by Arturo Calzona, 1000–37. Florence: L. S. Olschki, 2003.

Perrone Compagni, Vittoria. "Un'ipotesi non impossibile. Pomponazzi sulla generazione spontanea dell'uomo (1518)," *Bruniana & Campanelliana* 13, no. 1 (2007): 1000–13.

Peruzzi, Enrico. *La Nave di Ermete: La Cosmologia di Girolamo Fracastoro*. Florence: L. S. Olschki, 1995.

Peter of Abano. *Conciliator Controversiarum, Quae Inter Philosophos et Medicos Versantur*. Venice: Junta, 1565.

Peter of Abano. *Conciliator differentiarum philosophorum [et] medicorum*. Venice: Luceantonii de Giunta, 1520.

Petrescu, Lucian. "The Threefold Object of the Scientific Knowledge. Pseudo-Scotus and the Literature on the Meteorologica in Fourteenth-Century Paris." *Franciscan Studies* 72, no. 1 (October 22, 2016): 465–502.

Philastrius of Brescia. *De haeresibus liber*. Edited by Johannes Albertus Fabricius. Hamburg: Theodor Christopher Felginer, 1721.

Piccinni, Gabriella. In *Duccio: alle origini della pittura senese*, edited by Alessandro Bagnoli, 26–35. Milan: Silvana, 2003.

Piccoli, Gregorio. *La scienza dei cieli e dei corpi celesti e loro meravigliosa Posizione, Moto e Grandezza*. Verona: Vallarsi, 1741.

Piccoli, Gregorio. *Ragguaglio di una grotta, ove sono molte ossa di Belve diluviane nei Monti Veronesi*. Verona: Fratelli Merlo, 1739.

Piccolomini, Alessandro. *Della grandezza della terra et dell'acqua*. Venice: Appresso Giordano Ziletti, all'Insegna della Stella, 1557.

Piccolomini, Alessandro. *De la sfera del mondo. Libri quattro in lingua toscana... De le stelle fisse. Libro uno con le sue figure, e con le sue tavole*. Venice: al Segno del Pozzo, 1540.

Piccolomini, Alessandro. *La prima parte della filosofia naturale*. Rome: Valgrisi, 1551.

Pignatti, Franco. "Fausto (Fausto (Fausto da Longiano), Sebastiano," in *Dizionario Biografico Degli Italiani*, vol. XLV: 394–8. Rome: Treccani, 1995.

Pii, Eluggero. "Il pensiero politico di Scipione Maffei: dalla Repubblica di Roma alla Repubblica di Venezia." In *Scipione Maffei nell'Europa del Settecento: atti del convegno, Verona, 23–25 settembre 1996*, edited by Gian Paolo Romagnani, 93–117. Verona: Cierre Edizioni, 1998.

Pine, Martin. *Pietro Pomponazzi: Radical Philosopher of the Renaissance*. Padua: Antenore, 1986.

Pirillo, Diego. *Filosofia ed eresia nell'Inghilterra del tardo Cinquecento: Bruno, Sidney e i dissidenti religiosi italiani*. Rome: Edizioni di storia e letteratura, 2010.

Piscini, Angela. "Gelli, Giovan Battista," *Dizionario Biografico Degli Italiani*, vol. 53. Rome: Istituto della Enciclopedia Italiana, 2000.

Piva, Franco. "Cultura francese e censura a Venezia nel secondo settecento (ricerche storico-bibliografiche)." *Memorie dell'Istituto Veneto di Scienze, Lettere ed Arti* XXXVI, no. 3 (1973).

Pivati, Gianfrancesco. *Nuovo dizionario scientifico e curioso sacro-profano*. 10 vols. Venice: Milocco, 1746.

Pizzorusso, Ann. "Leonardo's Geology: The Authenticity of the 'Virgin of the Rocks.'" *Leonardo* 29, no. 3 (1996): 197–200.

Plastina, Sandra. "'Considering the Mutation of the Times and States and Humans': The Letters of Natural Philosophy of Camilla Erculiani." *Bruniana & Campanelliana* 20, no. 1 (2014): 145–56.

Plastina, Sandra. *Gli alunni di Crono: mito, linguaggio e storia in Francesco Patrizi da Cherso (1529–1597)*. Soveria Mannelli: Rubbettino, 1992.

Plato. *Timaeus*. Translated by Benjamin Jowett. Blacksburg, VA: Virginia Tech, 2001.

Pomian, Krzysztof. *Collectors and Curiosities: Paris and Venice, 1500–1800*. Translated by Elizabeth Wiles-Portier. Cambridge: Polity Press, 1990.

Pomponazzi, Pietro. *Trattato sull'immortalità dell'anima*. Edited by Vittoria Perrone Compagni. Florence: L. S. Olschki, 1999.

Pomponazzi, Pietro. *Tractatus de immortalitate animae*. Edited by William Henry Hay, (Haverford: Haverford College, 1938.

Pontano, Giovanni Gioviano *Pontani Opera. Vrania, siue de stellis libri quinque. Meteororum liber unus. De hortis hesperidum libri duo…* Venice: Aldo Manuzio, 1505.

Poole, William. "Francis Lodwick's Creation: Theology and Natural Philosophy in the Early Royal Society." *Journal of the History of Ideas* 66, no. 2 (2005): 245–63.

Poole, William. *The World Makers: Scientists of the Restoration and the Search for the Origins of the Earth*. Oxford: Peter Lang, 2010.

Popkin, Richard Henry. *Isaac La Peyrère (1596–1676): His Life, Work, and Influence*. Leiden: Brill, 1987.

Preti, Cesare. "Grandi, Jacopo." In *Dizionario Biografico Degli Italiani*, vol. 58. Rome: Treccani, 2002.

Prosperi, Adriano. *Il Concilio Di Trento: Una Introduzione Storica*. Turin: Einaudi, 2001.

Prosperi, Adriano, ed. *Livorno, 1606–1806: luogo di incontro tra popoli e culture.* Turin: U. Allemandi, 2009.

Pugliano, Valentina. "Natural History in the Apothecary's Shop." In *Worlds of Natural History*, edited by Helen Anne Curry, Nicholas Jardine, James Andrew Secord, and Emma C. Spary, 44–60. Cambridge UK: Cambridge University Press, 2018.

Puppi, Lionello, Ruggero Maschio, and Giacinto Cecchetto, eds. *Giovanni Rizzetti: scienziato e architetto (Castelfranco Veneto 1675–1751).* Castelfranco Veneto: Banca popolare di Castelfranco Veneto, 1996.

Quirini, Giovanni, and Jacopo Grandi. *J. Quirini de Festaceio fossilibus Musæi Septalliani, et J. Grandii de veritate diluvii Universalis, et testaceorum quæ procul a mari reperiuntur generatione, epistolæ.* Venice: Valsalva, 1676.

Ramazzini, Bernardino. *De fontium Mutinensium scaturigine.* Modena: Eredi Suliani, 1691.

Ramusio, Giovanni Battista. *Primo volume delle nauigationi et viaggi nel qual si contiene la descrittione dell'Africa, et del paese del Prete Ianni, con varii viaggi, dal mar Rosso a Calicut & infin all'isole Molucche, dove nascono le Spetiere et la navigatione attorno il mondo: li nomi de gli auttori, et le nauigationi, et i viaggi piu particolarmente si mostrano nel foglio seguente.* Venice: appresso gli heredi di Lucantonio Giunti, 1550.

Rao, Cesare. *Dell'origine de' monti.* Naples: Orazio Salviani, 1577.

Rao, Cesare. *I Meteori di Cesare Rao di Alessano città di terra d'Otranto. I quali contengono quanto intorno a tal materia si puo desiderare.* Venice: Giouanni Varisco & Compagni, 1582.

Rappaport, Rhoda. "Fontenelle Interprets the Earth's History." *Revue d'histoire Des Sciences* 44, no. 3/4 (1991): 281–300.

Rappaport, Rhoda. "Leibniz on Geology: A Newly Discovered Text." *Studia Leibnitiana* 29, no. 1 (1997): 6–11.

Rappaport, Rhoda. *When Geologists Were Historians, 1665–1750.* Ithaca: Cornell University Press, 1997.

Ray, Meredith K. *Daughters of Alchemy: Women and Scientific Culture in Early Modern Italy.* Cambridge, MA: Harvard University Press, 2015.

Resnick, Irven Michael, ed. *A Companion to Albert the Great: Theology, Philosophy, and the Sciences.* Leiden; Boston: Brill, 2013.

Restoro of Arezzo. *La composizione del mondo.* Edited by Alberto Morino. Lavis: La Finestra Editrice, 2007.

Restoro of Arezzo. *La composizione del mondo.* Edited by Alberto Morino. Parma: Fondazione Pietro Bembo, 1997.

Restoro of Arezzo. *La composizione del mondo: colle sue cascioni.* Edited by Alberto Morino. Florence: Presso l'Accademia della Crusca, 1976.

Rhodes, Frank H. T. "Darwin's Search for a Theory of the Earth: Symmetry, Simplicity and Speculation." *The British Journal for the History of Science* 24, no. 2 (June 1991): 193–229.

Riahi, Pari. *Ars et Ingenium: The Embodiment of Imagination in the Architectural Drawings of Francesco di Giorgio Martini.* Ph.D. Dissertation, McGill University, 2010.

Riccati, Jacopo, and Antonio Vallisneri. *Jacopo Riccati – Antonio Vallisneri. Carteggio (1719–1729).* Edited by Maria Luisa Soppelsa. Florence: Olschki, 1985.

Richardson, Brian. *Print Culture in Renaissance Italy: The Editor and the Vernacular Text, 1470–1600.* Cambridge: Cambridge University Press, 2003.

Richet, Pascal. *A Natural History of Time.* Chicago: University of Chicago Press, 2007.

Righi, Alessandro. *Il Conte Di Lilla e l'emigrazione Francese a Verona (1794–1796).* Perugia: Vincenzo Bartelli, 1909.

Righi, Alessandro. *Una loggia massonica a Verona nel 1792.* Verona: Franchini, 1912.

Sangalli, Maurizio. "Cesare Cremonini, la Compagnia di Gesù e la Repubblica di Venezia: eterodossia e protezione politica." In *Cesare Cremonini: aspetti del pensiero e scritti: atti del convegno di studio (Padova, 26–27 febbraio 1999),* edited by Ezio Riondato and Antonino Poppi, 207–18. Padua: Accademia galileiana di scienze, lettere ed arti in Padova, 2000.

Riva, Franco. *La dimestica stamperia del veronese conte Giuliari.* Florence: Sansoni, 1956.

Riva, Franco. "Le avventurose vicende dell'«Ittiolitologia Veronese» del mantovano Serafino Volta." *Civiltà Mantovana* 1, no. 5 (1966): 71–77.

Roda, Marica. "Pujati, Giuseppe Antonio." In *Dizionario Biografico Degli Italiani,* Vol. 85. Rome: Treccani, 2016.

Roger, Jacques. *Buffon: A Life in Natural History.* Edited by Pearce Williams L. Translated by Bonnefoi Sarah Lucille. Ithaca: Cornell University Press, 1997.

Rosenberg, Gary D., ed. *The Revolution in Geology from the Renaissance to the Enlightenment.* Boulder, CO: Geological Society of America, 2009.

Rossi, Paolo. *The Dark Abyss of Time: The History of the Earth & the History of Nations from Hooke to Vico.* Chicago: University of Chicago Press, 1984.

Rota, Orazio. *Dissertazione epistolare sopra i sistemi e le teorie de' due globi celeste e terracqueo che si stabiliscono da Mosé nella storia delle sei giornate della creazione del mondo al cap. primo della sacra Genesi.* Vicenza: Turra, 1789.

Rotta, Salvatore. *Il viaggio di Francesco Bianchini in Inghilterra. Contributo alla storia del newtonianesimo in Italia.* Brescia: Paideia, 1966.

Rudwick, M. J. S. *Bursting the Limits of Time: The Reconstruction of Geohistory in the Age of Revolution.* Chicago: University of Chicago Press, 2005.

Rudwick, M. J. S. *Earth's Deep History: How It Was Discovered and Why It Matters.* Chicago: The University of Chicago Press, 2014.

Rudwick, M. J. S. *The Great Devonian Controversy: The Shaping of Scientific Knowledge among Gentlemanly Specialists.* Chicago: University of Chicago Press, 1985.

Rudwick, M. J. S. *The Meaning of Fossils: Episodes in the History of Paleontology.* 2d rev. ed. New York: Science History Publications, 1976.

Rudwick, M. J. S. *Worlds before Adam: The Reconstruction of Geohistory in the Age of Reform.* Chicago: University of Chicago Press, 2008.

Russiliano, Tiberio. *Una reincarnazione di Pico ai tempi di Pomponazzi*. Edited by Paola Zambelli. Milan: Il Polifilo, 1994.

Rutkin, Darrel H. "Astrology and Magic." In *A Companion to Albert the Great*. Edited by Irven Resnick, 451–505. Leiden: Brill, 2013.

Rutkin, Darrel H. *Sapientia Astrologica: Astrology, Magic and Natural Knowledge, ca. 1250–1800: I. Medieval Structures (1250–1500): Conceptual, Institutional, Socio-Political, Theologico-Religious and Cultural*. Vol. 1. 3 vols. Cham: Springer International Publishing, 2019.

Sacks, Kenneth. *Diodorus Siculus and the First Century*. Princeton, NJ: Princeton University Press, 1990.

Sadrin, Paul. *Nicolas-Antoine Boulanger (1722–1759), ou, Avant nous le déluge*. Oxford: Voltaire Foundation, 1986.

Sainson, Katia. "Revolutions in Time: Chateaubriand on the Antiquity of the Earth." *French Forum* 30, no. 1 (2005): 47–63.

Sallis, John. *Chorology: On Beginning in Plato's Timaeus*. Bloomington: Indiana University Press, 1999.

Saraina, Torello. *Torelli Saraynae Veronensis . . . De origine et amplitudine ciuitatis Veronae. Eiusdem De viris illustribus antiquis Veronensibus. De his, qui potiti fuerunt dominio ciuitatis Veronae. De monumentis antiquis vrbis, & agri Veronensis De interpretatione litterarum antiquarum. Index praeterea huius operis in calce additus est*. Verona: ex officina Antonii Putelleti, 1540.

Sarpi, Paolo. *The History of the Council of Trent: Containing Eight Books in Which Besides the Ordinary Acts of the Council Are Declared Many Notable Occurrences Which Happened in Christendom, During the Space of Forty Years and More. And Particularly the Practices of the Court of Rome, to Hinder the Reformation of Their Errors, and to Maintain Their Greatness*. London: J. Macock, 1676.

Sasso, Gennaro. *Machiavelli e gli antichi, e altri saggi*, vol. 1. Milan: Riccardo Ricciardi, 1987.

Scafi, Alessandro. *Mapping Paradise: A History of Heaven on Earth*. London: British Library, 2006.

Schaffer, Simon. "Halley's Atheism and the End of the World." *Notes and Records of the Royal Society of London*. 32, no. 1 (July 1, 1977): 17–40.

Schechner, Sara. *Comets, Popular Culture, and the Birth of Modern Cosmology*. Princeton, NJ: Princeton University Press, 1999.

Schedel, Hartmann. *Liber chronicarum*. Nuremberg: Anton Koberger, 1493.

Scheuchzer, Johann Jackob. *Homo diluvii testis et Theoskopos*. Zurich: Joh. & Byrgkli, 1726.

Schmitt, Charles B. *Aristotle and the Renaissance*. Cambridge, MA: Harvard University Press, 1983.

Schoonheim, Pieter L., ed. *Aristotle's Meteorology in the Arabico-Latin Tradition: Critical Edition of the Texts, with Introduction and Indices*. Translated by Pieter L. Schoonheim. Leiden: Brill, 2000.

Schuessler, Rudolf. *The Debate on Probable Opinions in the Scholastic Tradition.* Leiden: Brill, 2019.

Scilla, Agostino. *La vana speculazione disingannata dal senso.* Edited by Paolo Rossi. Florence: Giunti, 1996.

Scilla, Agostino. *La vana speculazione disingannata dal senso: Lettera responsiua circa i corpi marini che petrificati si trouano in varij luoghi terrestri.* Naples: Andrea Colicchia, 1670.

Scott, Andrew C. "Federico Cesi and His Field Studies on the Origin of Fossils between 1610 and 1630." *Endeavour* 25, no. 3 (September 1, 2001): 93–103.

Scott, Andrew C., and David Freedberg. *Fossil Woods and Other Geological Specimens. The Paper Museum of Cassiano Dal Pozzo.* Turnhout; London: Harvey Miller, 2000.

Secord, James A. "The Discovery of a Vocation: Darwin's Early Geology." *The British Journal for the History of Science* 24, no. 2 (June 1991): 133–57.

Secord, James A. *Victorian Sensation: The Extraordinary Publication, Reception, and Secret Authorship of Vestiges of the Natural History of Creation.* Chicago: University of Chicago Press, 2000.

Séguin, Maria Susana. *Science et réligion dans la pensée française du XVIIIe siècle: le mythe du Déluge universel.* Paris: Champion, 2001.

Seneca, Lucius Annaeus. *Natural Questions.* Translated by Harry M. Hine. Chicago; London: University of Chicago Press, 2010.

Seripando, Girolamo. *Prediche del reuer.mo mons. Girolamo Seripando, arciuescouo di Salerno . . . sopra il simbolo de gli Apostoli, dichiarato co simboli del concilio Niceno, & di Santo Athanasio . . . .* Venice: Al segno della Salamandra, 1567.

Sgarbi, Marco and Maurizio Bertolotti, eds. *Pietro Pomponazzi: Tradizione e Dissenso. Atti del Congresso Internazionale di Studi su Pietro Pomponazzi, Mantova, 23–24 Ottobre 2008.* Florence: L. S. Olschki, 2010.

Sgarbi, Marco. "The Instatement of the Vernacular as Language of Culture. A New Aristotelian Paradigm in Sixteenth-Century Italy." *Intersezioni*, no. 3/2016 (2016): 317–42.

Sgarbi, Marco. "Aristotle and the People. Vernacular Philosophy in Renaissance Italy," *Renaissance & Reformation* 39, no. 3 (2016): 59–109.

Shapin, Steven. *A Social History of Truth: Civility and Science in Seventeenth-Century England.* Science and Its Conceptual Foundations. Chicago: University of Chicago Press, 1994.

Shapin, Steven, and Simon Schaffer. *Leviathan and the Air-Pump: Hobbes, Boyle, and the Experimental Life: Including a Translation of Thomas Hobbes, Dialogus Physicus de Natura Aeris by Simon Schaffer.* Princeton, NJ: Princeton University Press, 1985.

Shapiro, Adam R. *Trying Biology: The Scopes Trial, Textbooks, and the Antievolution Movement in American Schools.* Chicago: The University of Chicago Press, 2013.

Shell, Hanna Rose. "Ceramic Nature." In *Materials and Expertise in Early Modern Europe: Between Market and Laboratory*, edited by Ursula Klein and E. C. Spary, 50–70. Chicago: University of Chicago Press, 2010.

Siculus, Diodorus. *Diodorus Siculus: Library of History, Volume I, Books 1–2.34*. Translated by C. H. Oldfather. Cambridge, MA: Harvard University Press, 1933.

Siebert, Harald. "Kircher and His Critics. Censorial Practice and Pragmatic Disregard in the Society of Jesus." In *Athanasius Kircher: The Last Man Who Knew Everything*, edited by Paula Findlen, 79–103. New York: Routledge, 2004.

Silvestris, Bernard. *Cosmographia*. Edited by Peter Dronke. Leiden: Brill, 1978.

Silvestris, Bernard. *Cosmographie*. Edited and translated by Michel Lemoine. Paris: Cerf, 1998.

Silvestris, Bernard. *Poetic Works*. Edited and translated by Winthrop Wetherbee. Cambridge, MA: Harvard University Press, 2015.

Smith, Pamela H. *The Body of the Artisan: Art and Experience in the Scientific Revolution*. Chicago: University of Chicago Press, 2004.

Smith, Webster. "Observations on the Mona Lisa Landscape." *The Art Bulletin* 67, no. 2 (June 1, 1985): 183–99.

Smoller, Laura Ackerman. *History, Prophecy, and the Stars: The Christian Astrology of Pierre d'Ailly, 1350–1420*. Princeton, NJ: Princeton University Press, 1994.

Solinas, Francesco. "Il trattato del legno fossile di Francesco Stelluti e i quattro volumi della Natural History of Fossils nelle raccolte della Biblioteca Reale di Windsor." In *Il Museo cartaceo di Cassiano Dal Pozzo: Cassiano naturalista*, edited by Francis Haskell and David Freedberg, 84–94. Milan: Olivetti [Distribuzione A. Mondadori Arte], 1989.

Somfaj, Anna. "Calcidius' 'Commentary' on Plato's 'Timaeus' and Its Place in the Commentary Tradition: The Concept of 'Analogia' in Text and Diagrams." *Bulletin of the Institute of Classical Studies. Supplement*, no. 83 (2004): 203–20.

Sorabji, Richard, ed. *The Philosophy of the Commentators, 200–600 AD: Physics*. Vol. 2. 3 vols. Ithaca: Cornell University Press, 2005.

Spada, Giovanni Giacomo. *Catalogi Lapidum Veronensium Mantissa*. Verona: Dionigi Ramanzini, 1740.

Spada, Giovanni Giacomo. *Catalogus lapidum Veronensium... qui apud Joannem Jacobum Spadam Gretianam archibresbyterum asservantur*. Verona: Dionigi Ramanzini, 1739.

Spada, Giovanni Giacomo. *Dissertazione ove si prova, che li petrificati corpi marini, che nei monti adiacenti a Verona si trovano, non sono scherzi di natura, ne diluviani; ma antediluviani . . .* Verona: Dionigi Ramanzini, 1737.

Spada, Giovanni Giacomo. *Giovanni Giacomo Spada, Corporum lapidefactorum agri Veronensis catalogus quae apud Joan. Jacobum Spadam . . ., asservantur. Editio altera . . .* Verona: Dionigi Ramanzini, 1744.

Spada, Giovanni Giacomo. *Giunta alla dissertazione de' corpi marini petrificati ove si prova che sono antediluviani*. Verona: Dionigi Ramanzini, 1737.

Spada, Giovanni Giacomo. *Nuova invenzione di Ossa di Cervo petrificate*. Verona, 1737.

Spini, Giorgio. *Ricerca dei libertini: la teoria dell'impostura delle religioni nel Seicento italiano*. Florence: La nuova Italia, 1983.

Stabile. "Borri, Girolamo." In *Dizionario Biografico Degli Italiani*. Vol. 13. Rome: Treccani, 1971.

Steensen, Niels. *Elementorum myologiae specimen: seu, Musculi descriptio geometrica. Cui accedunt canis carchariae dissectum caput, et dissectus piscis ex canum genere . . .* Florence: Ex Typographia sub signo Stellae, 1667.

Steensen, Niels. *Nicolai Stenonis De Solido Intra Solidum Naturaliter Contento Dissertationis Prodromus*. Florence: Ex Typographia sub signo Stellae, 1669.

Stelluti, Francesco. *Trattato Del Legno Fossile Minerale Nuovamente Scoperto etc.* Rome: Mascardi, 1637.

Steno, Nicolaus. *Geological Papers*. Edited by Gustav Scherz. Odense: Odense University Press, 1969.

Stock, Brian. *Myth and Science in the Twelfth Century; a Study of Bernard Silvestris*. Princeton, NJ: Princeton University Press, 2015.

Strong, Donald. "The Triumph of Mona Lisa: Science and the Allegory of Time." In *Leonardo e l'età Della Ragione*, edited by Enrico Bellone and Paolo Rossi, 255–77. Milan: Scientia, 1982.

Sylla, Edith Dudley. "Ideo Quasi Mendicare Oportet Intellectum Humanum: The Role of Theology in John Buridan's Natural Philosophy." In *The Metaphysics and Natural Philosophy of John Buridan*, edited by Johannes. M. M. H. Thijssen and Jack Zupko, 221–45. Leiden: Brill, 2001.

Tanga, Mario. *Giacinto Cestoni, i rapporti con Redi e le scienze della vita nel XVII secolo*. Pisa: University of Pisa, 2008.

Tanner, Norman P., ed. *Decrees of the Ecumenical Councils*. 2 vols. London; Washington, DC: Sheed & Ward; Georgetown University Press, 1990.

Targhetta, Renata. "Ancora sulla massoneria veneta settecentesca, con qualche indugio a proposito di Verona." In *Tra conservazione e novità. Il mondo veneto innanzi alla rivoluzione del 1789*, 19–26. Verona: Accademia di Agricoltura, Scienze e Lettere, 1991.

Targhetta, Renata. *La massoneria veneta dalle origini alla chiusura delle logge (1729–1785)*. Udine: Del Bianco, 1988.

Tarrant, Neil. "Censoring Science in Sixteenth-Century Italy: Recent (and Not-So-Recent) Research." *History of Science* 52, no. 1 (March 1, 2014): 1–27.

Tarrant, Neil. "Disciplining the School of Athens: Censorship, Politics and Philosophy, Italy 1450–1600." Sussex: University of Sussex, 2009.

Tazzara, Corey. *The Free Port of Livorno and the Transformation of the Mediterranean World*. Oxford: Oxford University Press, 2017.

Testa, Simone. *Italian Academies and Their Networks, 1525–1700: From Local to Global*. New York: Palgrave Macmillan, 2015.

Asmussen, Tina, Pamela Long. "The Cultural and Material World of Mining in Early Modern Europe." *Renaissance Studies* 34, no. 1 (2020): 8–30

Thierry of Chartres. "Treatise of the Work of the Six Days." Translated by Katharine Park. https://www.academia.edu/31388090/Thierry_of_Chartres-Treatise_Six_D ays-trans._Park.pdf.

Thijssen, J. M. M. H. "The Buridan School Reassessed. John Buridan and Albert of Saxony." *Vivarium* 42, no. 1 (2004): 18–42.

Thijssen, Johannes. M. M. H. *Censure and Heresy at the University of Paris, 1200–1400.* Philadelphia: University of Pennsylvania Press, 1998.

Thijssen, Johannes M. M. H. "The Debate over the Nature of Motion: John Buridan, Nicole Oresme and Albert of Saxony. With an Edition of John Buridan's 'Quaestiones Super Libros Physicorum, Secundum Ultimam Lecturam', Book III, Q. 7." *Early Science and Medicine* 14, no. 1/3 (2009): 186–210.

Thijssen, Johannes. M. M. H., and Jack Zupko, eds. *The Metaphysics and Natural Philosophy of John Buridan.* Leiden; Boston: Brill, 2001.

Thomasset, Claude, and Joëlle Ducos. *Le temps qu'il fait au Moyen âge: phénomènes atmosphériques dans la littérature, la pensée scientifique et religieuse.* Paris: Presses Paris Sorbonne, 1998.

Thorndike, Lynn. *A History of Magic and Experimental Science: The First Thirteen Centuries of Our Era.* Vol. 1. 8 vols. New York: Columbia University Press, 1958.

Thorndike, Lynn. "The De Constitutione Mundi of John Michael Albert of Carrara." *Romanic Review* 17 (January 1, 1926): 193–216.

Tipaldo, Emilio de. *Biografia degli Italiani illustri nelle scienze: lettere ed arti del secolo XVIII. e de'contemporanei compilata da letterati Italiani di ogni provincia.* Vol. VIII. Venice: Alvisopoli, 1841.

Titelmans, Franciscus. *Compendium Naturalis Philosophiae. Libri duodecim de consideratione rerum naturalium earumque ad suum Creatorem reductione, Per fratrm Françisc. Titelmannum, etc.* Paris: Francisum Stephanum, 1547.

Toal, Ciaran. "Preaching at the British Association for the Advancement of Science: Sermons, Secularization and the Rhetoric of Conflict in the 1870s." *The British Journal for the History of Science* 45, no. 1 (2012): 75–95.

Tonti, Giacinto. *Augustiniana de rerum creatione sententia.* Padua: Giuseppe Corona, 1714.

Torrini, Maurizio. "Giuseppe Ferroni Gesuita e Galileiano." *Physis* 15 (1973): 411–23.

Toulmin, Stephen, and June Goodfield. *The Discovery of Time.* New York: Harper & Row, 1965.

Trentafonte, Franco. *Giurisdizionalismo, illuminismo e massoneria nel tramonto della Repubblica Veneta.* Venice: Deputazione Editrice, 1984.

Trivellato, Francesca. *The Familiarity of Strangers: The Sephardic Diaspora, Livorno, and Cross-Cultural Trade in the Early Modern Period.* New Haven: Yale University Press, 2009.

Turetta, Laura. "Bibliografia delle Opere a Stampa di e su Pietro d'Abano." *Medicina nei Secoli* 20, no. 2 (2008): 659–734.

Turnbull, Herbert W., ed. *The Correspondence of Isaac Newton*. Vol. 3: 1688–1693. Cambridge: Cambridge University Press, 1961.

Turner, Frank M. "The Late Victorian Conflict of Science and Religion as an Event in Nineteenth-Century Intellectual and Cultural History." In *Science and Religion: New Historical Perspectives*, edited by Thomas Dixon, Geoffrey N. Cantor, and Stephen Pumfrey, 87–109. New York: Cambridge University Press, 2010.

Uglietti, Francesco. *Un erudito veronese alle soglie del settecento. Mons. Francesco Bianchini 1662–1729*. Verona: Biblioteca Capitolare di Verona, 1986.

Ulvioni, Paolo. *Atene sulle lagune: Bernardo Trevisan e la cultura veneziana tra Sei e Settecento*. Venice: Ateneo veneto, 2000.

Ulvioni, Paolo. "La filosofia morale di Scipione Maffei." In *Scipione Maffei nell'Europa del Settecento: atti del convegno, Verona, 23–25 settembre 1996*, edited by Gian Paolo Romagnani, 147–63. Verona: Cierre Edizioni, 1998.

Ulvioni, Paolo. *Riformar il mondo: il pensiero civile di Scipione Maffei. Con una nuova edizione del "Consiglio Politico."* Alessandria: Edizioni dell'Orso, 2008.

Vaccari, Ezio. *Giovanni Arduino (1714–1795): il contributo di uno scienziato veneto al dibattito settecentesco sulle scienze della terra*. Florence: Olschki, 1993.

Vaccaro, Giulio. "Questo libretto che t'ho volgarizzato e chiosato. La Traduzione Nel Medioevo." In *I traduttori come mediatori interculturali*, edited by Sergio Portelli and Bart Van Den Bossche, 11–19. Florence: Cesati, 2016.

Vai, Gian Battista. "I viaggi di Leonardo lungo le valli romagnole: riflessi di geologia nei quadri, disegni e codici." In *Leonardo, Machiavelli, Cesare Borgia – Arte, Storia e Scienza in Romagna, 1500–1503*, 37–47. Rome: De Luca, 2003.

Valerio, Adriana. "La Verità luogo teologico in Bellarmino." In *Bellarmino e la Controriforma: Atti del Simposio Internazionale di Studi, Sora, 15–18 Ottobre, 1986*, edited by Romeo De Maio, Agostino Borromeo, Luigi Giulia, Georg Lutz, and Aldo Mazzacane, 51–87. Sora, Italy: Centro di Studi Sorani "Vincenzo Patriarca," 1990.

Vallisneri, Antonio. *De' corpi marini, che su' monti si trovano; della loro origine; e dello stato del mondo avanti'l diluvio, nel diluvio, e dopo il diluvio: lettere critiche di Antonio Vallisneri*. Venice: Domenico Lovisa, 1721.

Vallisneri, Antonio. *Epistolario 1714–1729*. Edited by Dario Generali. Florence: Olschki, 2006.

Vallisneri, Antonio. *Istoria della generazione*. Edited by Maria Teresa Monti. Vol. 1. 2 vols. Florence: Olschki, 2009.

Vallisneri, Antonio. "Lezione Accademica intorno all'ordine della progressione, e della connessione, che hanno insieme tutte le cose create, etc." In *Istoria della generazione*, edited by Maria Teresa Monti, 2: 511–26. Florence: Olschki, 2009.

Vallisneri, Antonio. "Specimen geographiae physicae." *Galleria di Minerva* VI (1708): 17.

Varchi, Benedetto. *La via della dottrina: le lezioni accademiche di Benedetto Varchi*. Edited by Annalisa Andreoni. Pisa: ETS, 2012.

Varenius, Bernhard. *Geographia generalis, in qua affectiones generales telluris explicantur. Autore Bernh.* Varenio: Apud Ludovicum Elzevirium, 1650.

Vasiliki, Grigoropoulous. "Steno's Critique of Descartes and Louis de La Forge's Response." In *Steno and the Philosophers*, edited by Raphaële Andrault and Mogens Lærke, 113–37. Boston: Brill, 2018.

Vasoli, Cesare. *Francesco Patrizi Da Cherso*. Rome: Bulzoni, 1989.

Vecce, Carlo. *La biblioteca perduta: i libri di Leonardo*. Rome: Salerno editrice, 2017.

Ventrice, Pasquale. *La discussione sulle maree tra astronomia, meccanica e filosofia nella cultura veneto-padovana del Cinquecento*. Venice: Istituto veneto di scienze, lettere ed arti, 1989.

Venturi, Franco. *Settecento Riformatore*. Vol. V:2. VII vols. Turin: G. Einaudi, 1990.

Verardi, Donato. "Rao, Cesare." In *Encyclopedia of Renaissance Philosophy*, 1–3. Cham: Springer, 2015.

Vermij, Rienk. "A Science of Signs. Aristotelian Meteorology in Reformation Germany." *Early Science and Medicine* 15, no. 6 (October 1, 2010): 648–74.

Vermij, Rienk. "Stevin's Physical Geography: The World as a Chemical Furnace." In *Rethinking Stevin, Stevin Rethinking. Constructions of a Dutch Polymath*, edited by Rienk Vermij, C.A. Davids, Fokko Jan Dijksterhuis, and Ida H. Stamhuis, 106–22. Leiden: Brill, 2020.

Vermij, Rienk. "Subterranean Fire. Changing Theories of the Earth During the Renaissance." *Early Science and Medicine* 3, no. 4 (1998): 323–47.

Vismara Chiappa, Paola. *Oltre l'usura: la Chiesa moderna e il prestito a interesse*. Catanzaro: Rubbettino, 2004.

Vogel, Klaus Anselm. *Sphaera Terrae – Das Mittelalterliche Bild Der Erde Und Die Kosmographische Revolution*. Ph.D. Dissertation, University of Gottingen, 1995.

Volta, Giovanni Serafino. *Ittiolitologia veronese del Museo Bozziano ora annesso a quello del Conte Giovambattista Gazola e di altri gabinetti di fossili veronesi: Con la versione latina*. Vol. 1. 2 vols. Verona: Stamperia Giuliari, 1796.

Volta, Serafino. *Degl'impietrimenti del territorio veronese ed in particolare dei pesci fossili del celebre Monte Bolca*. Mantua: 1789.

Voltaire. *A Pocket Philosophical Dictionary*. Translated by John Fletcher. Oxford: Oxford University Press, 2011.

Voltaire. *Candide or, Optimism: The Robert M. Adams Translation, Backgrounds, Criticism*. Edited by Nicholas Cronk. Translated by Robert M. Adams. New York: W. W. Norton, 2016.

Voltaire. "Dissertation envoyée par l'auteur en italien, à l'Académie de Bologne et traduite par lui-même en français sur le changements arrivés dans notre globe et sur les pétrifications qu'on prétend en être encore les témoignages." In *Oeuvres complètes de Voltaire*, XXXI:375–89. Basel: Jean-Jacques Tourneisen, 1786.

Voltaire. *Elementi della filosofia del Neuton esposti dal signor di Voltaire tradotti dal francese*. Venice: Giam-Maria Lazzaroni, 1741.

Voltaire. *Eléments De La Philosophie De Newton*. Edited by Robert Walters and W.H. Barber. Oxford: The Voltaire Foundation, 1992.

Voltaire. *Principj fisici tratti dagli Elementi di fisica Nevtoniana dell'insigne m.r de Voltaire, e combinati a dovere dal C.A.G.D.C.* Lucca: Filippo Maria Benedini, 1754.

Waddell, Mark A. *Jesuit Science and the End of Nature's Secrets.* London: Routledge, 2016.

Waddell, Mark A. "The World, As It Might Be: Iconography and Probabilism in the Mundus Subterraneus of Athanasius Kircher." *Centaurus* 48, no. 1 (January 1, 2006): 3–22.

Wallace, William. "Natural Philosophy: Traditional Natural Philosophy." In *The Cambridge History of Renaissance Philosophy*, edited by C. B. Schmitt, Quentin Skinner, Eckhard Kessler, and Jill Kraye, 199–235. Cambridge: Cambridge University Press, 1988.

Walsh, James J. *The Popes and Science; the History of the Papal Relations to Science During the Middle Ages and down to Our Own Time.* New York: Fordham University Press, 1908.

Wheeler, Stephen M. "Imago Mundi: Another View of the Creation in Ovid's Metamorphoses." *The American Journal of Philology* 116, no. 1 (1995): 95–121.

White, Andrew Dickson. *A History of the Warfare of Science with Theology in Christendom.* New York: D. Appleton, 1896.

White, Andrew Dickson. *A History of the Warfare of Science with Theology in Christendom.* New York: D. Appleton, 1901.

William of Conches. *A Dialogue on Natural Philosophy (Dragmaticon Philosophiae).* Edited by Italo Ronca and Matthew Curr. Notre Dame: University of Notre Dame Press, 1997.

William of Conches. *Philosophia mundi: Ausgabe des. 1. Buchs von Wilhelm von Conches' "Philosophia" mit Anhang, Übersetzung und Anmerkungen.* Edited by Gregor Maurach. Pretoria: University of South Africa, 1974.

William of Conches. *Philosophicarum et astronomicarum institutionum, Guilielmi Hirsaugiensis olim abbatis, libri tres.* Basel: Heinrich Petrus, 1531.

William of Moerbeke. *Meteorologica. Translatio Guillelmi de Morbeka.* Edited by Gudrun Vuillemin-Diem. 2 vols. Turnhout: Brepols, 2008.

Williams, Arnold. *The Common Expositor: An Account of the Commentaries on Genesis, 1527–1633.* Chapel Hill: University of North Carolina Press, 1948.

Wilson, Malcolm. *Structure and Method in Aristotle's Meteorologica: A More Disorderly Nature.* Cambridge: Cambridge University Press, 2013.

Wissink, Jozef, ed. *The Eternity of the World in the Thought of Thomas Aquinas and His Contemporaries.* Leiden; New York: Brill, 1990.

Witcombe, Christopher L. C. E. *Copyright in the Renaissance: Prints and the Privilegio in Sixteenth-Century Venice and Rome.* Leiden; Boston: Brill, 2004.

Witt, Ronald. "What Did Giovannino Read and Write? Literacy in Early Renaissance Florence." *I Tatti Studies in the Italian Renaissance* 6 (1995): 83–114.

Wolf, Kenneth B., ed. *Chronicon, Isidore of Seville, c. 616.* Medieval Texts in Translation, 2008. https://canilup.googlepages.com.

Woodward, John. *An Essay toward a Natural History of the Earth: And Terrestrial Bodies, Especially Minerals: As Also of the Sea, Rivers, and Springs: With an Account of the Universal Deluge: And of the Effects That It Had upon the Earth.* London: Printed for R. Wilkin, 1695.

Woodward, John. *Geografia fisica, ovvero saggio intorno alla storia naturale della terra, del sig. Woodward ... con la giunta dell'apologia del saggio contra le osservazioni del dottor Camerario; e d'un trattato de' fossili d'ogni spezie, divisi metodicamente in varie classi.* Venice: Giambatista Pasquali, 1739.

Woodward, John. *Specimen geographiae physicae Quo agitur de terra, et Corporibus Terrestribus Speciatim mineralibus: nec non Mari, Fluminibus, & fontibus: Accedit diluvii universalis Effectrumque ejus in Terra descriptio.* Translated by Johann Jackob Scheuchzer. Zurich: Typis D. Gessneri, 1704.

Wootton, David. *The Invention of Science: A New History of the Scientific Revolution.* London: Allen Lane, 2015.

Yushi, Ito. "Earth Science in the Scientific Revolution 1600–1728." Melbourne: University of Melbourne, 1985.

Yushi, Ito. "Hooke's Cyclic Theory of the Earth in the Context of Seventeenth Century England." *The British Journal for the History of Science* 21, no. 3 (1988): 295–314.

Zannichelli, Giovanni Girolamo. *Enumeratio rerum naturalium quae in Museo Zannichelliano asservantur.* Venice: Antonio Bortoli, 1736.

Zannichelli, Giovanni Girolamo. *Lithographia duorum montium veronensium. Unius nempe vulgo dicti di Boniolo; et alterius di Zoppica.* Venice: Giuseppe Corona, 1721.

Ziggelaar, August. "The Age of Earth in Niels Stensen's Geology." In *The Revolution in Geology from the Renaissance to the Enlightenment,* edited by Gary D. Rosenberg, 135–42. Boulder, CO: Geological Society of America, 2009.

Zuccolo, Vitale. *Dialogo delle cose meteorologiche. Di D. Vitale Zuccolo padoano ... In cui si dichiarano tutte le cose marauigliose, che si generano nell'aere, & alcune mirabili proprietà de' fonti, fiumi, e mari, secondo la dottrina d'Aristotele con le opinioni d'altri illustri scrittori.* Venice: Paolo Megietti, 1590.

Zupko, Jack. *John Buridan: Portrait of a Fourteenth-Century Arts Master.* Notre Dame: University of Notre Dame Press, 2003.

Index

For the benefit of digital users, indexed terms that span two pages (e.g., 52–53) may, on occasion, appear on only one of those pages.

Note: Page numbers followed by *f* indicate a figure on the corresponding page.